Growth Hormone Secretagogues
in Clinical Practice

Growth Hormone Secretagogues
in Clinical Practice

edited by

Barry B. Bercu

University of South Florida College of Medicine
Tampa, and
All Children's Hospital
St. Petersburg, Florida

Richard F. Walker

University of South Florida
Tampa, Florida

CRC Press
Taylor & Francis Group
Boca Raton London New York

CRC Press is an imprint of the
Taylor & Francis Group, an **informa** business

The editors acknowledge the generous support of Pharmacia & Upjohn, Inc., for an unrestricted educational grant.

CRC Press
Taylor & Francis Group
6000 Broken Sound Parkway NW, Suite 300
Boca Raton, FL 33487-2742

First issued in paperback 2019

ISBN-13: 978-0-8247-9832-1 (hbk)
ISBN-13: 978-0-367-40048-4 (pbk)

**Visit the Taylor & Francis Web site at
http://www.taylorandfrancis.com**

**and the CRC Press Web site at
http://www.crcpress.com**

Preface

The symposium on which this book is based, held in Tampa, Florida, was the second in a series devoted to achieving a better understanding of growth hormone (GH) secretagogues and the mechanisms by which they stimulate pituitary release of somatotropin. These objectives were achieved through the presentation of recent findings and the discussion of contemporary issues in this important area of endocrinology.

Historically, net GH release from the pituitary was thought to represent the balance of single inhibitory and stimulatory influences. Although a stimulatory agent, growth hormone-releasing factor (GRF) had been postulated in the early 1960s but it was not until 1982 that growth hormone-releasing hormone (GHRH) was isolated and characterized. That discovery served as the basis for a regulatory system in which net GH secretion represented a balance of stimulatory and inhibitory signals from two different hypothalamic peptides—GHRH and somatostatin, respectively. Subsequent analysis of the GHRH molecule revealed that the structural requirements for GH-stimulating activity resided in the first 29 amino acids of the peptide.

Shortly after GHRH was characterized and identified, synthetic peptides capable of stimulating GH secretion in vitro and in vivo were derived from met-enkephalin. It was known that other molecules such as arginine, morphine, clinidine, and levodopa could also provoke GH secretion, but these increased GH release by altering the secretion of endogenous GHRH or somatostatin. Thus, they were not true GH secretagogues per se, and they were incapable of directly altering somatotroph function in vitro when hypothalamic peptides were absent. In contrast, the opioid-derived peptides were capable of directly stimulating GH release from cultured pituitary tissue and cells, even though none of them shared structural homology with the active portion of the stimulatory GHRH molecule (i.e., amino acids, 1–29). Furthermore, binding and functional

studies demonstrated that GH-releasing activity and opioidergic character di-
verged as GHRP efficacy and potency increased. The new molecules, which
were called GH-releasing peptides (GHRPs) to distinguish them from the naturally occurring hypothalamic peptide, GHRH, seemed to be xenobiotic analogs
of another, still-unidentified GH secretagogue.

However, the acceptance of another endogenous secretagogue for GH that
complemented GHRH function was not immediately forthcoming. The extremely
low activity of the early analogs, coupled with a poor understanding of their
relationship to endogenous GHRH and how it created differences in in vitro
activity, caused initial findings to be viewed skeptically and, occasionally,
dismissed as contamination artifacts of natural GHRH within the in vitro testing system.

Despite initial skepticism, Dr. Cyril Y. Bowers, who headed the team that
synthesized GHRP and discovered its potential as a GH secretatogue, persevered
in his efforts to have the scientific community recognize the importance of his
findings. After a decade of work demonstrating the efficacy of GHRP through
the publication of many important papers, his labors finally bore fruit.

One of us (AVS) has been fortunate to share a close friendship with Dr.
Bowers for more than 30 years and was privileged to collaborate with him during
the 1960s and early 1970s. This collaboration led to the isolation, elucidation
of structure, and synthesis of thyrotropin-releasing hormone (TRH), the first
hypothalamic hormone to be completely identified. It also placed the projects
involving other hypothalamic hormones on a solid foundation.

Through his vision and untiring devotion to the advancement of knowledge in the science of endocrinology, Dr. Bowers has earned the respect and
admiration of his colleagues. Accordingly, the symposium was conceived as a
tribute to Dr. Bowers and dedicated in his honor.

Appropriately, the meeting began with Dr. Bower's historical perspective
on the discovery and development of GHRPs as therapeutic agents. He explained
that the prototype molecule of the GHRP-type GH secretagogues, growth hormone-releasing hexapeptide (GHRP-6, His-D-Trp-Ala-Trp-D-Phe-Lys-NH$_2$), was
used for pharmacological analysis of this new family of potential drug substances. Although much was learned about their activity, there was little interest in clinical application of GHRPs until oral bioavailability was later demonstrated in rats, dogs, monkeys, and finally, humans, leading to a concerted effort
by several pharmaceutical companies to develop better orally bioavailable
molecules. Those efforts culminated in the discovery of orally active, nonpeptidyl, and truncated peptide GH secretagogues. Despite the apparent structural diversity of the xenobiotic molecules, they all shared the pharmacodynamic
and mechanistic characteristics of GHRP-6, which were distinct from those for
GHRH. Furthermore, the GH secretagogue characteristics of GHRPs were not
shared by the other molecules that were capable of indirectly stimulating GH

secretion, such as TRH, arginine, clonidine, morphine, and others. Although GHRP had certain constraints on its ability to stimulate GH secretion, it was a reliable secretagogue that produced unique patterns of GH secretion that were distinctly different from those produced by GHRH. Subsequent comparative analysis of GHRH and GHRP efficacies as GH secretagogues revealed that they acted through clearly dissimilar signal transduction mechanisms.

Following Dr. Bower's interesting and thorough review of GHRP evolution, contemporary issues related to the broad scope of GH secretagogue action were presented for discussion. The topics discussed during this first session included an analysis of structural requirements for GH secretagogue efficacy, the use of mathematical models for predicting the differential values of individual stimuli for GH secretion from the pituitary, and the presentation of a unifying mechanism by which peptidyl and nonpeptidyl GH secretagogues of the GHRP class elicited their action. Structural analyses of the xenobiotic GHRPs indicated that small-molecule mimetics of the prototypic GHRP-6 could be useful for overcoming many of the potential disadvantages associated with the peptides, which are poorly absorbed and rapidly metabolized, and thus suffer from low bioavailability. In an attempt to circumvent these problems, a new series of nonpeptidyl GHRPs was derived from a molecule-modeling analysis of reported small-molecule mimetics. These included the benzolactam L-692,429 and the acyclic dipeptides L-162,752 and MK-0677. From a satisfactory conformational overlay, a series of new compounds containing substituted tetrahydroisoquinoline and isoindoline rings was synthesized. These nonpeptidyl mimetics of GHRP were proposed as potentially useful tools to investigate the utility and mechanisms of action of non-GHRH GH secretagogues.

Data were presented within this context to show that all of the non–GHRH-type secretagogues used a common receptor type that had binding and dissociation characteristics different from those identified for GHRH-related secretagogues. Furthermore, phosphotidyl inositide second messengers were shown to mediate intracellular transduction of the GHRP message, compared with cyclic nucleotides that were used by GHRH. As an extension of the concept that unique binding sites subserve GHRP activity, the results of a study providing the molecular characterization of the new receptor were presented. By utilizing *Xenopus* oocyte expression cloning, a swine pituitary cDNA was isolated that encoded a protein with a high-affinity binding site for MK-0677, a nonpeptidyl GHRP. From its pharmacology for GHRP binding, the investigators concluded that the cDNA encoded a swine GHRP receptor. Similar receptors were also identified for human and rat GHRP cDNA clones. The GHRP receptor represents a G–protein-coupled receptor that is most closely related to the neurotension receptor (35% amino acid sequence identity) and is encoded by a single highly conserved gene of diverse vertebrate species. This receptor is expressed in pituitary tissues, as well as in areas of the brain that include the hypothala-

mus, hippocampus, and other sites. The central message of this study was that
the presence of a GHRP receptor implies the existence of a naturally occurring
ligand that may be part of the GH secretion regulatory complex.

The initial session was followed by one in experimental pharmacology and
involved topics such as agonists of growth hormone-releasing hormone, interactions of the exogenous GH secretagogues, evidence for extrahypophyseal sites
of action for the GHRP-like molecules, and an analysis of the physiological state
as it affects GHRH and GHRP activity. The presentations were wide and varied. It was recognized that such efforts, as described in the foregoing, had been
devoted to the design, synthesis, and evaluation of the different families of GH
secretagogues. However, suitable screening systems for the rapid evaluation of
numerous novel compounds representing potential candidates were lacking.
Thus, one laboratory reported the results of its efforts to develop a simple
screening system based on polyclonal antibodies. For this purpose, a pentapeptide (Gly-Ala-Asn-Ala-Gly) was bound to the COOH or NH-terminus of GHRP-
6, and the resulting undecapeptides, termed P1 and P2, respectively, were
further conjugated with bovine serum albumin to serve as antigens for the
immunization of rabbits and preparation of antisera.

Another presentation focusing on GHRH reported that many agonistic
analogs intended for potential clinical and veterinary applications had been
synthesized by various laboratories. Analogs with enhanced biological activities and, thus, greater potential therapeutic usefulness were shown to result from
derivatization of functional groups or exchanges of amino acids. Many different analogs were identified, and their potential role as therapeutic agents suggested. Similarly, a clinical need for antagonistic analogs of GHRH was reported. Among proposed applications for these molecules were acromegaly,
diabetic retinopathy, diabetic nephropathy (glomerulosclerosis), and in cancer
treatment GHRH antagonists could be given alone or in combination with
superactive somatostatin analogs to suppress GH and insulin-like growth factor
I (IGF-1) levels in patients with neoplasms potentially dependent on IGF-1.
GHRH antagonists could also be used for inhibiting growth of tumors that do
not express somatostatin receptors, such as osteosarcomas or pancreatic cancers. The main point of this presentation was to summarize the work of one of
us (AVS) in synthesis of these antagonists and also to review oncological investigations performed in various cancer models with the new antitumor agents.
Potential clinical applications of antagonists in treating IGF-1–dependent cancers were discussed in light of increasing evidence of the involvement of growth
factors IGF-I and IGF-II involvement in the progression and metastases of
various malignancies.

Consistent with the discovery of GHRP receptors in the hypothalamus,
the more pronounced effect of GHRP in vivo compared with in vitro, and
GHRH and GHRP synergy in vivo, was the direct demonstration of GHRP

activity in brain structures through enhancement of c-*fos* expression and electrical activity in hypothalamic neurons. GHRP was reported to behave like a hypothalamic neurohormone that released GH in two phases characterized by early hypersensitivity and late hyposensitivity. These responses were typical of receptors activated by neurohormones, and the investigator provided evidence for extrahypophyseal site of action of non-GHRH GH secretagogues. Further direct evidence was provided by the use of electrophysiological tests. Neurosecretory neurons in the arcuate nucleus were identified antidromically, and their electrical characteristics were recorded. Physiological evidence for the effects of GHRP on behavior was provided in a series of tests involving voluntary and forced exercise. The data showed that GHRP has an effect on the ability to perform a behavioral task; however, it was unclear whether this improvement resulted from enhanced physical capacity, attitude, or a combination of these neurophysiological factors. Finally, action of GH secretagogues in the brain were discussed by determination of the activity of these substances on hypothalamic hormones in hypothalamic incubation systems. GHRP stimulated arginine vasopressin release, but had no stimulatory effect on corticotropin-releasing hormone. In addition, release of posterior pituitary hormones was inhibited by small synthetic molecules capable of stimulating GH release from the anterior pituitary.

After the basic aspects of GH secretagogue endocrinology had been discussed, the meeting turned to topics of clinical application for GH secretagogues. Within this context, investigators presented a novel diagnostic test for evaluating pituitary function in slowly growing children and aging adults. The objective was to show that differential secretory responses to GHRH and GHRP could be used to allow appropriate selection of therapeutic entities, ranging from GHRH and GHRP alone, the combination of GHRH, or recombinant GH, when patients lack a pituitary mechanism for GH production or secretion. In another study, the presence of a significant GH response to GHRP in at least a subset of patients suggested the possibility of anabolic benefits from multiple doses given over time. Another study reported different effects of GHRH and GHRP on sleep, indicating that the two secretagogues have different activities here. GHRH was effective in producing slow-wave sleep, whereas GHRP was associated with rapid eye movement (REM) sleep. Finally, reports were presented on interactions of GH secretagogues with other hormones and endogenous substances, ranging from sex steroids to interferons.

In conclusion, this book provides the reader with information based on presentations from the Second International Symposium on GH Secretagogues. Through those details provided by pioneers in the field, one can view the origins of a new area of GH research involving the xenobiotic GH secretagogues. Dr. Bowers is recognized for his significant contributions to this research and, accordingly, we thank him, the scientific committee, session chairs, speakers, and poster presenters for their outstanding contributions to the symposium and to the publication of this book. We are especially grateful to Rolf Gunnarsson, Vice President for Medical Affairs, and to Barbara Lippe, Senior Medical Director, Peptide Hormones, Pharmacia & Upjohn, Inc., without whose support this project would not have been possible. We appreciate the educational grant provided by Pharmacia & Upjohn as evidence of the company's commitment to research and support of science and medicine.

Barry B. Bercu
Richard F. Walker

Andrew V. Schally
(contributor and advisor)

Contents

Contents

Contributors

B. Håkan Ahlman, M.D., Ph.D. Professor, Department of Surgery, Sahlgrenska University Hospital, Göteborg, Sweden

Gianluca Aimaretti, M.D. Division of Endocrinology, Department of Clinical Physiopathology, University of Turin, Turin, Italy

Antonino Alberti, M.D. Department of Pediatrics, OASI Institute for Research in Mental Retardation and Brain Aging, Troina, Italy

Thomas Anderson Department of Biological Sciences, The Royal Danish School of Pharmacy, Copenhagen, Denmark

Emanuela Arvat, M.D. Consultant, Department of Internal Medicine, Division of Endocrinology, University of Turin, Turin, Italy

Mark A. Bach, M.D., Ph.D. Director, Clinical Research/Endocrinology and Metabolism, Merck Research Laboratories, Merck and Co., Inc., Rahway, New Jersey

Alex R. T. Bailey, Ph.D. Department of Physiology, University Medical School, Edinburgh, Scotland

Gerhard Baumann, M.D. Professor of Medicine, Department of Medicine, Northwestern University Medical School, Chicago, Illinois

Jaele Bellone, M.D. Division of Endocrinology, Department of Clinical Physiopathology, University of Turin, Turin, Italy

Bengt-Åke Bengtsson, M.D., Ph.D.   Assistant Professor, Department of Medicine, Endocrine Division, Sahlgrenska University Hospital, Göteborg, Sweden

Barry B. Bercu, M.D.   Professor of Pediatrics, Pharmacology and Therapeutics, University of South Florida College of Medicine, Tampa, and All Children's Hospital, St. Petersburg, Florida

Michael Besser, M.D., D.Sc., F.R.C.P.   Professor of Medicine, Department of Endocrinology, St. Bartholomew's Hospital, London, England

Muny Franklin Boghen   Scientific Medical Advisor, Pharmacia & Upjohn, Milan, Italy

Roger Bouillon, M.D., Ph.D.   Professor, Department of Endocrinology, Katholieke Universiteit Leuven, Leuven, Belgium

Cyril Y. Bowers, M.D.   Professor of Medicine, Department of Medicine, Tulane University Medical Center, New Orleans, Louisiana

Fabio Broglio, M.D.   Trainer in Endocrinology, Department of Internal Medicine, University of Turin, Turin, Italy

David Brown, B.Sc., M.Sc., D.I.C.   Laboratory of Biomathematics, The Babraham Institute, Cambridge, England

Franco Camanni, M.D.   Chief, Division of Endocrinology, Department of Internal Medicine, University of Turin, Turin, Italy

Marco Cappa, M.D.   Division of Pediatrics, Ospedale Pediatrico Bambino Gesù, IRCCS, Rome, Italy

Philip A. Carpino, Ph.D.   Senior Research Scientist, Department of Cardiovascular and Metabolic Diseases, Pfizer Central Research, Groton, Connecticut

Eva Maria Carro, B.Sc.   Research Fellow, Department of Physiology, University of Santiago de Compostela, Santiago de Compostela, Spain

Felipe F. Casanueva, M.D., Ph.D.   Professor, Department of Medicine, University of Santiago de Compostela, Santiago de Compostela, Spain

Chen Chen, M.D., Ph.D.   Research Fellow (Australian NH and MRC), Department of Neuroendocrinology, Prince Henry's Institute of Medical Research, Melbourne, Victoria, Australia

Kristin L. Chidsey-Frink Assistant Scientist, Department of Cardiovascular and Metabolic Diseases, Pfizer Central Research, Groton, Connecticut

Iain J. Clarke, Ph.D. Associate Professor, Prince Henry's Institute of Medical Research, Melbourne, Victoria, Australia

Fabio Colabucci Pediatric Clinic, Catholic University, Rome, Italy

Annamaria Colao, Ph.D. Department of Endocrinology and Clinical and Molecular Oncology, Università Federico II, Naples, Italy

Ana Maria Comaru-Schally, M.D. Professor of Medicine, Department of Medicine, Tulane University School of Medicine, and Veterans Affairs Medical Center, New Orleans, Louisiana

Alfredo Costa, M.D. Research Associate, Laboratory of Neuroendocrinology, Institute of Endocrinology "C. Mondino," University of Pavia, Pavia, Italy

Paul DaSilva Jardine, Ph.D. Manager, Department of Cardiovascular and Metabolic Diseases, Pfizer Central Research, Groton, Connecticut

Romano Deghenghi, M.D., Ph.D. President, Europeptides, Argenteuil, France

Laura De Marinis, M.D. Researcher, Institute of Endocrinology, Catholic University School of Medicine, Rome, Italy

Francis de Zegher, M.D., Ph.D. Professor, Department of Pediatrics, University of Leuven, Leuven, Belgium

Frank M. DiCapua, Ph.D. Senior Research Scientist, Department of Computational Chemistry, Pfizer Central Research, Groton, Connecticut

Suzanne L. Dickson, B.Sc., Ph.D. Department of Physiology, University of Cambridge, Cambridge, England

Carlos Dieguez, M.D. Professor of Physiology, University of Santiago de Compostela, Santiago de Compostela, Spain

Lidia Di Vito, M.D. Department of Internal Medicine, University of Turin, Turin, Italy

Patrick du Souich, M.D., Ph.D. Professor, Department of Pharmacology, Faculty of Medicine, University of Montreal, Montreal, Quebec, Canada

Antonella Faedda, M.D., Ph.D.   Pediatric Endocrinology Service, Ospedale Regionale per le Microcitemie, Cagliari, Italy

Giovanni Farello, M.D.   Pediatric Clinic, Università di L'Aquila, L'Aquila, Italy

Scott D. Feighner, Ph.D.   Senior Research Fellow, Department of Biochemistry and Physiology, Merck Research Laboratories, Merck and Co., Inc., Rahway, New Jersey

Bjarne Fjalland, Ph.D.   Associate Professor and Head of Department of Biological Sciences, The Royal Danish School of Pharmacy, Copenhagen, Denmark

Mary L. Forsling, Ph.D., D.Sc.   Professor, Department of Physiology, United Medical and Dental Schools, St. Thomas's Campus, London, England

Ralf-Michael Frieboes, M.D.   Clinical Institute, Max Planck Institute of Psychiatry, Munich, Germany

Lawrence A. Frohman, M.D.   Professor and Head, Department of Medicine, University of Illinois at Chicago, Chicago, Illinois

Giovanni Gasbarrini, M.D.   Professor of Internal Medicine, Department of Internal Medicine, Catholic University School of Medicine, Rome, Italy

Ezio Ghigo, M.D.   Associate Professor of Endocrinology, Department of Internal Medicine, University of Turin, Turin, Italy

Laura Gianotti, M.D.   Department of Internal Medicine, Division of Endocrinology, University of Turin, Turin, Italy

David A. Griffith, Ph.D.   Research Scientist, Animal Health Discovery, Pfizer Central Research, Groton, Connecticut

Ashley Grossman, M.D., B.Sc., F.R.C.P.   Professor of Neuroendocrinology, Department of Endocrinology, St. Bartholomew's Hospital, London, England

William A. Hada, Jr., M.S.   Scientist, Department of Medicinal Chemistry, Pfizer Central Research, Groton, Connecticut

Andrew D. Howard, Ph.D.   Senior Research Fellow, Biochemistry and Physiology, Merck Research Laboratories, Merck and Co., Inc., Rahway, New Jersey

John K. Inthavongsay Associate Scientist, Department of Medicinal Chemistry, Pfizer Central Research, Groton, Connecticut

John-Olov Jansson, M.D., Ph.D. Associate Professor, Research Center for Endocrinology and Metabolism, Sahlgrenska University Hospital, Göteborg, Sweden

Peter B. Johansen, Ph.D. Principal Research Scientist, Department of Growth Hormone Biology, Novo Nordisk A/S, Copenhagen, Denmark

Jelena Joksimović, Ph.D. Senior Member, Head of the Department of Molecular Biology and Biochemistry, Institute for Biological Research, Belgrade, Yugoslavia

Rhonda D. Kineman, Ph.D. Research Assistant Professor, Department of Medicine, University of Illinois at Chicago, Chicago, Illinois

Márta Korbonits, M.D. Department of Endocrinology, St. Bartholomew's Hospital, London, England

Magdolna Kovacs, M.D., Ph.D. Associate Professor, Section of Experimental Medicine, Department of Medicine, Tulane University School of Medicine, and Endocrine, Polypeptide and Cancer Institute, Veterans Affairs Medical Center, New Orleans, Louisiana

Zvi Laron, M.D. Irene and Nicholas Marsh Chair in Endocrinology and Diabetes, Endocrinology and Diabetes Research Unit, Schneider Children's Medical Center, Tel Aviv, Israel

Alfonso Leal-Cerro, M.D. Consultant in Clinical Endocrinology, Endocrinology Division, Hospital Virgen del Rocio, Sevilla, Spain

Bruce A. Lefker, Ph.D. Senior Research Investigator, Department of Cardiovascular and Metabolic Diseases, Pfizer Central Research, Groton, Connecticut

Gareth Leng, Ph.D. Professor of Experimental Physiology, Department of Physiology, University Medical School, Edinburgh, Scotland

Sharon K. Lewis Associate Scientist, Animal Health Discovery Pharmaceuticals, Pfizer Central Research, Groton, Connecticut

Anders Lindahl, M.D., Ph.D. Associate Professor, Institute of Laboratory Medicine, Sahlgrenska University Hospital, Göteborg, Sweden

John A. Little, B.Sc., Ph.D., C.Chem., M.R.S.C.   Department of Chemical Endocrinology, St. Bartholomew's Hospital, London, England

Sandro Loche, M.D.   Pediatric Endocrinology Service, Ospedale Regionale per le Microcitemie, Cagliari, Italy

Gaetano Lombardi, M.D., Ph.D.   Professor, Department of Endocrinology and Molecular and Clinical Oncology, Università Federico II, Naples, Italy

Simon M. Luckman, Ph.D.   Department of Neurobiology, The Babraham Institute, Cambridge, England

Hiralal G. Maheshwari, M.B.B.S., Ph.D.   Senior Research Associate, Department of Medicine, Northwestern University Medical School, Chicago, Illinois

Antonio Mancini, M.D.   Researcher, Institute of Endocrinology, Catholic University School of Medicine, Rome, Italy

F. Michael Mangano, M.S.   Scientist, Department of Medicinal Chemistry, Pfizer Central Research, Groton, Connecticut

Sylvie Marleau, Ph.D.   Assistant Professor, Faculty of Pharmacy, University of Montreal, Montreal, Quebec, Canada

Bartolomeo Merola, M.D.   Department of Endocrinology and Molecular and Clinical Oncology, Università Federico II, Naples, Italy

Dragan Micić, M.D., Ph.D.   Professor, Department of Endocrinology, Institute of Endocrinology, Diabetes, and Diseases of Metabolism, Belgrade, Yugoslavia

Giampiero Muccioli, Ph.D.   Associate Professor of Pharmacology, Department of Anatomy, Pharmacology, and Forensic Medicine, University of Turin, Turin, Italy

Harald Murck, M.D.   Clinical Institute, Max Planck Institute of Psychiatry, Munich, Germany

Marianne C. Murray, Ph.D.   Manager, Animal Health Discovery Pharmaceuticals, Pfizer Central Research, Groton, Connecticut

Pierluigi Navarra, M.D.   Associate Professor, Institute of Pharmacology, Catholic University Medical School, Rome, Italy

Magnus H. L. Nilsson, Dr. Med. Sc.   Project Manager, Pharmacia & Upjohn, Stockholm, Sweden

Michael Nilsson, M.D., Ph.D.   Department of Anatomy and Cell Biology, Institute of Neurobiology, Sahlgrenska University Hospital, Göteborg, Sweden

Ola Nilsson, M.D., Ph.D.   Associate Professor, Department of Pathology, Sahlgrenska University Hospital, Göteborg, Sweden

OiCheng Ng, M.S.   Scientist, Department of Cardiovascular and Metabolic Disease, Pfizer Central Research, Groton, Connecticut

Huy Ong, Ph.D.   Professor, Department of Pharmacy, University of Montreal, Montreal, Quebec, Canada

Jane G. Owens, D.V.M., Ph.D.   Research Scientist, Animal Health Safety and Metabolism, Pfizer Central Research, Groton, Connecticut

Lydia C. Pan, Ph.D.   Senior Research Investigator, Department of Cardiovascular and Metabolic Diseases, Pfizer Central Research, Groton, Connecticut

Roberto Peinó, M.D.   Consultant in Clinical Endocrinology, Department of Medicine, University of Santiago de Compostela, Santiago de Compostela, Spain

Vojislav Pejović, M.Sc.   Research Assistant, Center for Chemistry, Institute for Chemistry, Technology, and Metallurgy, Belgrade, Yugoslavia

Christine M. Pirie, M.S.   Associate Scientist, Department of Cardiovascular and Metabolic Diseases, Pfizer Central Research, Groton, Connecticut

Manuel Pombo Arias, M.D.   Professor, Department of Paediatrics, University of Santiago de Compostela, Santiago de Compostela, Spain

Vera Popović, M.D., Ph.D.   Professor, Department of Endocrinology, Institute of Endocrinology, Diabetes, and Diseases of Metabolism, Belgrade, Yugoslavia

Caterina Proto   Biologist, Laboratory of Radioimmunology, OASI Institute for Research in Mental Retardation and Brain Aging, Troina, Italy

Lucia Puglisi, M.D.   Fellow, Department of Internal Medicine, Catholic University School of Medicine, Rome, Italy

Letizia Ragusa, M.D.   Department of Pediatrics, OASI Institute for Research in Mental Retardation and Brain Aging, Troina, Italy

Gian Lodovico Rapaccini, M.D.   Researcher, Department of Internal Medicine, Catholic University School of Medicine, Rome, Italy

Corrado Romano, M.D.   Head of the Department of Pediatrics, OASI Institute for Research in Mental Retardation and Brain Aging, Troina, Italy

Colin R. Rose, M.S.   Assistant Scientist, Department of Medicinal Chemistry, Pfizer Central Research, Groton, Connecticut

Marie Roumi, M.Sc.   Faculty of Pharmacy, University of Montreal, Montreal, Quebec, Canada

Nancy I. Ryan   Assistant Scientist, Animal Health Research, Pfizer Central Research, Groton, Connecticut

John R. Schafer   Assistant Scientist II, Department of Cancer, Immunology, and Infectious Diseases, Pfizer Central Research, Groton, Connecticut

Andrew V. Schally, Ph.D., D.Sc.H.C., M.D.H.C.   Professor, Department of Medicine, Tulane University School of Medicine, and Chief, Endocrine, Polypeptide, and Cancer Institute, Veterans Affairs Medical Center, New Orleans, Louisiana

Eva Sjögren-Jansson, M.Sc.   Research Assistant, Institute of Laboratory Medicine, Sahlgrenska University Hospital, Göteborg, Sweden

Roy G. Smith, Ph.D.   Vice President, Department of Biochemistry and Physiology, Merck Research Laboratories, Merck and Co., Inc., Rahway, New Jersey

Vukić Šoškić, Ph.D.   Senior Research Associate, Department of Molecular Biology and Biochemistry, Institute for Biological Research, Belgrade, Yugoslavia

Axel Steiger, M.D.   Clinical Institute, Max Planck Institute of Psychiatry, Munich, Germany

Elinor A. Stephens, B.Sc., M.Sc.   Laboratory of Biomathematics, The Babraham Institute, Cambridge, England

Johan Svensson, M.D.   Specialist in Endocrinology, Department of Internal Medicine, Research Center for Endocrinology, Sahlgrenska University Hospital, Göteborg, Sweden

David D. Thompson, Ph.D.   Assistant Director, Metabolic Diseases, Pfizer Central Research, Groton, Connecticut

Katalin Toth, Ph.D.   Senior Researcher, Endocrine, Polypeptide, and Cancer Institute, Section of Experimental Medicine, Department of Medicine, Tulane University School of Medicine and Veterans Affairs Medical Center, New Orleans, Louisiana

Peter J. Trainer, M.D., B.Sc., M.R.C.P.   Senior Lecturer in Endocrinology, Department of Endocrinology, St. Bartholomew's Hospital, London, England

Giuseppe Tringali, Ph.D.   Research Fellow, Institute of Pharmacology, Catholic University Medical School, Rome, Italy

John L. Tucker   Assistant Scientist, Department of Process Research and Development, Pfizer Central Research, Groton, Connecticut

Maria Rosa Valetto, M.D.   Department of Clinical Pathophysiology, Division of Endocrinology, University of Turin, Turin, Italy

Domenico Valle, M.D.   Fellow, Department of Internal Medicine, Catholic University School of Medicine, Rome, Italy

Greet Van den Berghe, M.D., Ph.D.   Professor, Department of Intensive Care Medicine, University of Leuven, Leuven, Belgium

Lex H. T. Van der Ploeg, Ph.D.   Senior Director, Department of Genetics and Molecular Biology, Merck Research Laboratories, Merck and Co., Inc., Rahway, New Jersey

Johannes D. Veldhuis, M.D.   Professor, Division of Endocrinology and Metabolism, Department of Internal Medicine, University of Virginia, Charlottesville, Virginia

Richard F. Walker, Ph.D.   Director, Division of Pharmaceutical Studies and Division of Compliance Services, and Associate Professor of Biochemistry and Molecular Biology, University of South Florida, Tampa, Florida

Bo Wängberg, M.D., Ph.D.   Associate Professor, Department of Surgery, Sahlgrenska University Hospital, Göteborg, Sweden

Ann S. Wright   Associate Scientist, Department of Cardiovascular and Metabolic Diseases, Pfizer Central Research, Groton, Connecticut

Danxing Wu, M.D., Ph.D. Postdoctoral Fellow, Department of Neuroendo-crinology, Prince Henry's Institute of Medical Research, Melbourne, Victoria, Australia

Michael P. Zawistoski, M.A. Senior Scientist, Department of Cardiovascular and Metabolic Diseases, Pfizer Central Research, Groton, Connecticut

Growth Hormone Secretagogues

in Clinical Practice

1

Synergistic Release of Growth Hormone by GHRP and GHRH: Scope and Implication

Cyril Y. Bowers
Tulane University Medical Center, New Orleans, Louisiana

I. INTRODUCTION

The initial growth hormone-releasing peptide (GHRP), Tyr-D-Trp-Gly-Phe-Met-NH$_2$ was developed in 1976 (1), and since then the GHRPs have been greatly expanded at many different levels including accomplishments of both potentially practical and theoretical value. There are now three major chemical classes of GHRP (i.e., peptides, partial peptides, and nonpeptides or peptidomimetics; 2,3). Peptide GHRPs, which consist of four major types, were established between 1977 and 1980. Despite the broad range of chemistry among the GHRPs (2–12), almost all of them appear to act on the same receptor and activate the same intracellular signal transduction pathway (13,14). By 1984, the biological action of GHRP was considered sufficiently and significantly different from native GH-releasing hormone (GHRH) to propose that it may reflect the activity of a new hypothalamic hypophysiotropic hormone (15). The activity of GHRP in multiple animal species (15) and, subsequently, in humans in 1990, again strongly supported this hypothesis (16). In 1996, the seminal achievement of cloning the GHRP–GH secretagogue receptor was accomplished (17). It is a new seven-transmembrane domain G–protein-coupled receptor (18). Genomic analysis supports the presence of a single highly conserved gene in human, chimpanzee, swine, bovine, rat, and mouse genomic DNA. Two immediate objectives of primary importance include isolation of the putative native GHRP–like hormone and further elucidation of how GHRP releases GH. The more long-term primary objectives are the following: (1) to determine what role a putative GHRP system might have in the physiological regulation of GH secretion; (2) to determine whether the putative GHRP-like hormone might play a role in the pathophysiological secretion of GH; and (3) to determine the possible diagnostic and therapeutic value of GHRP in various disorders of GH secretion in humans.

The GHRP acts at two anatomical sites—the hypothalamus and the pituitary—to release GH (19–24). The pituitary, but not the hypothalamic action of GHRP is well established. It releases GH by activation of the phospholipase C–protein kinase C pathway, which is different from the adenylate cyclase cAMP protein kinase A pathway activated by GHRH (25–27). Evidence indicates that crosstalk between these intracellular pathways does occur, probably from inositol triphosphate (IP_3) to cAMP, but apparently not cAMP to IP_3. The combination of GHRP with GHRH does synergistically raise cAMP levels in vitro; however, the effect of these two peptides on GH release is mainly additive, or only slightly synergistic (19,20).

A seemingly important point is that, even though there are sufficient favorable molecular interactions of GHRP plus GHRH on the pituitary to believe, and reasonably conclude, that not infrequently dramatic synergistic effects in vivo might be explained by a direct pituitary action of these two peptides in combination, this conclusion is not supported by other results (20). By using a coupled in vitro–in vivo approach, it has been possible to demonstrate that the scope of the relation between GHRP and GHRH on GH release can be both independent and dependent, additive and synergistic, as well as permissive.

II. RESULTS AND DISCUSSION

Table 1 shows some of our earlier results that demonstrate how effectively GHRP-6 plus GHRH synergistically releases GH in monkeys and cows. In humans, the complementary effect of these two combined peptides has been even more dramatic, regardless of age and sex. Not only have the results on synergism in animals and humans indicated new dimensions in the secretion of GH that were not previously apparent and appreciated, but they seem to indicate new practical diagnostic and therapeutic aspects of GHRP. A major challenge will be to determine the more fundamental mechanisms of this synergism and whether they may play a role in the physiological regulation of GH secretion or, even possibly, the pathophysiological secretion of GH.

Table 1 Synergism of GHRP-6 and GHRH

	Dose (μg/kg)	GH(ng/mL ± SEM) +20 min	
		Male	Female
Monkeys			
Control	—	2 ± 1.6	6 ± 5
GHRP-6	5	1 ± 0.6	1 ± 0.5
GHRH[a]	5	6 ± 2	2 ± 1
GHRP-6 + GHRH[a]	5 + 5	21 ± 8	26 ± 9
	Dose (μg/kg)	GH(ng/mL ± SEM) +15 min	
Nonlactating holstein cows			
Control	—	0.17 ± 0.19	
Ala[1]GHRP-6	3	8.6 ± 2.5	
GHRH[b]	3	5.7 ± 0.58	
Ala[1]GHRP-6 + GHRH[b]	3 + 3	88.0 ± 19.0	

[a]GHRH = 1–44NH$_2$; mean of 6 ± SEM
[b]GHRH = Nle^{27}GHRH(1–29)NH$_2$; mean BW = 543 kg; mean of 4 ± SEM

An aspect of synergism that needs more investigation concerns how opiates, GHRP, and GHRH interact to release GH synergistically in rats. Table 2 records results that demonstrate the synergistic release of GH induced by the heptapeptide dermorphin or the tetrapeptide dermorphin analog, Tyr-D-Arg-Phe-Gly-NH$_2$ in combination with GHRP(s) or GHRH or both (28). Each pair of these different classes of GH secretagogues synergistically releases GH, and when all three are administered together, this effect is even greater. Even though the GH-releasing activity of the opiates and GHRP is dependent on the presence of endogenous GHRH, GHRH appears to play a permissive role, rather than being responsible for mediating the release of GH. The synergism produced by these three types of GH secretagogues demonstrates that each of them induces synergism by an independent action, even though they may also have overlapping actions on GH release. Previous results in rats support that the opiates, but not GHRP or GHRH, inhibit the release of somatostatin (somatotropin release-inhibiting factor; SRIF) from the hypothalamus (19,20). Because exogenous GHRH has a marked in vivo effect on elevating both pituitary cAMP levels and GH release in parallel, the concomitant changes of GH release and pituitary cAMP levels were used as an index for the action of GHRH (19,20). A parallel rise of GH release and pituitary cAMP was considered to indicate an effect caused by endogenous GHRH. It is readily apparent

Table 2 Concomitant In Vivo Effects of Opiates + GHRP + GHRH on GH Release and Pituitary cAMP Levels in Rats[a]

Peptide	Dose (μg iv)	Serum GH (ng/mL ± SEM) C	Serum GH (ng/mL ± SEM) E	Pit. cAMP (pmol/mg protein ± SEM) C	Pit. cAMP (pmol/mg protein ± SEM) E
A. Opiate, GHRP, GHRH					
DM	30	12 ± 2	166 ± 44	6 ± 1	11 ± 3
	30	14 ± 1	161 ± 14	10 ± 1	8 ± 1
DM-A	30	13 ± 1	88 ± 21	6 ± 1	4 ± 1
	30	14 ± 1	154 ± 1	10 ± 1	9 ± 1
	100	13 ± 1	270 ± 45	6 ± 1	8 ± 2
GHRP-6	10	12 ± 2	166 ± 62	6 ± 1	25 ± 5
Ala¹GHRP-6	1	13 ± 1	46 ± 7	6 ± 1	10 ± 3
GHRP-1	10	13 ± 1	156 ± 46	6 ± 1	7 ± 1
	10	14 ± 1	209 ± 30	6 ± 1	26 ± 2
GHRH	1	13 ± 1	73 ± 12	6 ± 1	41 ± 8
	10	12 ± 2	94 ± 22	6 ± 1	159 ± 19
	10	14 ± 1	98 ± 16	10 ± 1	160 ± 14
B. Opiate + GHRP, opiate + GHRH, GHRP + GHRH					
DM + GHRP-6	30 + 10	12 ± 2	1817 ± 309	6 ± 1	28 ± 5
DM-A + Ala¹GHRP-6	30 + 10	13 ± 1	1995 ± 241	6 ± 1	19 ± 6
	100 + 1	13 ± 1	1221 ± 243	6 ± 1	5 ± 6
DM-A + GHRP-1	30 + 10	14 ± 1	921 ± 159	10 ± 1	39 ± 3
DM + GHRH	30 + 10	14 ± 1	1157 ± 123	10 ± 1	260 ± 16
DM-A + GHRH	100 + 1	13 ± 1	3912 ± 614	6 ± 1	57 ± 6
	30 + 10	14 ± 1	1574 ± 303	10 ± 1	256 ± 29

GHRP-6 + GHRH	10 + 10	12 ± 2	1853 ± 271	6 ± 1	193 ± 13
GHRP-1 + GHRH	10 + 10	14 ± 1	1012 ± 158	10 ± 1	217 ± 35
Ala¹GHRP-6 + GHRH	1 + 1	13 ± 1	814 ± 107	6 ± 1	62 ± 10
	10 + 10	13 ± 1	2109 ± 369	6 ± 1	104 ± 16
C. Opiate + GHRP + GHRH					
DM + GHRP-6 + GHRH	30 + 10 + 10	12 ± 2	3480 ± 580	6 ± 1	267 ± 33
DM + GHRP-1 + GHRH	30 + 10 + 10	14 ± 1	4003 ± 317	10 ± 1	301 ± 35
DM-A + GHRP-1 + GHRH	100 + 1 + 1	13 ± 1	4750 ± 322	6 ± 1	62 ± 10
	30 + 10 + 10	13 ± 1	3344 ± 357	6 ± 1	104 ± 16
	1 + 1 + 1	8 ± 1	1966 ± 224	—	—
	30 + 1 + 1	8 ± 1	3794 ± 284	—	—
	100 + 1 + 1	8 ± 1	3639 ± 339	—	—
	10 + 10 + 10	8 ± 1	3488 ± 523	—	—
DM – A + Ala¹GHRP-6 + GHRH	30 + 10 + 10	14 ± 1	3769 ± 318	10 ± 1	479 ± 59
	1 + 1 + 1	8 ± 1	947 ± 192	—	—
	30 + 1 + 1	8 ± 1	3218 ± 434	—	—
	10 + 10 + 10	8 ± 1	3154 ± 384	—	—

Dermorphin (DM), TyrDAlaPheGlyTyrProSerNH$_2$; Dermorphin analog (DM – A), TyrDAlaPheGlyTyrProGlyNH$_2$; GHRP-6, HisDTrpAlaTrpDPheGlyNH$_2$; GHRP-1, AlaHisDβNalAlaTrpDPheLysNH$_2$; Ala¹GHRP-6, AlaHisDTrpAlaTrpDPheLysNH$_2$; GHRH, Nle²⁷GHRH(1-29)NH$_2$.

[a]Untreated 26-day-old female rats, GH and pituitary cAMP determined 10 min after iv bolus peptide(s).

from the results in Table 2 that the opiates released GH without elevating pituitary cAMP. In contrast to the opiates, GHRP sometimes induced a small rise of pituitary cAMP, but this was much less than that induced by exogenous GHRH, even though both peptides released essentially the same amount of GH. GH release induced by the opiate plus GHRP was marked, without raising or only slightly raising pituitary cAMP levels. A conclusion from these studies is that the in vivo release of GH induced by the opiates or GHRPs does not result from the release of endogenous GHRH; however, from passive GHRH immunoneutralization studies in rats, the action of both types of GH secretagogues (opiates and GHRPs) on GH release depends on the presence of endogenous GHRH (20).

The possibility that opiates release GH in part by stimulating the release of the putative endogenous GHRP–like hormone from the hypothalamus was also investigated. In this study, a GHRP antagonist (substance P analog), which inhibits the GH response of GHRP, but not GHRH was administered together with the opiate plus GHRP as well as the opiate plus GHRH (29). As recorded in Table 1, the antagonist definitely inhibited the opiate plus GHRP-induced GH release, indicating it was inhibiting either the effect of exogenous GHRP or the putative endogenous GHRP-like hormone possibly released by the opiate. From the lack of effect of the antagonist on the release of GH induced by the opiate plus GHRH (see Table 2), it was concluded that the antagonist inhibited the exogenous GHRP-2, rather than the putative endogenous GHRP–like hormone. Undoubtedly, inhibition of SRIF release by the opiates explains how they augment the GH action of GHRP and GHRH. However, still uncertain is whether the singular action of the opiate to inhibit SRIF release can entirely explain how an opiate mediates the synergistic release of GH when administered in combination with GHRP, GHRH, or both. If the opiates are acting by inhibition of SRIF release, presumably the action of endogenous GHRH would be enhanced which, in turn, would be expected to increase both GH and cAMP. Additionally, if the opiate plus GHRP synergistically releases GH because of its inhibition of SRIF release, this would imply that GHRP is not acting as a functional SRIF antagonist to induce synergism. Future studies with these three different classes of GH secretagogues may reveal new insight into the regulation of GH secretion. Because there are various types of opiates, it will be necessary to determine which types might release GH by different mechanisms and, thereby, interact differently with GHRP or GHRH to release GH.

A series of studies were performed in the same nine normal young men and the same seven normal young women at three different doses of GHRH 1–44NH$_2$ and four doses of GHRP-2 alone and in various combinations. These types of dose–response studies are important to fully reveal

Table 3 Release of GH in Rats: Opiate (A) + GHRP-2/GHRH (B) with and Without Antagonist (C)

Dosage (μg iv)			Serum GH (ng/mL ± SEM)	Dosage (μg iv)			Serum GH (ng/mL ± SEM)
A +	B⁻ +	C		A +	B +	C	
I.							
—	—	—	26 ± 3	—	—	—	26 ± 3
100	0.1	—	1525 ± 194	100	0.1	100	182 ± 46
100	0.3	—	1170 ± 668	100	0.3	100	223 ± 53
100	1.0	—	1992 ± 802	100	1.0	100	797 ± 337
100	3.0	—	2670 ± 381	100	3.0	100	2567 ± 510
100	10.0	—	3267 ± 389	100	10.0	100	3720 ± 1158
II.							
—	—	—	26 ± 3	—	—	—	26 ± 3
100	0.1	—	773 ± 36	100	0.1	100	1085 ± 342
100	0.3	—	1667 ± 197	100	0.3	100	2478 ± 500
100	1.0	—	1958 ± 219	100	1.0	100	2356 ± 712
100	3.0	—	2130 ± 302	100	3.0	100	2684 ± 24
100	10.0	—	1810 ± 436	100	10.0	100	2820 ± 547

I. TyrDArgPheGlyNH$_2$; B, GHRP-2; C, [DArg^{1}DPhe^{5}DTrp^{7,9}Leu^{11}]-substance P GHRP antagonist.
n = Four, 26-day-old female rats, GH at + 10 min.
II. TyrDArgPheGlyNH$_2$; B, Nle^{27}GHRH(1-29)NH$_2$; C, [DArg^{1}DPhe^{5}DTrp^{7,9}Leu^{11}]-substance P GHRP antagonist.
n = Four, 26-day-old female rats, GH at + 10 min.

the synergistic GH response of GHRP-2 plus GHRH and to better appreciate the action of GHRP-2 on GH release. At 1 µg/kg, GHRP-2 released the same amount of GH in normal young men and women; however, at three lower doses, more GH was released in the women than in the men. Added to this complexity is that at the two lower doses of GHRH (0.1 and 0.3 µg/kg) the same amount of GH was released in the men and women, whereas at the high dose, 1 µg/kg, a much larger amount of GH was released in the women than in the men. Because of these dose-related differences, it is not difficult to envision that special results may be induced by certain ratios of GHRP to GHRH administered together. Relative to the synergistic response, the GH response was greater in women at all dose combinations of GHRP-2 plus GHRH in which synergism was induced. The dose combination of 0.3 + 0.3 µg and 1.0 + 1.0, but not 0.1 + 0.1 µg/kg, induced a synergistic GH response in both the men and women. The area under the curve (AUC) of the GH responses to 0.1 + 0.1, 0.3 + 0.3, and 1.0 + 1.0 µg/kg, respectively, for the men and women were 1,146, 5,242, and 10,800 µg/L over 4 h for the men and 3,210, 6,652 and 16,477 µg/L over 4 h for the women.

As recorded in Figures 1 and 2, the combination of 0.1 + 1.0 µg/kg GHRP-2 plus GHRH induced synergism in normal young men, but not when the doses were reversed and 1.0 + 0.1 µg/kg GHRP-2 plus GHRH was administered. In normal young women both 0.1 + 1.0 µg/kg and 1.0 + 0.1 µg/kg GHRP-2 plus GHRH released GH synergistically. The combination dose of 1.0 + 0.3 µg/kg GHRP-2 plus GHRH synergistically released GH in both the men and women. These various dose combination studies were performed not only to gain more insight into the related actions of these two peptides on GH release, but also to determine the optimum dose ratios for administering the peptides together to induce the synergistic release of GH. It is apparent that GH can be more optimally released when relatively higher doses of GHRH (1 µg/kg) are administered in combination with relatively lower doses of GHRP-2 (0.1 and 0.3 µg/kg). Until there is a better understanding of how the combination of GHRP-2 plus GHRH releases GH synergistically, it is not possible to speculate why 0.1 + 1.0, but not 1.0 + 0.1 µg/kg GHRP-2 plus GHRH synergistically releases GH in men, whereas in women, both combinations were synergistic.

To emphasize the unusual synergistic release of GH induced by GHRP-2 plus GHRH in certain normal individuals, select examples of these studies are shown in Figures 3–6. At the dose of 0.1 + 0.1 µg/kg (see Figs. 5 and 6) synergism could be demonstrated in selected individuals, but not when determined as a group. Of possible future importance will be to perform more detailed studies on normal subjects with unusual GH responses

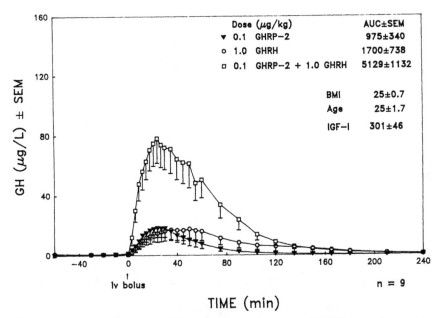

Figure 1 Synergistic release of GH induced by a low dose of GHRP-2 combined with a high dose of GHRH in normal young men. Units for figures are AUC, μg/L over 4 h; IGF, ng/mL.

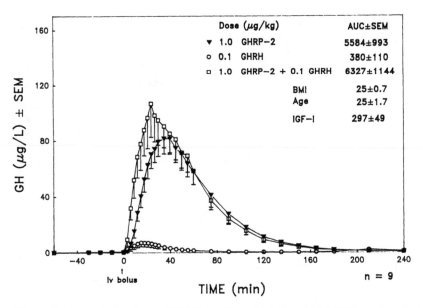

Figure 2 Synergistic release of GH induced by a high dose of GHRP-2 combined with a low dose of GHRH in normal young men.

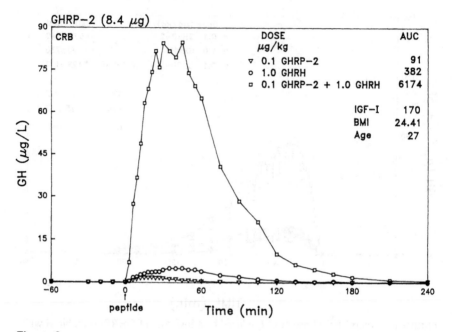

Figure 3

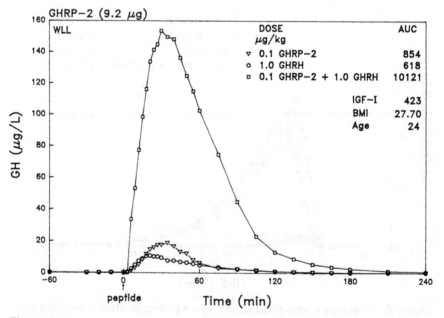

Figure 4

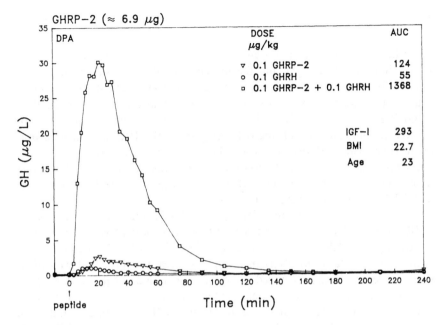

Figure 5

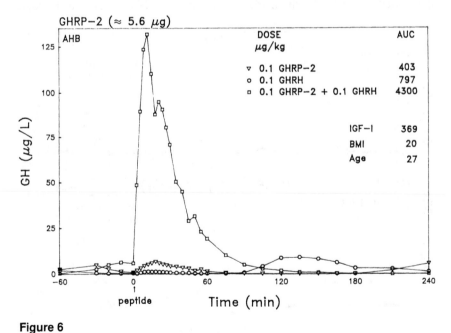

Figure 6

Figure 3–6 Unusual synergistic GH responses induced by GHRP-2 plus GHRH in select normal young subjects.

to further elucidate and understand the mechanisms involved in GH release and to impart new insight into the dimensions of neuroendocrine regulation of GH secretion. When some of these subjects were studied in more detail, one finding in common was that the GH response to the 1μg/kg dose of GHRH was low (< 1000 AUC), whereas administration of this dose of GHRH in combination with the 0.1 μg/kg dose of GHRP-2 induced a dramatic synergistic GH response (see Figs. 3 and 4). This latter result does not support that the low GH response to a maximal dose of GHRH is due to an increase of SRIF secretion. If SRIF had been increased, the low dose of GHRP-2 would not have induced synergism.

Further insight into the GHRP–GHRH relation was revealed when GHRP-6 or GHRH was continuously infused into normal young men (Fig. 7). During the GHRP-6 prolonged 24 or 36-h infusion by Huhn and Thorner et al. (30) and Jaffe and Barkan et al. (31), respectively, both sensitization and desensitization occurred. Increase in amplitude of the normal spontaneous GH pulses as well as enhancement of the GH response to intravenous (iv) bolus GHRH at the end of the GHRP-6 infusion indicated sensitization. Evidence of desensitization was demonstrated at the end of the GHRP-6 infusion by the marked decrease in the GH response to iv bolus

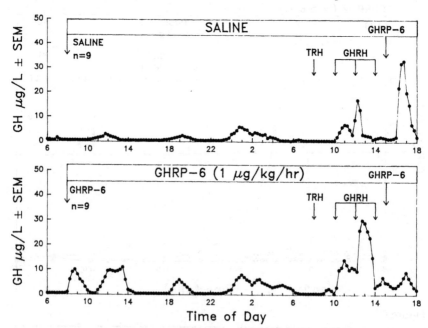

Figure 7 Effect of continuous 36-h infusion of saline or GHRP-6 in normal young men on the spontaneous pulsatile secretion of GH (From Ref. 31.)

GHRP-6. Infusion of GHRH for 8 h by Robinson and Barkan et al. (32) produced the opposite result, in that at the end of the GHRH infusion, the iv bolus GHRH GH response was markedly decreased, whereas the iv bolus GHRP GH response was increased. These results again clearly emphasize the independent actions of these two peptides on GH release. Whether the infusion of GHRP sensitizes the action of GHRH at each spontaneous GH secretory pulse at the time when GHRH is released is a provocative possibility. GHRP infusion did not appear to act by an effect on SRIF release or action because the SRIF-dependent low basal interpulse levels of GH, as well as the frequency of the GH pulse, remained essentially unchanged during the infusion. Because there was no effect on the frequency of the GH pulse, a direct pacemaker role for GHRP seems unlikely.

It is very possible that the GHRP plus GHRH synergistic release of GH in humans is the primary result of the hypothalamic action of GHRP, with the pituitary action being secondary. From a theoretical viewpoint, a seemingly valuable and meaningful finding is the dose-dependency of the effect of GHRP on the synergistic release of GH.

In normal young men, very low doses of GHRP-2 plus high doses of GHRH induced synergism, indicating that the increased release of endogenous GHRH is not producing the synergism. At the high dose of GHRP-2, large amounts of GH are released, suggesting endogenous GHRH release may be increased and synergism induced. Examples of these low- and high-dose effects of GHRP-2 on GH release are shown in Figures 8–11. Results in Figures 8 and 9 show that the very low, subthreshold dose of GHRP-2 ($\approx$2 µg or 0.03 µg/kg) in combination with a maximal dose of GHRH(1–44)NH$_2$ ($\approx$70 µg or 1 µg/kg), released GH synergistically. The results in Figure 10 show that the GH response to the high 10µg/kg GHRP-2 dose in comparison to the 1 µg/kg dose was much greater. As recorded in Figure 11, the high dose of GHRP-2 alone ($\approx$700 µg or 10 µg/kg) released essentially the same amount of GH that was released by GHRP-2 plus GHRH ($\approx$70 µg or 1 µg/kg of each peptide) administered together.

Notable is that the very low dose ($\approx$2 µg) of GHRP-2 probably does not induce synergistic release of GH by a pituitary action, but rather, a hypothalamic action. The effectiveness of a subthreshold dose and the mainly additive GH-releasing effects in vitro of GHRP plus GHRH both indirectly support that synergism is mediated by a hypothalamic action. Administration of high-dose GHRH in combination with low-dose GHRP-2 obviates the role of endogenous GHRH in the production of synergism. Because of this finding and the lack of experimental data to support that GHRP inhibits release of SRIF from the hypothalamus (i.e., agents known to inhibit SRIF release also enhance the GH response of GHRP), it has been concluded that GHRP does not induce synergism by releasing endogenous

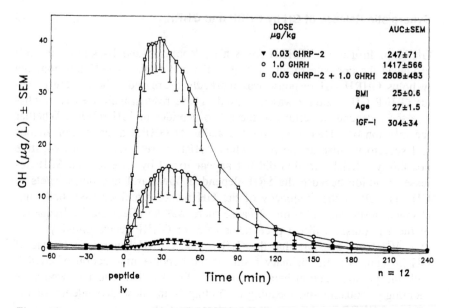

Figure 8

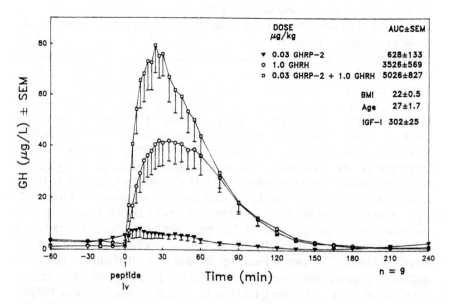

Figure 9

Figures 8–9 Effect of a very low dose of GHRP-2 (0.03 μg/kg) combined with a high dose of GHRH (1.0 μg/kg) on the synergistic release of GH in normal young men and women.

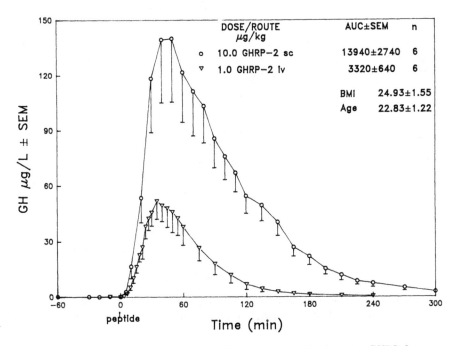

Figure 10 Comparative effects on the GH release to 1 μg/kg iv bolus GHRP-2 versus 10 μg/kg sc GHRP-2 in the same six normal young men.

GHRH, except at a high dose, or by inhibiting SRIF release. It is hypothesized that GHRP-2 acts on the median eminence of the hypothalamus to release the putative U-factor, or unknown factor. The synergistic release of GH, in turn, is mediated by the dual complementary pituitary action of U-factor plus GHRH. To what degree U-factor might play a physiological role versus only a pharmacological role is unknown.

In rat studies, the effective dose (ID_{50}) of SRIF to inhibit the GH response of GHRP-2 plus GHRH was much greater than for GHRP-2 or GHRH alone. These same studies were performed on cultured dispersed rat pituitary cells but the ID_{50} of SRIF for GHRP-2, GHRH, and GHRP-2 plus GHRH were all about the same in these in vitro studies (20). Because GHRP plus GHRH appeared to have a special effect in attenuating the SRIF inhibitory action on GH release in rats, a similar study was performed in normal young men by determining the effect of the SRIF agonist octreotide, to inhibit the GH responses to GHRP-2 and GHRH alone and together. In this study, 100 μg octreotide was administered subcutaneously (sc) 3 h before iv bolus 1 μg/kg GHRP-2, 1 μg/kg GHRP-2 plus GHRH or 10 μg/kg sc GHRP-2 (Fig. 12). Despite the much larger GH responses in the lat-

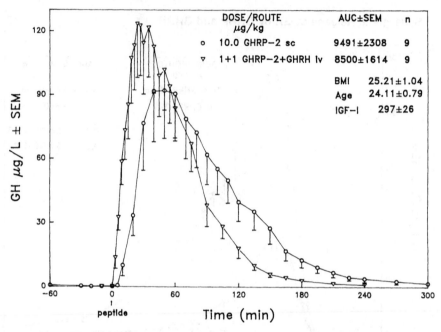

Figure 11 Comparative effects on the GH release to 1 µg/kg sc GHRP-2 versus 1 + 1 µg/kg iv bolus GHRP-2 + GHRH in the same nine normal young men.

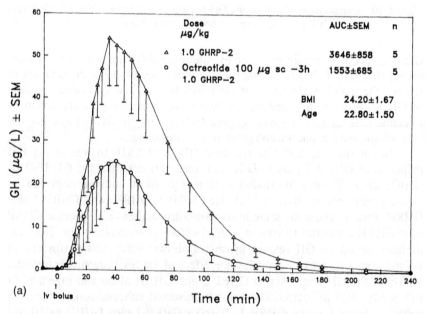

Figure 12 Inhibitory effect of octreotide on the GH responses to (a) 1 µg/kg iv bolus GHRP-2; (b) 10 µg/kg sc GHRP-2; and (c) 1 + 1 µg/kg iv bolus GHRP-2 plus GHRH.

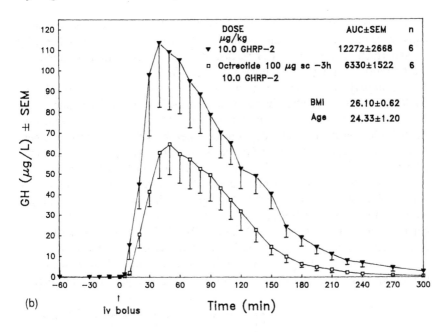

(b)

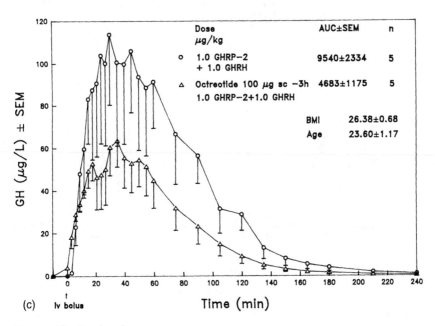

(c)

Figure 12 Continued

ter two groups, octreotide inhibited the GH response of all three groups equally, about 50%. Thus, the combined effect of GHRP-2 plus GHRH by its synergistic action(s) on GH release did not disproportionally attenuate the octreotide pituitary inhibitory action on GH release.

Comparison of the GH response to 1 µg/kg GHRP-2, GHRH, or GHRP-2 plus GHRH after iv bolus administration to normal younger and older men and women is shown in Figure 13. The GH responses were substantially less in the normal older subjects in each of the three groups. There was a disproportionally greater GH decrease in the women than in the men administered GHRH or GHRP-2 plus GHRH but not GHRP-2 alone.

In an attempt to learn more about the synergistic release of GH in normal older subjects and possibly develop more insight into the pathophysiology of the decreased GH secretion in these subjects, GH responses were compared after administration of 1 µg/kg GHRH, 0.1 and 1.0 µg/kg GHRP-2, as well as 1 and 0.1 µg/kg GHRH plus GHRP-2. In Table 4, the peak GH response to the 1 µg/kg dose of GHRH was only 2.7 µg/L while the 0.1 µg/kg dose of GHRP-2 was 0.7 µg/L. When 0.1 plus 1 µg/kg GHRP-

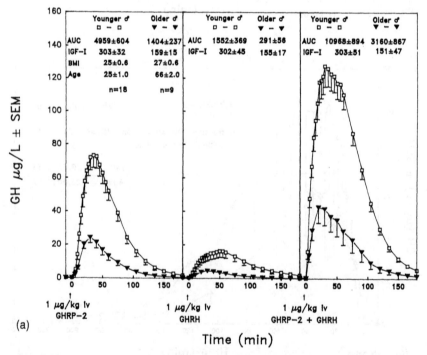

(a)

Figure 13 Comparative effects on the GH response to GHRP-2, GHRH, and GHRP-2 plus GHRH in normal younger and older (a) men and (b) women.

2 plus GHRH was administered, GH rose to a peak of 44 µg/L. GHRP-2 alone at 1 µg/kg induced a peak GH response of 47 µg/L. Results of another elderly subject in whom the same type of study was performed are shown in Table 5. Again, the same pattern of GH response was obtained. Three noteworthy findings from these studies are as follows. First, pituitary GH release is markedly impaired to even a maximal 1 µg/kg dose of GHRH. Second, GH release is substantially increased by the combined 0.1 plus 1.0 µg/kg dose of GHRP-2 plus GHRH. Third, the GH response to the 1 µg/kg dose of GHRP-2 alone is relatively high.

From the foregoing results the following conclusions have evolved about the pathophysiology of decreased GH secretion in some elderly subjects. The substantial release of GH by the 1 µg/kg dose of GHRP-2 alone indicates endogenous secretion of GHRH is occurring in elderly subjects with decreased GH secretion because the action of GHRP-2 on GH release in vivo depends on the presence of endogenous GHRH secretion, at least in a permissive way. An impaired GHRH pituitary GH response is probably the major abnormality, but this is a secondary, rather than a primary, pituitary abnormality. The insensitivity of the pituitary to the GH-releas-

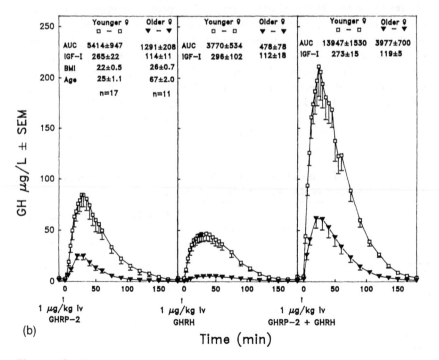

Figure 13 Continued

Table 4 Peak GH Concentrations in an Elderly 66-Year-Old Man with a Postulated Deficiency of the Putative GHRP-L[a] Hormone

Peptide	Dose/route (μg/kg iv bolus, sc[a])	Peak GH (μg/L)	AUC GH(μg/L·4 h)
GHRH	1.0	2.7	183
GHRP-2	0.1[a]	0.7	221
GHRH + GHRP-2	1.0 + 0.1	44.1	2436
GHRP-2	1.0	47.6	2540

IGF-I = 85 μg/L
BMI = 21.4
[a]GHRP-L, GHRP-like hormone

ing action of GHRH is reversed by GHRP-2 plus GHRH, with a low minimally effective or subthreshold GH-releasing dose of GHRP-2. Thus, the insensitivity of the GHRH action appears to be a functional abnormality that GHRP-2 can overcome. The high GHRP-2 sensitivity is against either an increased release or action of SRIF as an explanation for the decreased GH response to GHRH. Collectively, it has been concluded from results when using these two peptides that the decreased GH secretion in some elderly subjects may result from a deficiency of the putative native GHRP-like hormone. If so, this would be another reason for believing that a putative GHRP-like system exists that may be involved in the physiological regulation of GH release. Also, these and other studies indicate the hypothalamic action of GHRP that mediates the GHRP-2 plus GHRH synergism involves neither release of GHRH nor inhibition of SRIF release from the hypothalamus. Accordingly, GHRP-2 has been postulated to act on the

Table 5 Peak GH Concentration in an Elderly 69-Year-Old Woman with a Postulated Deficiency of the Putative GHRP-L[a] Hormone

Peptide	Dose/route (μg/kg) iv bolus	Peak GH (μg/L)	AUC GH (μg/L· 4 h)
GHRH	1.0	6.2	428
GHRP-2	0.1	4.2	185
GHRH + GHRP-2	1.0 + 0.1	33.0	1275
GHRP-2	1.0	37.3	2290

IGF-I = 111 μg/L
BMI = 25.8
[a]GHRP-L, GHRP-like hormone.

hypothalamus to release the putative U-factor or unknown factor. U-factor together with endogenous GHRH is envisioned to act on the pituitary to synergistically release GH.

Because of the complexity of the GH system, a new approach is needed to distinguish low secretion of GH associated with aging per se from the pathological decreased secretion of GH owing to a specific hormonal deficiency. Our present hypothesis is that the putative GHRP-like hormone is the primary hormone deficiency, and a decreased GHRH or excess SRIF secretion plays a secondary role. Currently, two diagnostic-linked indices of GH release, designated as a quantitative and a qualitative index, are being evaluated. For estimation of the quantitative index of GH secretion, the GH response to a 1μg/kg iv bolus GHRH and GHRP-2 has been selected. The 1 μg/kg GHRP-2 GH response assesses the pituitary capacity to release GH. However, because 1 plus 1μg/kg iv bolus GHRP-2 plus GHRH or 10 μg/ kg sc GHRP-2 alone induces such a large amount of GH release in normal younger men and women, eventually one of these doses may be considered more optimal to assess the maximal capacity of the pituitary to release GH. To what degree the pituitary GH capacity to release GH will parallel and determine the efficacy of a neuroendocrine therapeutic approach will need special consideration. The secretory status of endogenous GHRH, SRIF, and the pituitary somatotrophs alone and collectively will probably significantly dictate the type and design of neuroendocrine GHRP-2 therapeutic approach.

For the qualitative index of GH release, tentatively the 0.1 plus 1.0 μg GHRP-2 plus GHRH iv bolus has been initially selected to evaluate this index. How this may change in the future will depend on the range of the results and further understanding the pathophysiology of GH secretion in older men and women, as well as understanding the full spectrum of the in vivo actions of GHRP in humans. Additionally, it will be necessary to understand the degree to which endogenous GHRH and SRIF may be secondarily responsible for the decreased GH secretion. There is the possibility that GHRH or SRIF may play a primary, rather than a secondary, role in pathological decreased secretion of GH. The qualitative index is presently envisioned as an all-or-none indicator of the pathological decreased secretion of GH caused by a specific hormonal deficiency. Currently, if the quantitative index is abnormally low (i.e., the GHRH GH response is less than 6 μg/L to a 1μg/kg iv bolus GHRH) and the qualitative index reveals the synergistic release of GH (i.e., by the GH response to iv bolus 0.1 plus 1.0 μg/kg GHRP-2 plus GHRH), a decreased GH secretion would be considered to be pathological due to an abnormal functioning hypothalamic–pituitary neuroendocrine unit.

III. CONCLUSION

The finding that the synergistic release of GH induced by GHRP plus GHRH
can be elicited, regardless of age, sex, or species, underscores its broad
scope and possible basic importance. The uniqueness and unusualness of
this effect, especially in humans, imparts new dimensions on the release
of GH and appears to reflect possibly new hypothalamic regulators of GH
secretion. At this stage, the effect of the two peptides is obviously a pharmacological phenomenon. It will be a major challenge to elucidate just how
this occurs and the mechanisms involved to determine whether synergism
may also be involved in the physiological regulation of GH secretion. If
these relations and mechanisms eventually are found to be involved in the
physiological regulation of GH, this again, would imply that a putative
native GHRP-like hormone and system exists.

ACKNOWLEDGMENTS

This work was supported in part by National Institutes of Health grants AM-
06164 and PHS RR05096 (GCRC). I am greatly indebted to all of the fol-
lowing people: G. Burch, A.V. Schally, K. Folkers, G.A. Reynolds, F.
Momany, T. Badger, O. Sartor, M.O. Thorner, A. Barkan, C. Jaffe, C.
Pihoker, G. Merriam, F. Cassorla, V. Mericq, A. Tuilpakov, A. Bulatov,
Z. Laron, G. Van den Berghe, F. de Zegher, E. Adams, J. Veldhuis, Y.,
Wakiyama, H. Shibuya, N. Kajikawa, K. Uchida, B. Gonen, F. Wagner,
as well as to Kaken, Wyeth Ayerst, and BioNebraska, and many other
unnamed persons. In addition, special appreciation is expressed to Dr.
Granda-Ayala, to the technicians, and to the fellows of the Endocrinology
and Metabolism Section of the Department of Medicine, and to Robin
Alexander for typing the manuscript.

REFERENCES

1. Bowers CY, Chang JK, Fong TTW. A synthetic pentapeptide which specifically
    releases GH, in vitro. 59th Annual Meeting of the Endocrine Society. Chicago,
    IL, June 8-10, 1977.
2. Bowers CY. Xenobiotic growth hormone secretagogues. In: Bercu B, Walker R,
    eds. Growth Hormone Secretagogues. New York: Springer-Verlag, 1996:9-28.
3. Smith RG, Cheng K, Schoen WR, Pong SS, Hickey G, Jacks T, Butler B, Chan
    WWS, Chaung LYP, Judith F, Taylor J, Wyvratt MJ, Fisher MH. A novel
    nonpeptidyl growth hormone secretagogue. Science 1993; 260:1640-1643.

4. Bowers CY, Chang J, Momany F, Folkers K. Effects of the enkephalins and enkephalin analogs on release of pituitary hormones in vitro. In: MacIntyre I, Szelke M, eds. Molecular Endocrinology. Amsterdam. New York: Elsevier/ North-Holland Biomedical Press 1977:287–292.

5. Bowers CY, Momany F, Chang D, Hong A, Chang K. Structure–activity relationships of a synthetic pentapeptide that specifically releases GH in vitro. Endocrinology 1981; 106:663–667.

6. Bowers CY, Reynolds GA, Chang D, Hong A, Chang K, Momany F. A Study on the regulation of GH release from the pituitary of rats, in vitro. Endocrinology 1981; 108:1070–1079.

7. Bowers CY, Momany F, Reynolds GA. In vitro and in vivo activity of a small synthetic peptide with potent GH releasing activity. 64th Annual Meeting of the Endocrine Society. San Francisco, CA, June 16–18, 1982.

8. Momany FA, Bowers CY, Reynolds GA, Chang D, Hong A, Newlander K. Design, synthesis and biological activity of peptides which release growth hormone, in vitro. Endocrinology 1981; 108:31–39.

9. Momany F, Bowers CY, Reynolds GA, Hong A, Newlander K. Conformational energy studies and in vitro activity data on active GH releasing peptides. Endocrinology 1984;114:1531–1536.

10. Momany FA, Bowers CY. Computer-assisted modeling of xenobiotic growth hormone secretagogues. In: Bercu B, Walker R, eds. Growth Hormone Secretagogues. New York: Springer-Verlag, 1996:73–83.

11. Deghenghi, R. Growth hormone releasing peptides. In: Bercu B, Walker R, eds. Growth Hormone Secretagogues. New York: Springer-Verlag, 1996:85–102.

12. Elias, KA, Ingle GS, Burnier JP, Hammonds RG, McDowell RS, Rawson TE, Somers TC, Stanley MS, Cronin MJ. In vitro characterization of four novel classes of growth hormone releasing peptide. Endocrinology 1995; 136:5694–5699.

13. Wu D, Chen C, Katoh K, Zhang J, Clarke IJ. Effect of a new generation of growth hormone-releasing peptide (GHRP-2) on growth hormone secretion from ovine pituitary cells can be abolished by a specific GRF receptor antagonist. J Endocrinol 1994; 140:R9–R13.

14. Wu D, Chen C, Zhang J, Bowers CY, Clarke IJ. The effect of growth hormone-releasing peptide-6 (GHRP-6) and GHRP-2 on intracellular adenosine 3′,5′-monophosphate (cAMP) levels in ovine and rat somatotrophs. J Endocrinol 1996; 148:197–205.

15. Bowers CY, Momany F, Reynolds GA, Hong A. On the in vitro and in vivo activity of a new synthetic hexapeptide that acts on the pituitary to specifically release growth hormone. Endocrinology 1984; 114:1537–1545.

16. Bowers CY, Reynolds GA, Durham D, Barrera CM, Pezzoli SS, Thorner MO. Growth hormone releasing peptide stimulates GH release in normal men and acts synergistically with GH-releasing hormone. J Clin Endocrinol Metab 1990; 70:975–982.

17. Howard AD, Feighner SD, Cully DF, Arena JP, Liberator PA, Rosenblum CI, Hameline M, Hreniuk DL, Palyha OC, Anderson J, Paress PS, Diaz C, Chou M,

Liu KK, McKee, KK, Pong SS, Chaung LY, Elbrecht A, Dashkevicz M, Heavens R, Rigby M, Sirinathsinghji DJS, Dean DC, Melillo DG, Patchett AA, Nargund R, Griffin PR, DeMartino JA, Gupta SK, Schaeffer JM, Smith RG, Van der Ploeg LHT. A receptor in pituitary and hypothalamus that functions in growth hormone release. Science 1996; 273:974–977.

18. Pong SS, Chaung LYP, Dean DC, Nargund RP, Patchett AA, Smith RG. Identification of a new G-protein–linked receptor for growth hormone secretagogues. Mol Endocrinol 1996; 10:57–61.

19. Bowers CY, Sartor AO, Reynolds GA, Badger TM. On the actions of the growth hormone releasing hexapeptide GHRP. Endocrinology 1991; 128:2027–2035.

20. Bowers CY, Veeraragavan K, Sethumadhavan K. Atypical growth hormone releasing peptides. In: Bercu B, Walker R, eds. Growth Hormone II: Basic and Clinical Aspects. New York: Springer-Verlag 1994:203–222.

21. Dickson SL. Evidence for a central site and mechanism of action of growth hormone releasing peptide (GHRP-6). In: Bercu B, Walker R, eds. Growth Hormone II: Basic and Clinical Aspects. New York: Springer-Verlag 1994:237–251.

22. Dickson SL, Luckman SM. Induction of c-*fos* mRNA in NPY and GRF neurons in the rat arcuate nucleus following systemic injection of the growth hormone secretagogue, GHRP-6. Endocrinology 1997; 138:771–777.

23. Robinson ICAF. Regulation of growth hormone output: the GRF signal. In: Bercu B, Walker R, eds. Growth Hormone II: Basic and Clinical Aspects. New York: Springer-Verlag 1994:47–65.

24. Fairhall KM, Mynett A, Thomas GB, Robinson ICAF. Central and peripheral effects of peptide and nonpeptide GH secretagogues on GH release in vivo. In: Bercu B, Walker R, eds. Growth Hormone Secretagogues. New York: Springer-Verlag, 1996:219–236.

25. Adams EF, Lei T, Buchfelder M, Bowers CY, Fahlbusch R. Protein kinase C-dependent growth hormone releasing peptides stimulate cAMP production by human pituitary somatotrophinomas expressing GSP oncogenes: evidence for cross-talk between transduction pathways. Mol Endocrinol 1996; 10:432–438.

26. Cheng K, Chan WWS, Barreto A, Convey EM, Smith RG. The synergistic effects of His-D-Trp-Ala-Trp-D-Phe-Lys-NH$_2$ on growth hormone (GH) releasing factor-stimulated GH release and intracellular adenosine 3′,5′-monophosphate accumulation in rat primary pituitary cell culture. Endocrinology 1989; 124:2791–2798.

27. Smith RG, Pong SS, Hickey G, Jacks T, Cheng K, Leonard R. Modulation of pulsatile GH release through a novel receptor in hypothalamus and pituitary gland. Recent Prog Horm Res 1996; 51:261–286.

28. Reynolds GA, Momany FA, Bowers CY. Synthetic tetrapeptides that release GH synergistically in combination with GHRP and GHRH. 73rd Annual Meeting of the Endocrine Society. Washington, DC, June 19–22, 1991.

29. Bitar KG, Bowers CY, Coy DH. Effects of substance P-/bombesin antagonist on the release of growth hormone by GHRP and GHRH. Biochem Biophys Res Commun 1991; 180:156–161.

30. Huhn WC, Hartman ML, Pezzoli SS, Thorner MO. 24-h growth hormone (GH)-

releasing peptide (GHRP) infusion enhances pulsatile GH secretion and specifically attenuates the response to a subsequent GHRP bolus. J Clin Endocrinol Metab 1993; 76:1201–1208.

31. Jaffe CA, Ho J, Demott-Friberg R, Bowers CY, Barkan AL. Effects of a prolonged growth hormone (GH)-releasing peptide infusion on pulsatile GH secretion in normal men. J Clin Endocrinol Metab 1993; 77:1641–1647.

32. Robinson BM, Friberg RD, Bowers CY, Barkan AL. Acute growth hormone (GH) response to GH-releasing hexapeptide in humans is independent of endogenous GH-releasing hormone. J Clin Endocrinol Metab 1992; 75:1121–1124.

2

Structural Requirements of Growth Hormone Secretagogues

Romano Deghenghi
Europeptides, Argenteuil, France

I. INTRODUCTION

The discovery of growth hormone-releasing peptides (GHRPs), progenitors of more recent growth hormone secretagogues (GHSs) that are not limited to peptidyl structures, was a fortuitous and serendipitous event, for which we are indebted to Bowers and his collaborators (1). The existence of the hypothetical endogenous ligand (Bowers' U-factor) has virtually been confirmed by the isolation and cloning of a receptor from pituitary and hypothalamic tissues that function in growth hormone (GH) release (2). More recently, strong but unconfirmed evidence has come to light, indicating the existence of several peripheral receptors that bind some, but not all, secretagogues (3,4) and cloud the issue of determining the structural requirements of GHS by opening a Pandora's box of hitherto unsuspected properties attributable to the endogenous ligand that GHSs are supposed to mimic.

II. THE BOWERS–MOMANY LEGACY

Apart from discovering the GHRP series of secretagogues, Bowers and Momany have laid the foundation that advanced our understanding of the biology of these products and have established the basic relation between their structure and their GH-releasing properties (5). It is noteworthy that

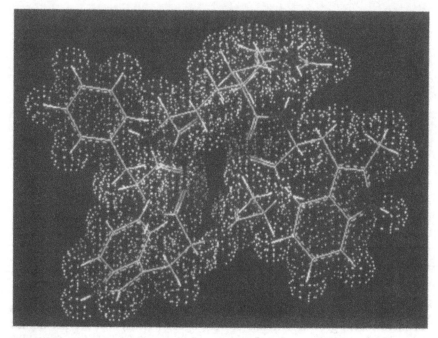

Figure 1 Structure of hexarelin.

it took 4 years of work to arrive at the first GHRP that was active in vivo (the hexapeptide, GHRP-6). The in vivo activity was maintained with the heptapeptide GHRP-1, but there was obviously no motivation to lengthen the chain and explore the activity of say nona- or decapeptides, which would have been more expensive and, presumably, metabolically less stable. No explanation was offered for why peptides smaller than GHRP-6 or GHRP-2 were devoid of in vivo activity, even if they were potent releasers of GH in vitro. Possible explanations include appropriate solubility and metabolic stability. Momany (5) has clearly pointed out that nuclear magnetic resonance (NMR) studies on GHRP-6 and related analogs indicate the formation of an energetically favorable folded structure, and our computer simulation of the hexapeptide hexarelin* (Fig. 1) likewise shows a β-turn, almost cyclic structure, with the carbonyls of the peptide bonds on the inside, reminiscent of a musk-ox configuration. The peptide bonds are thus less accessible to peptidases, resulting in metabolically stable secretagogues, in addition to the inherent stability imparted by the D-configuration of some substituents.

---

*Hexarelin is a trademark for examorelin (I.N.N.).

Experimentally the metabolic stability of GHRP-6 (SK&F 110679) or of hexarelin has been confirmed at least in the rat from which more than 50% of these peptides can be recovered unchanged in the bile following their subcutaneous (sc) administration. This observation prompted the SK&F group to observe that GHRP-6 "was not designed with metabolic stability in mind [but] it is tempting to speculate that the structural features that are important for receptor binding and pharmacological activity of these peptides may also confer metabolic stability, protecting them from degradation by peptidases" (6). We propose the term *impervious peptides* to describe the metabolic stability characteristic of this series of secretagogues.

III. MORE RECENT GHRPs

Hexarelin (a 2-Me-D-Trp analog of GHRP-6) was developed to take advantage of the increased chemical stability of 2-Me-Trp (Mrp) versus the unsubstituted Trp (7). In humans, hexarelin is equipotent to GHRP-2 (8), and in vitro-binding studies by a group at Merck (9) have shown increased binding and higher biological potency of hexarelin than those of GHRP-6. We have continued our synthetic efforts, maintaining our privileged Mrp substituent as a constant substituent, and have prepared a series of downsized hexarelin analogs (Table 1) with satisfactory GH-releasing properties when given sc to normal infant rats (10). This animal model has given reproducible results in our hands, but it has limited usefulness owing to its high sensitivity. Some of our most active analogs have thus been retested in dogs by parenteral and oral routes to confirm the initial screening assay in the infant rat (Fig. 2). Preliminary results with the pentapeptide EP 51216 indicate oral activity in humans (normal volunteers) without cortisol stimulation (11). The Genentech group has disclosed (12) a series of potent GHRPs, including the cyclic analog G-7203 and a Novo Nordisk group has published the structures of analogs with an aminomethylene-substituted peptide bond (13).

IV. NONPEPTIDE SECRETAGOGUES

With few exceptions, peptides are generally not active when given orally. Even impervious peptides resistant to metabolic degradation are poorly absorbed in the gastrointestinal tract either because they are intensely solvated through hydrogen bonding with the aqueous medium, or because of their size. GHRP-6, GHRP-2, and hexarelin are active orally, but their

Table 1 Activity of Peptides (300 μg/kg, sc) in the 10-Day Rat Model
(GH at 15 min)

Inactive (not different from controls)	
L-164,080	(Aib-D-Trp-D-HomoPhe-OEt)
EP 50887	(TXM-D-Mrp-D-βNal-Phe-Lys-NH$_2$)
EP 51322	(GAB-D-Mrp-D-βNal-NH$_2$
EP 51343	(Aib-D-Ser(Bzl)-D-Mrp-NH$_2$)
EP 60021	(D-Mrp-D-Mrp-NH$_2$)
EP 60022	(GAB-D-Mrp-D-Mrp-NH$_2$)
EP 60260	(D-Mrp-D-Mrp-Phe-NH$_2$)
Weakly Active (GH range 30–40 ng/mL)	
EP 51321	(GAB-D-Mrp-D-βNal-OEt)
EP 60261	(D-Mrp-D-Mrp-Mrp-NH$_2$)
EP 60274	(GAB-D-Mrp-Mrp-NH$_2$)
EP 60275	(D-Mrp-Mrp-NH$_2$)
Active (GH range 50–90 ng/mL)	
EP 60761	(GAB-D-Mrp-D-Mrp-D-Mrp-Lys-NH$_2$)
EP 41616	(IMA-D-Mrp-D-Trp-Phe-Lys-NH$_2$)
EP 41617	(IMA-D-Mrp-D-βNal-Phe-Lys-NH$_2$)
EP 51390	(Aib-D-Mrp-Mrp-NH$_2$)
MK 677	(cf. Ref. 16)
Very active (GH range 100–150 ng/mL)	
Hexarelin	(His-D-Mrp-Ala-Trp-D-Phe-Lys-NH$_2$)
GHRP-2	(D-Ala-D-β-Nal-Ala-Trp-D-Phe-Lys-NH$_2$)
G 7509	(INIP-D-β-Nal-D-Trp-Phe-Lys-NH$_2$)
G 7039	(INIP-D-β-Nal-D-βNal-Phe-Lys-NH$_2$)
EP 41614	(INIP-D-Mrp-D-Trp-Phe-Lys-NH$_2$)
EP 41615	(INIP-D-Mrp-D-βNal-Phe-Lys-NH$_2$)
EP 50477	(GAB-D-Mrp-D-Trp-Phe-Lys-NH$_2$)
EP 50886	(TXM-D-Mrp-D-Trp-Phe-Lys-NH$_2$)
EP 51215	(GAB-D-Mrp-D-Mrp-Phe-Lys-NH$_2$)
Most active (GH range 160–200 ng/mL)	
EP 50885	(GAB-D-Mrp-D-β-Nal-Phe-Lys-NH$_2$)
EP 51216	(GAB-D-Mrp-D-Mrp-Mrp-Lys-NH$_2$)
EP 51389	(Aib-D-Mrp-D-Mrp-NH$_2$)

INIP, isonipecotinyl; IMA, imidazolylacetyl; GAB, γ-aminobutyryl; TXM, tranexamyl; 4-(aminomethyl)-cyclohexanecarbonyl; Mrp, 2-methyl-Trp; Aib, α-aminoisobutyryl; β–Nal, β-(2-naphthyl)alanine.

absorption rarely exceeds 1% of the administered dose. The concern here is a potential lack of reproducibility, rather than the actual amount, which, owing to the high potency of these agents, does not represent an economic obstacle. Indeed, other classes of useful drugs, such as the bisphos-

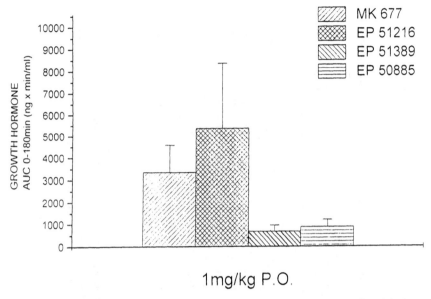

Figure 2 The most active peptides in the infant rat model were tested in adult dogs.

phonates, are notoriously poorly absorbed by the oral route (14). Nevertheless, a reproducible oral absorption accompanied by a reasonable potency, is an attractive feature, and this has prompted the now well-known effort of Merck to develop nonpeptide secretagogues such as L-692,429, L-692,585, and L-700,653 (15).

The more recent L-163,191 (MK 677) has a hybrid structure being two-thirds peptidic and having the spiroindoline "privileged" structure mimicking a tryptophan. MK 677 has been considered a peptidomimetic of GHRP-6, and its labeled [^{35}S]sulfonamide radioligand provided the key breakthrough in identifying saturable, specific, and high-affinity–binding sites in hypothalamic and pituitary tissues (2,16). The GHRP-6-based earlier radioligands were not considered suitable to identify specific-binding sites for GHSs (9). What are we trying to mimic? GHRP-6, or the putative natural ligand(s)? Can we anticipate the structure of such ligands based on present available information? One is tempted to assume a similar hydrophobic domain common to both the putative ligand and some of the most active GHRPs. It is also not a coincidence that, as the Genentech group has pointed out (17), the binding of human GH (hGH) to its receptor is concentrated on two tryptophans, (Trp-104 and Trp-169) of the extracellular binding protein, that fit in a hydrophobic cleft of hGH. Clearly, studies on the possible interaction of GHRPs and GH are urgently needed.

Table 2 Inhibition of ^{125}I-Tyr-Ala-Hexarelin Binding to Human Heart Membranes[a] by Hexarelin, EP-50885, and EP-51389

	Human Heart					
	Specific binding of ^{125}I-Tyr-Ala-hexarelin/0.1 mg membrane protein expressed as % of total radioactivity added					
	Hexarelin		EP-50885		EP-51389	
Peptide concentration (nM)	Subject 1	Subject 2	Subject 1	Subject 2	Subject 1	Subject 2
0	21.0 (0)	23.9 (0)	20.3 (0)	24.0 (0)	25.1 (0)	24.6 (0)
0.1	20.1 (4)	22.8 (5)	20.1 (1)	23.8 (1)	25.1 (0)	24.5 (0)
1	16.9 (20)	17.3 (28)	19.8 (3)	23.5 (2)	24.9 (1)	24.5 (0)
10	10.4 (51)	10.5 (56)	16.5 (19)	19.2 (20)	24.9 (1)	24.0 (3)
100	1.7 (92)	2.4 (90)	10.2 (50)	11.2 (53)	24.7 (2)	24.1 (2)
1000	0 (100)	0 (100)	4.0 (80)	4.0 (83)	24.4 (3)	24.0 (3)
IC$_{50}$ value (nM)	10.6	8.3	62.7	57	Inactive	Inactive
Mean IC$_{50}$ value (nM)	9.5		59.8		—	
Comparative IC$_{50}$ value (Hexarelin = 1)	1		6		—	

[a]Tissue membranes (0.1 mg protein) obtained from two different adult male subjects were incubated in triplicate with a subsaturating concentration (34 pM, about 48,000 cpm) of ^{125}I-Tyr-Ala-hexarelin for 40 min at 0°C in the absence and in the presence of increasing concentrations of the indicated unlabeled peptides. The value in parentheses represents the % of inhibition of ^{125}I-Tyr-Ala-hexarelin specifically bound.

V. RECEPTORS FOR GH SECRETAGOGUES: HOW MANY?

The GH secretagogues are reasonably specific in their biological action, at least in normal subjects, but they do liberate, in a dose-related fashion, other hormones, such as corticotropin (adrenocorticotropic hormone; ACTH), cortisol, and prolactin; also an effect on appetite and sleep has been noted (18,19). In certain pathological situations, some of these effects are greatly enhanced, such as cortisol release in Cushing's syndrome (20). GH was recently found promising in dilated cardiomyopathy patients (21), a finding that prompted a pharmacological evaluation of GHRPs in GH-deficient rats. Hexarelin prevents the aggravation of ischemic reperfusion damage in hearts of such rats (22), and heart tissues were examined for binding properties of hexarelin, a related pentapeptide EP 50885, and the tripeptide EP 51389.

Surprisingly, the tripeptide does not bind to heart tissues (as measured by displacement of ^{125}I-Tyr-Ala-hexarelin), although binding was observed in hypothalamic tissues (23; Table 2). If confirmed, these findings will presumably be helpful in the development of organ-specific secretagogues.

VI. CONCLUSIONS

Bowers and Momany have laid the foundation for several exciting studies involving peptide and nonpeptide synthetic secretagogues. Future developments will include the enlargement of the GHS receptor family, the isolation and characterization of the putative endogenous ligands and the discovery of additional organ-specific secretagogues. It is possible that this new family of secretagogues, by acting on peripheral receptors, will elicit even greater attention of endocrinologists and pharmacologists in their quest for novel therapeutic indications.

ACKNOWLEDGMENTS

I am deeply indebted to Professors Eugenio Müller, Vittorio Locatelli, and co-workers at the University of Milan for most of the animal work done with the novel peptides described in the foregoing. I acknowledge the outstanding contributions from Professor Gianpiero Muccioli, University of Turin and of Professor Huy Ong, University of Montreal, for their important binding studies in human and animal tissues. My colleagues at Europeptides in France, François Boutignon, Hélène Touchet, Sandrine David, and Edith Barré have given much of their time and ability to our project. I am particularly indebted to Professors Ezio Ghigo and Franco Camanni, and their team at the University of Turin for their innovative, competent, and enthusiastic contributions for both basic and clinical aspects of this project.

REFERENCES

1. Bowers CY. Xenobiotic growth hormone secretagogues: growth hormone releasing peptides. In: Bercu BB, Walker RF, eds. Growth Hormone Secretagogues. New York: Springer-Verlag, 1996:9–28.
2. Howard AD, et al. A receptor in pituitary and hypothalamus that functions in growth hormone release. Science 1996; 273:974–923.
3. Muccioli G, et al. Presence of specific receptors for hexarelin in human brain and pituitary gland. 20th International Symposium on Growth Hormone Growth Factors. Sept 29–30. Endocrinol Metab 1995. Abstr El. p. 110.
4. Muccioli G. Personal communication, Nov 26, 1996.
5. Momany FA, Bowers CY. Computer-assisted modelling of xenobiotic growth hormone secretagogues. In: Bercu BB, Walker RF, eds. Growth Hormone Secretagogues. New York: Springer-Verlag, 1996:73–83.
6. Davis CB, et al. Disposition of growth hormone-releasing peptide (SK&F 110679) in rat and dog following intravenous or subcutaneous administration. Drug Metab Dispos 1994; 22:90–98.
7. Deghenghi R. Growth hormone releasing peptides. In: Bercu BB, Walker RF, eds. Growth Hormone Secretagogues. New York: Springer-Verlag, 1996:85–102.
8. Arvat E, et al. Effects of GHRP-2 and hexarelin, two synthetic GH-releasing peptides, on GH, prolactin, ACTH and cortisol levels in man. Comparison with the effects of GHRH, TRH and hCRH. Peptides 1997; 18:885–891.
9. Pong SS, et al. Identification of a new G-protein-linked receptor for growth hormone secretagogues. Mol Endocrinol 1996; 10:57–61.
10. Deghenghi R, Boutignon F, Luoni M, Grilli R, Guidi M, Locatelli V. Characterization of the activity of new growth hormone secretagogues in the infant rat. 10th International Congress of Endocrinology, June 12–15, 1996, San Francisco, CA. Abstr P1–581.
11. Deghenghi R, et al. Small impervious peptides as GH secretagogues. 2nd International Symposium on Growth Hormone Secretagogues, Feb 13–16, 1997, Tampa, FL.
12. Elias KA, et al. In vitro characterization of four novel classes of growth hormone-releasing peptides. Endocrinology 1995; 136:5694–5699.
13. Langeland Johansen N, Hansen Sehested B, Klitgaard H, Ankersen M. Structure activity relationship of GHRP analogs. 10th International Congress of Endocrinology, June 12–15, 1996, San Francisco, CA. Abstr P1–586.
14. Lin JH. Pharmacokinetic properties of bisphosphonates. Bone 1996; 18:75–85.

15. Chen MH, et al. Analogs of the orally active growth hormone secretagogue L-162,752. Biorg Med Chem Lett 1996; 6:2163–2168.
16. Dean DC, et al. Development of a high specific activity sulfur-35-labelled sulfonamide radioligand that allowed the identification of a new growth hormone secretagogue receptor. J Med Chem 1996; 39:1767–1770.
17. Clackson T, Wells JA. A hot spot of binding energy in a hormone–receptor interface. Science 1995; 26:383–386.
18. Frieboes RM, Murck H, Maier P, Schier T, Holsboer F, Steiger A. Growth hormone releasing peptide-6 stimulates sleep, growth hormone, ACTH and cortisol release in normal man. Neuroendocrinology 1995; 61:584–589.
19. Locke W. Kirgis HD, Bowers CY, Abdoh AA. Intracerebroventricular growth-hormone-releasing peptide-6 stimulates eating without affecting plasma growth hormone responses in rats. Life Sci 1995; 56:1347–1352.
20. Ghigo E, Arvat E, Ramunni J, Colao AM, Gianotti L, Deghenghi R, Lombardi G, Camanni F. ACTH- and cortisol-releasing effect of hexarelin, a synthetic GHRP, in normal subjects and in patients with Cushing's syndrome. J Clin Endocrinol Metab 1997; 82(8):2439–2444.
21. Fazio S, et al. A preliminary study of growth hormone in the treatment of dilated cardiomyopathy. N Engl J Med 1996; 334:809–814.
22. De Gennaro Colonna V, Rossoni G, Bernareggi M, Müller E, Berti F. Hexarelin prevents the aggravation of ischemic-reperfusion damage in hearts from rats with selective growth hormone deficiency. Eur J Pharmacol 1997. In press.
23. Muccioli G. University of Turin, Italy. Personal communication, Dec. 18, 1996.

3

Unifying Mechanism of a Non-GHRH Growth Hormone Secretagogue's Action

Roy G. Smith
Merck Research Laboratories, Merck and Co., Inc., Rahway, New Jersey

I. INTRODUCTION

Pulsatile GH release from the pituitary gland is thought to be regulated by episodic changes in two hypothalamic hormones, growth hormone-releasing hormone (GHRH) and the inhibitory hormone, somatostatin. The discovery of synthetic growth hormone-releasing peptides (GHRPs) by Bowers and co-workers (1–5) that also stimulate GH release was an important fundamental observation. It demanded continued evaluation of whether the physiology of GH release was governed by more than just GHRH and somatostatin and whether GHRPs were mimicking an unknown natural hormone that also played a role. With this in mind, in 1987, the Merck group focused on this issue for two reasons. First, understanding this pathway would likely lead to new strategies that could be exploited for the treatment of GH deficiencies. Second, because GHRPs were small molecules, they might be appropriate templates for the design of nonpeptide analogs. Peptidomimetics have an important advantage over peptides because their structures are more readily modified to allow optimization of oral bioavailability and pharmacokinetic properties. Extensive investigations of the biological properties of GHRPs and peptidomimetics lead one to conclude that these molecules mimic a natural hormone. Indeed, most recently, the cloning and sequencing of the receptor for these novel GH secretagogues (GHSs) showed the receptor to be a highly conserved G–protein-coupled receptor that apparently does not belong to any of the known families of G–protein-coupled receptors. Because of its apparent pivotal role in modulating

GH release across a variety of species, identification of the natural hormone mimicked by this class of GHSs now becomes critically important to our understanding of the physiology of GH release.

II. MECHANISM OF ACTION IN SOMATOTROPHS

The early studies of Cheng et al. demonstrated that GHRP-6, in contrast to GHRH, appeared to signal through the phospholipase C (PLC) pathway (6). They showed that the effects of GHRP-6 were mimicked, at least partially, by activators of protein kinase C (PKC) and that the effects of GHRP-6 were antagonized by a protein kinase C inhibitor (6). Moreover, prolonged exposure of rat anterior pituitary cells to an activating phorbol ester almost completely desensitized the cells to GHRP-6, without affecting their response to GHRH. These data were consistent with suggestions that the receptors for GHRH and GHRP-6 were distinct and that the GHRP-6 receptor was G–protein-coupled and transduced its signal through phospholipase C. From these conclusions and from a consideration of the structural features, the Merck group searched for a pharmacophore that would mimic either GHRP-6 or GHRH by screening a library of nonpeptidyl "privileged structures" that were known to bind to G–protein-coupled receptors (7). With this approach, the first potent nonpeptide GH secretagogue, L-692,429, that increased GH release from rat pituitary cells was discovered (8). As shown in Figure 1 the (S)-enantiomer of L-692,429, L-692,428,

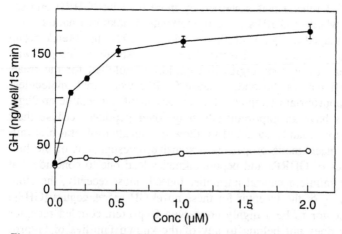

Figure 1   Dose-dependent increase in GH secretion from rat pituitary cells in response to L-692,429 but not L-692,428 demonstrating the stereoselectivity of the response. (From Ref. 8.)

was inactive, demonstrating the stereoselectivity of the response (8). Additional studies based on mechanism of action and the use of selective antagonist demonstrated that L-692,429 was a peptidomimetic of GHRP-6, rather than a mimetic of GHRH (8). Through a combination of medicinal chemistry and selection of privileged structures, other GHRP-6 peptidomimetics were identified (Fig. 2); because of its high oral bioavailability, MK-0677 was selected as a clinical candidate (9).

The conclusions of the earlier mechanistic studies of Cheng et al. suggesting that the GHRP-6 signal was transduced through phospholipase C (6) were supported by more direct evidence. Herrington and Hille (10) and Bresson-Bepoldin and Dufy-Barbone (11) reported that GHRP-6 stimulated Ca^{2+} release from intracellular stores. Adams et al. (12) showed that both GHRP-6 and the nonpeptide mimetic L-692,429 (8) caused increased in inositol triphosphate (IP_3) and Mau et al. demonstrated translocation of protein kinase C (13). These observations strengthened the notion that GHRP-6 and L-692,429 interacted with a G-protein-coupled receptor that

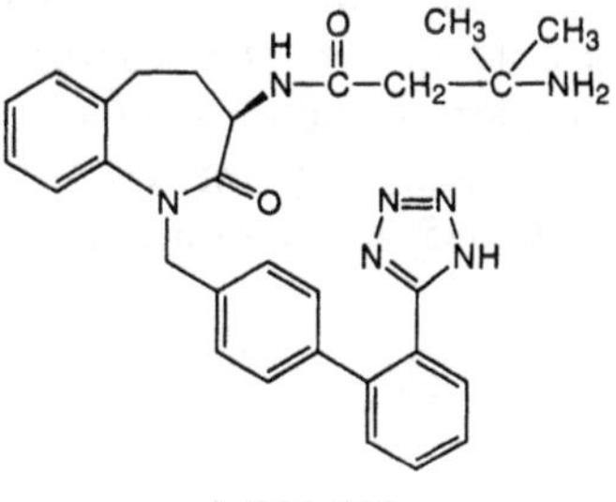

L-692,429

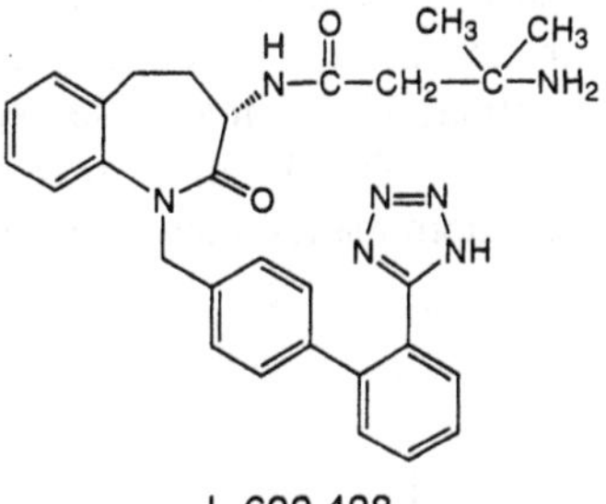

L-692,428

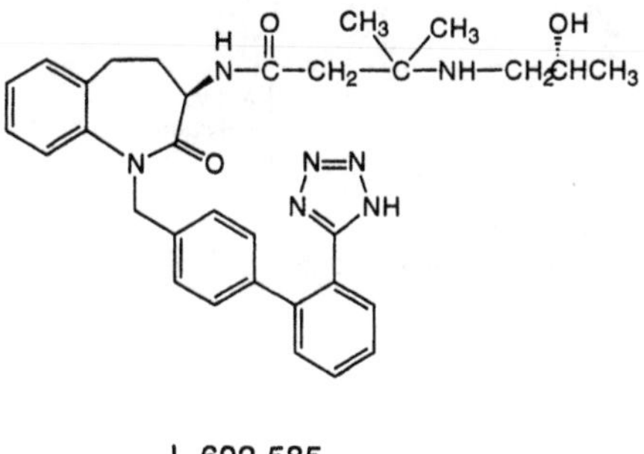

L-692,585

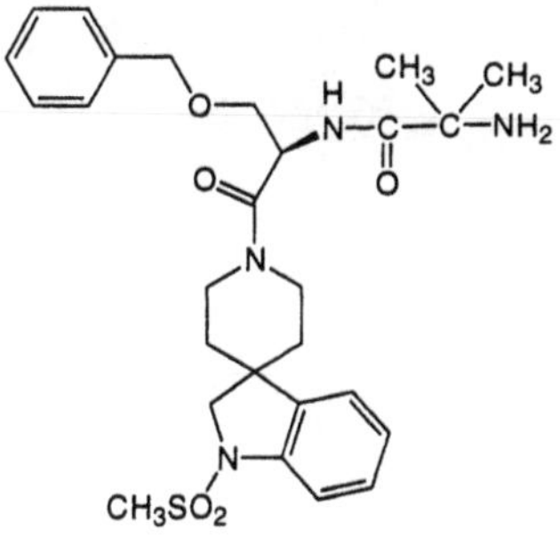

MK-0677

Figure 2 Structural variants of mimetics of GHRP-6 that with the exception of L-692,428, the *S* = enantiomer of L-692,429, are potent GH secretagogues.

signaled through phospholipase C, liberating the second messengers IP_3 and diacylglycerol (Fig. 3). Studies with the clinical candidate MK-0677 showed it behaved similarly (14,15). Figure 4 contrasts the distinct signal transduction pathways activated by GHRH and MK-0677; GHRH transduces its signal through cAMP and MK-0677 by redistribution of intracellular Ca^{2+} following phospholipase C activation (15).

Characterization of a high-affinity–binding site in the pituitary gland by ligand-binding assays was problematic in our hands owing to relatively low-affinity, high-capacity binding of radiolabeled GHRP-6. In our search for peptidomimetics of GHRP-6 that had high oral bioavailability (> 50%) we identified MK-0677 (14). To characterize the receptor Dean et al. (16) synthesized radiolabeled MK-0677 by substituting [^{35}S] in place of [^{32}S]. This radiolabeled secretagogue had very high specific activity (700–1100 Ci/mol) and was ideally suited for investigations designed to identify high-affinity binding to pituitary membranes. Pong et al. (17) demonstrated, by Scatchard plot analysis, high-affinity (K_d = 140 pM), limited-capacity binding (B_{max} = 6 fmol/mg protein) to pig pituitary gland membranes (Fig. 5a) and to rat anterior pituitary membranes (K_d = 200 pM; B_{max} = 2 fmol/mg protein). The low concentrations of these binding sites is remarkable and is about 1% of that observed for other receptors found on pituitary membranes. The binding was highly selective, and only the GHRPs and peptidomimetics such as L-692,429 and L-692,585 that were active in caus-

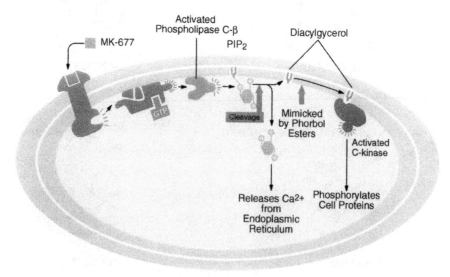

Figure 3 Mechanism of action of MK677 on somatotrophs illustrating the signal transduction pathway. (From Ref. 15.)

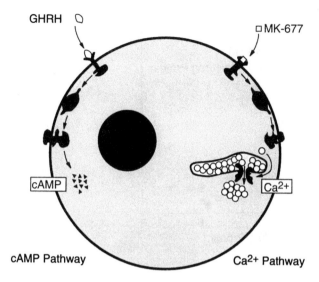

Figure 4 Distinction between the initial steps in the signal transduction pathways in rat somatotrophs following binding of GHRH and MK-0677 to their respective G–protein-coupled receptors. GHRH causes increases in intracellular cAMP whereas MK-0677 has no affect on cAMP, but causes the release of Ca^{2+} from intracellular stores. (From Ref. 15.)

ing GH release-displaced [^{35}S]MK-0677 binding to anterior pituitary membranes. Their efficacy in displacing MK-0677 binding correlated well with concentrations required to stimulate GH release (17). Figure 5b illustrates that the peptide GHRP-6 is a competitive inhibitor of [^{35}S]MK-0677 binding (15). Binding was Mg^{2+}-dependent and was markedly inhibited by nonhydrolyzable analogs of GTP, but not of ATP (17). These properties further supported the results of the foregoing studies on the signal transduction pathway, illustrating that GHRP-6, L-692,429, and MK-0677 probably mediated their effects on GH release through a receptor belonging to the superfamily of serpentine receptors that couple to G proteins. To determine whether binding sites could be found in other rat tissues, membranes were prepared from the hypothalamus, posterior pituitary, liver, and pons. High-affinity [^{35}S]MK-0677 binding was detected only in the hypothalamus, and the concentration of binding sites (8 fmol/mg protein) was higher than that measured in the anterior pituitary membranes.

Influx of extracellular Ca^{2+} is a common signal for the release of hormones stored in secretory granules. GHRH-induced GH release is blocked by Ca^{2+} channel blockers. Therefore, it was important to investigate the role of ion channels that might be involved in the action of GHRP-

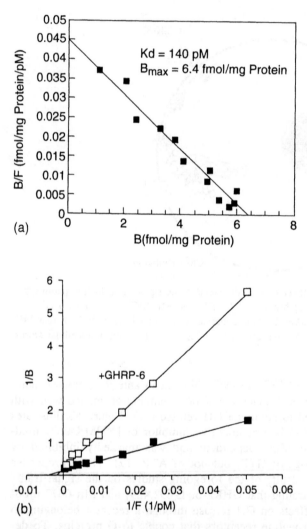

Figure 5   (a) Scatchard plot analysis of binding of [^{35}S]MK-0677 to pig pituitary membranes demonstrating a single class of high affinity binding sites. (b) Double reciprocal plot of [^{35}S]MK-0677 binding to pit pituitary membranes in the presence of GHRP-6 demonstrating that GHRP-6 binds to the same receptor and is a competitive inhibitor of [^{35}S]MK-0677 binding. (From Refs. 15 and 17.)

6, L-692,429, and MK-0677. The role of Ca^{2+} in the signaling pathway was demonstrated using fura-2 to monitor changes in fluorescence in somatotrophs following treatment with GHRP-6, L-692,429, and MK-0677 (8,9,15). Use of selective inhibitors showed that the effects on fura-2 fluorescence and GH release were dependent on activation of L-type Ca^{2+} channels on somatotrophs. These results also revealed that, although the receptors, the signal transduction pathways, and the second messengers activated by the GHSs were different from those activated by GHRH, the pathways converge so that the pivotal event associated with GH release is influx of Ca^{2+} through L-type Ca^{2+} channels.

Cheng et al. demonstrated that the dose–response curve for the inhibitory effects of somatostatin on GHRH-stimulated growth hormone was shifted to the right by GHRP-6 (6). Similarly, Blake and Smith illustrated that GHRP-6 behaved as a functional antagonist of somatostatin (18). Because somatostatin acts by inhibiting increases in cAMP and causes hyperpolarization of the plasma membrane by opening K^+ channels and inhibiting Ca^{2+} uptake, it was speculated that the GHRP-6 and peptidomimetics functionally antagonize somatostatin by inhibiting K^+ channels and opening Ca^{2+} channels. Indeed, K^+ antagonists, such as tetramethylammonium and 4-aminopyridine, in common with the GHSs that bind to the MK-0677-binding site, augmented GH release induced by GHRH. However, they did not increase GH release in the presence of saturating concentrations of GHRP-6 (9). Identical observations were made when the pituitary cells were depolarized by treatment with the Na^+ channel agonist veratridine (9). Direct evidence for the depolarizing properties of the L-692,429 and MK-0677 was shown by use of the membrane–potential-sensitive dye bisoxanol (8,14). To directly investigate the effects on K^+ channels, electrophysiological studies were performed on somatotrophs using both perforated patch and on-cell single-channel recording. The GHSs blocked K^+ currents, resulting in depolarization and electrical spiking to enhance Ca^{2+} entry through voltage-gated channels (9,19–22). Interestingly, modulation of these channels was not observed when the somatrotrophs were dialyzed using the whole-cell voltage clamp configuration, consistent with the involvement of a soluble second messenger. Although, depolarizing agents were able to mimic the effects of the GHRP-6 and peptidomimetics on amplifying the effects of GHRH-induced GH release, in comparison with the GHSs, when used alone, the depolarizing agents were far less effective, illustrating the additional importance of activation of the phospholipase C pathway (9).

III. CLONING OF THE GHS-R

A strategy for expression cloning of the receptor was provided by the demonstration that GHRP-6, L-692,429, and MK-0677 acted through a receptor expressed in the pituitary gland and that the receptor signaled through the phospholipase C pathway (15). Poly (A^+) RNA isolated from pig pituitaries was injected into xenopus oocytes and, after incubation, for 2–3 days the addition of MK-0677 caused activation of a Ca^{2+} activated Cl current (23). However, this signal was not reproducible enough to allow the effective cloning of a rare cDNA from a pituitary cDNA library. Two modifications were made to the expression-cloning strategy. To provide a more robust signal for the detection of changes in Ca^{2+}, aequorin bioluminescence was used, and to optimize receptor G protein coupling, various G protein partners were expressed. A robust MK-0677-inducible aequorin bioluminescent signal was reproducibly observed when cRNA encoding aequorin and cRNA encoding $G_{\alpha11}$ were coinjected with RNA derived from a pig pituitary cDNA library (23). By using this bioluminescent assay and fractionating cDNA pools, a single cDNA clone was isolated that encoded a protein that bound [^{35}S]MK-0677 with high affinity. Displacement of [^{35}S]MK-0677 binding correlated with the biological activity of MK-0677, GHRP-2, and GHRP-6, but not with the biological activity of GHRH or somatostatin (15,17).

A long (GHS-R1a) and a short (GHS-R1b) form of the GH secretagogue receptor (GHS-R) was isolated from a pig pituitary cDNA library. The predicted amino acid sequence of GHS-R1a was consistent with that of a new G–protein-coupled receptor (GPC-R) having appropriate residues for seven transmembrane-spanning domains and an ERY signature motif (23). Notably, based on the nucleotide and predicted amino acid sequence, the GHS-R1a appeared to be the first member of a new family of GPC-Rs. Subsequent cloning of the human and rat homologues showed that the receptor was highly conserved across these species (23,24). For example, the human and rat receptors share 95% identify at the amino acid level. Because the GHRPs, L-692,429, and MK-0677 class of GHSs cause GH release in a variety of species, including chickens and humans, and because the receptor is highly conserved, it is likely that a natural hormone for the GHS-R exists.

The GHS-R1b was isolated from both pig and human cDNA libraries, and compared with GHS-R1a, GHS-R1b lacked transmembrane domains 6 and 7. Inspection of the genomic sequence of GHS-R confirmed that this second form of the receptor was derived from the same gene as GHS-R1a. GHS-R-1b when expressed in COS7 and in HEK293 cells did not bind [^{35}S]MK-0677 and lacked functional activity (23). The potential regulatory role of heteromeric complexes of GHS-R1a and 1b are currently under investigation.

IV. EFFECTS OF GHS IN THE CENTRAL NERVOUS SYSTEM

The studies designed to determine the mechanism of action of GHRPs and the peptidomimetics focused largely on the pituitary gland, and it is undeniable that these molecules have profound effects on stimulating GH release from somatotrophs (1,5,8,15,25). At this level, in addition to being very effective when administered alone, they synergize with GHRH and functionally antagonize somatostatin (26). However, it is also clear that pituitary cells become refractory to continued stimulation in vitro, yet paradoxically, constant infusion of GHRP-6 and benzolactam peptidomimetics into animals and humans increases the amplitude of the GH pulsatile profile over a 24-h period, without complete tachyphylaxis (15,18,27–30). These observations led investigators to speculate that the GHRPs and mimetics may also act in the hypothalamus at centers involved in the control of GHRH and somatostatin release (31–40). In support of such speculation, in guinea-pigs and sheep, lower doses of GHRPs and peptidomimetics are more effective as GHSs when administered centrally, rather than peripherally (39,41). Also, experiments in hypothalamic stalk-sectioned pigs showed that a peptidomimetic GHS required an intact hypothalamic–pituitary axis for optimal GH release (42). Similarly, human subjects with hypothalamic–pituitary stalk disconnection fail to respond to hexarelin* (43,44).

Speculation on a central role for GHRP-6 led to elegant electrophysiology studies showing that intravenous administration of GHRP-6 activated arcuate neurons that project to the median eminence (35). Activation of arcuate neurons has also been observed following systemic treatment with the peptidomimetics (34,45). Similarly, increased expression of c-*fos* activity in arcuate neurons was demonstrated following systemic administration of GHRP-6, L-692,429, L-692,585, and MK-0677 to rats, mice, lit/lit and dw/dw mice (46,47). Activation of c-*fos* in arcuate neurons was not caused by GH because the dwarf rodents fail to release GH in response to GHSs. Dual localization studies showed the c-*fos* expression occurred in GHRH and neuropeptide Y (NPY)-containing arcuate neurons (36). The activation of GHRH-containing neurons are consistent with increased GHRH and GHRH pulses measured in hypothalamic–pituitary portal vessels of sheep following systemic treatment with GHRPs (48,49). The effects on NPY-containing neurons are consistent with the reports that ad-

*Hexarelin is a trademark for examorelin (I.N.N.).

ministration of GHRPs increases feed intake (50). Alternatively, or in addition to, activation of NPY neurons might be involved in signaling somatostatin release from neurons in the periventricular nucleus (51).

When the effects of GHRP-6 were investigated by recording electrical activity in arcuate neurons in rat hypothalamic slices, two different effects were observed (34). In secretory arcuate neurons, GHRP-6 caused excitation; however, in certain nonsecretory neurons, electrical activity was suppressed. These results provide a basis for speculating that the neurons in which activity is suppressed are somatostatin-containing neurons; hence, GHRP-6 is acting locally to suppress somatostatin release. Because these neurons are closely associated with GHRH-containing neurons that are excited by GHRP-6, perhaps GHRP-6 controls GHRH release indirectly by relieving somatostatin tone on GHRH-containing neurons. Combinations of indirect and direct stimulation and functional antagonism of somatostain in GHRH neurons are also conceivable, given the properties of the GHRPs and their mimetics and provide a means to amplify the functional effects of GHRP-6 and MK-0677 in the hypothalamus.

Consistent with in vitro GH-release data and the electrophysiology and c-*fos* activation studies, in situ hybridization, using a radiolabeled rat cDNA probe, shows that the GHS-R receptor is expressed in the anterior pituitary gland, in the arcuate nucleus, and in the ventromedial hypothalamus (52). A similar distribution was observed when a human GHS-R cDNA probe was used for in situ hybridization to rhesus monkey brain (23). These results are persuasive that this class of GHSs act directly on arcuate neurons in rodents and primates and are consistent with their effects on neurons involved in the control of GHRH, NPY, and somatostatin release. The receptor is also expressed in regions of the brain other than those associated with the regulation of GH release. Remarkably, expression is clearly evident in the dentate gyrus, CA2 and CA3 regions of the hippocampus, and several nuclei within the brain stem (52). Expression in these locations suggest that the receptor might have a role in learning and memory functions.

V. PROLONGED TREATMENT OF DOGS WITH GHS WITH MK-0677, L-692,429, AND L-692.585

Pharmacokinetic, pharmacodynamic, and oral bioavailability measurements confirmed that MK-0677 was a viable candidate for once-daily oral dosing (14,53). To establish whether the effects on GH release could be sustained, dogs were treated with MK-0677 (1 mg/kg) for 4 days. On days 1 and 4 blood was collected before treatment and at 15-min intervals for 8 h following treatment. MK-0677 treatment resulted in sustained amplification of the pulsatile profile of GH, but on day 4 the amplitude of the GH peaks were markedly attenuated compared with day 1 (15,54). However, continued treatment with MK-0677 for up to 14 days did not further decrease the GH peak amplitude. During treatment, serum insulin-like growth factors-1 (IGF-1) levels increased by approximately 120% and were sustained. In direct contrast to MK-0677, once-daily intravenous administration of the shorter-acting secretagogues L-692,429 and L-692,585, for up to 14 days did not result in tachyphylaxis of the GH response. The sustained increase in IGF-1 levels observed with MK-0677 was also not evident with L-692,429 and L-692,585. When MK-0677 was dosed on alternate days the GH response to repeated dosing was not attenuated. Because this dosing regimen allowed IGF-1 to return to basal levels before the next dose of MK-0677, it was speculated that tachyphylaxis was associated with increases in IGF-1, rather than MK-0677-receptor desensitization. Because of the unavailability of sufficient quantities of IGF-1, this hypothesis was addressed indirectly. In a 4-day study in dogs, MK-0677 treatment was substituted for porcine GH on days 2 and 3. On day 4, after GH had returned to basal levels, but IGF-1 levels were still elevated, the dogs were challenged with MK-0677. Similar to that observed in dogs that were treated for 4 days with MK-0677, the response to MK-0677 on day 4 compared with day 1 was clearly attenuated. These results argue that reduced responsiveness to repeated MK-0677 treatment was not necessarily because of desensitization of MK-0677 receptors. Most likely tachyphylaxis was associated with reduced responsiveness of the GH–GHRH axis caused by sustained increases in IGF-1 (54).

In addition to stimulating GH release, GHRP-6, L-692,429, and L-692,585 caused a corticotropin (adrenocorticotropic hormone; ACTH)-mediated increase in cortisol (55,56). This effect is apparently mediated at the hypothalamic level because these secretagogues do not stimulate ACTH release from cultured pituitary cells. MK-0677 also increased cortisol levels when given briefly; however, during repeated daily oral administration to dogs the effect on cortisol disappears (54,57). Similar to that observed with GH, on alternate-day treatment with MK-0677, there is no down-regulation of the cortisol response (54). Thus, attenuation of the cortisol and GH responses are both associated with elevations in serum IGF-1. Interestingly, the magnitude of the stimulatory effect of MK-0677 on GH release and IGF-1 is apparently limited by negative feedback associated with increases in IGF-1 concentrations acting on the hypothalamus and pituitary gland, thereby preventing MK-677 hyperstimulation of the GH/IGF-1 axis (Fig. 6).

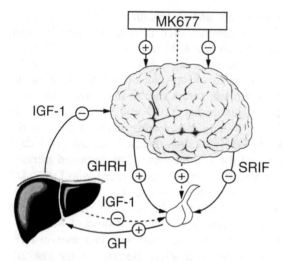

Figure 6 Effects of MK-0677 on GH secretion: MK-0677 acts directly on the pituitary gland to stimulate GH release by amplifying the stimulatory action of GHRH and antagonizing somatostatin (SRIF). In the hypothalamus, MK-0677 stimulates arcuate neurons containing GHRH and reduces somatostatin tone. The GH levels act on the liver to increase IGF-1 that, in turn, feeds back negatively on the hypothalamus and pituitary gland to reduce the response to MK-0677, thus preventing hyperstimulation of ther GH to IGF-1 axis. Not illustrated is the brief GH-mediated negative-feedback pathway that regulates self-entrainment of GH pulsatility (70). (From Ref. 15.)

VI. REGULATION OF GH PULSATILITY

Growth hormone secretion is pulsatile in all species studied (58–62). Episodic secretion is biologically significant because GH replacement is more efficient given in a pulsatile manner versus constant infusion (63). Pulsatility is likely to be dependent on sequential cyclic changes in GHRH and somatostatin (67). Experiments using GHRH antibodies and GHRH antagonists show that GH pulses are dependent on the release of GHRH (64,65). There is also evidence pointing to pulsatile GH release being related to increased somatostatin secretion during GH troughs (66–68). GHRH, GHRP-6, or the peptidomimetics are positive regulators that amplify and sustain pulsatile GH release in humans (28,30,69). However, administration of somatostatin or GH suppresses this pulsatility (70–72); thus, GH and somatostatin are the negative regulators that interrupt GH release.

The periodicity of GH release appears to be entrained by GH because when GH is given exogenously to rats at intervals of 3 h, approximately in phase with their endogenous GH pulses, the exogenous and endogenous

GH peaks become entrained (70). When the exogenous pulses are repeated more frequently the regular endogenous pulsatility disappears (70). Entrainment of the endogenous pulses can also be accomplished by administering a GHRH analog (73) or L-692,585 instead of GH at 3-h intervals (39,41). These episodic rhythms at 3- to 3.5-h intervals are probably best explained by GH-mediated feedback that increases somatostatin tone at the hypothalamus and pituitary gland, resulting in attenuation of GHRH and GH release, respectively. The suggestion that GH acts on the hypothalamus is consistent with experiments showing that GH does not inhibit its own release from rat pituitary cells (74). Moreover, introduction of GH into the third ventricle prevents endogenous GH release (75). Physiologically, negative feedback in the hypothalamus might occur by retrograde transport of GH through the portal vessels (76,77). The precise mechanism of GH-mediated negative feedback has yet to be elucidated, for although it is thought that GH directly causes release of somatostatin from neurons in the periventricular nucleus, indirect affects mediated by release of NPY from arcuate neurons must also be considered (51). The role of the GHRPs and peptidomimetics on these pathways remains to be elucidated.

An important problem to be resolved is whether pulsatility is controlled at the hypothalamic level, at the level of the pituitary, or both. That hypothalamic–pituitary stalk disconnection results in loss of normal GH pulsatility (78) suggests pulsatility is controlled centrally by changes in GHRH, somatostatin, or another hypothalamic factor, such as the natural ligand for the GHS-R. Physiologically, the natural ligand for the GHS-R probably plays a crucial role in modulating pulsatile GH release by acting at both the hypothalamic and pituitary levels. It could be argued that the natural ligand plays only a permissive role in fine-tuning the amplitude of GH release; however, given the biological properties of its mimetics, it seems more likely that the ligand plays a more dominant role. For example, in contrast with GHRH, when ligands for the GHS-R are administered to animals, GH release is always stimulated. This characteristic is probably because this class of secretagogues are functional somatostatin antagonists; therefore, in contrast with GHRH, they are able to overcome the inhibitory effects of high somatostatin tone on GHRH and GH release. Because the pulsatile GH-release profile appears to be entrained by GH itself (70), by stimulating GH release these small molecules have the capacity to initiate and thus reset the cycle of GH pulsatility.

Once the cycle is initiated, oscillations in GH and somatostatin, and perhaps the natural ligand, maintaining the periodicity. Clearly, these arguments also support an important role for somatostatin, which appears to be the crucial mediator of GH short-loop, negative feedback that entrains GH pulsatility (67,68,72,79–81). Studies of the effects of somatostatin on the desensitization and resensitization of pituitary cells to GHRP-6 and peptidomimetics show that somatostatin in a dose-dependent manner can inhibit desensitization of pituitary cells (Cheng K, Smith RG, unpublished observations). Similarly, perifusion studies show that a short pulse of somatostatin superimposed on a constant infusion of GHRH with GHRP-6 induces GH pulsatility by preventing desensitization (Blake AD, Smith RG, unpublished observations). These experiments suggest an interdependence of somatostatin and the GHSs that is mediated through their individual receptors. Mathematical-modeling studies provide a rationale for how changes in somatostatin tone or concentrations of somatostatin can modulate pulsatility, even in the continued presence of the GHSs (82).

REFERENCES

1. Bowers CY, Momany FA, Reynolds GA, Hong A. On the in vitro and in vivo activity of a new synthetic hexapeptide that acts on the pituitary to specifically release growth hormone. Endocrinology 1984; 114:1537–1545.
2. Momany FA, Bowers CY, Reynolds GA, Hong A, Newlander K. Conformational energy studies and in vitro and in vivo activity data on growth hormone-releasing peptides. Endocrinology 1984; 114:1531–1536.
3. Momany Fa, Bowers CY, Reynolds GA, Chang D, Hong A, Newlander K. Design, synthesis and biological activity of peptides which release growth hormone, in vitro. Endocrinology 1981; 108:31–39.
4. Bowers CY. GH-releasing peptides—structure and kinetics. J Pediatr Endocrinol 1993; 6:21–31.
5. Bowers CY. Xenobiotic growth hormone secretagogues: growth hormone releasing peptides, In: Bercu BB, Walker RF, eds. Growth Hormone Secretagogues. New York: Springer-Verlag, 1996:9–28.
6. Cheng K, Chan WW-S, Butler BS, Barreto A, Smith RG. Evidence for a role of protein kinase-C in His-DTrp-Ala-Trp-DPhe-Lys-NH$_2$-induced growth hormone release from rat primary pituitary cells. Endocrinology 1991; 129:3337–3342.
7. Evans BE, Rittle KE, Bock MG, et al. Methods for drug discovery: development of potent, selective orally effective cholecystokinin antagonists. J Med Chem 1988; 31:2235–2246.
8. Smith RG, Cheng K, Schoen WR, et al. A nonpeptidyl growth hormone secretagogue. Science 1993; 260:1640–1643.
9. Smith RG, Cheng K, Pong S-S, et al. Mechanism of action of GHRP-6 and nonpeptidyl growth hormone secretagogues, In: Bercu BB, Walker RF, eds. Growth Hormone Secretagogues. New York: Springer-Verlag, 1996:147–163.
10. Herrington J, Hille B. Growth hormone-releasing hexapeptide elevates intracellular calcium in rat somatotrophs by two mechanisms. Endocrinology 1994; 135:1100–1108.

11. Bresson-Bepoldin L, Dufy-Barbe L. GHRP-6 induces a biphasic calcium response in rat pituitary somatotrophs. Cell Calcium 1994; 15:247–258.

12. Adams EF, Petersen B, Lei T, Buchfelder M, Fahlbusch R. The growth hormone secretagogue, L-692,429, induces phosphatidylinositol hydrolysis and hormone secretion by human pituitary tumors. Biochem Biophys Res Commun 1995; 208:555–561.

13. Mau SE, Witt MR, Bjerrum OJ, Saermark T, Vilhardt H. Growth hormone releasing hexapeptide (GHRP-6) activates the inositol (1,4,5)-trisphosphate/diacylglycerol pathway in rat anterior pituitary cells. J Recept Signal Transduction Res 1995; 15:311–323.

14. Patchett AA, Nargund RP, Tata JR, et al. The design and biological activities of L-163,191 (MK-0677): a potent orally active growth hormone secretagogue. Proc Natl Acad Sci USA 1995; 92:7001–7005.

15. Smith RG, Pong S-S, Hickey GJ, et al. Modulation of pulsatile GH release through a novel receptor in hypothalamus and pituitary gland. Rec Prog Horm Res 1996; 51:261–286.

16. Dean DC, Nargund RP, Pong S-S, et al. Development of a high specific activity sulfur-35-labeled sulfonamide radioligand that allowed the identification of a new growth hormone secretagogue receptor. J Med Chem 1996; 39:1767–1770.

17. Pong S-S, Chaung L-YP, Dean DC, Nargund RP, Patchett AA, Smith RG. Identification of a new G-protein-linked receptor for growth hormone secretagogues. Mol Endocrinol 1996; 10:57–61.

18. Blake AD, smith RG. Desensitization studies using perifused rat pituitary cells show that growth hormone releasing hormone and His-D-Trp-Ala-Trp-D-Phe-Lys-NH$_2$ stimulates growth hormone release through distinct receptor sites. J Endocrinol 1991; 129:11–19.

19. Pong S-S, Chaung L-YP, Smith RG, Ertel EA, Smith MM, Cohen CJ. Role of calcium channels in growth hormone secretion induced by GHRP-s (His-D-Trp-Ala-Trp-D-Phe-Lys-NH$_2$) and other secretagogues in rat somatotrophs (abstr). Proceedings of the 74th Annual Meeting of The Endocrine Society 1992:255.

20. McGurk JF, Pong S-S, Chaung L-YP, Gall M, Butler BS, Arena JP. Growth hormone secretagogues modulate potassium currents in rat somatotrophs (abstr). Soc Neurosci 1993; 19:1559.

21. Pong S-S, Chaung L-YP, Leonard RJ. The involvement of ions in the activity of a novel growth hormone secretagogue L-692,429 in rat pituitary cell culture (abstr). Proceedings of the 75th Annual Meeting of The Endocrine Society 1993; 172.

22. Leonard RJ, Chaung L-YP, Pong S-S. Ionic conductances of identified rat somatotroph cells studies by perforated patch recording are modulated by growth hormone secretagogues (abstr). Biophys J 1991; 59:254a.

23. Howard AD, Feighner SD, Cully DF, et al. A receptor in pituitary and hypothalamus that functions in growth hormone release. Science 1996; 273:974–977.

24. McKee KK, Palyha OC, Feighner SD, et al. Molecular analysis of growth hormone secretagogue receptors (GHS-Rs): cloning of rat pituitary and hypothalamic GHS-R type 1a cDNAs. Mol Endocrinol 1997; 11:415–423.

25. Bowers CY. [Editorial]: On a peptidomimetic growth hormone-releasing peptide. J Clin Endocrinol Metab 1994; 79:940–942.

26. Cheng K, Chan WW-S, Butler BS, Barreto A, Smith RG. The synergistic effects of His-D-Trp-Ala-Trp-D-Phe-Lys-NH$_2$ on GRF stimulated growth hormone release and intracellular cAMP accumulation in rat primary pituitary cell cultures. Endocrinology 1989; 124:2791–2797.

27. Bowers CY, Reynolds GA, Durham D, Barrera CM, Pezzoli SS, Thorner MO. Growth hormone (GH)-releasing peptide stimulates GH release in normal man and acts synergistically with GH-releasing hormone. J Clin Endocrinol Metab 1990; 70:975–982.

28. Chapman IM, Hartman ML, Pezzoli SS, Thorner MO. Enhancement of pulsatile growth hormone secretion by continuous infusion of a growth hormone-releasing peptide mimetic, L-692,429, in older adults—a clinical research center study. J Clin Endocrinol Metab 1996; 81:2874–2880.

29. Chapman IM, Bach MA, Van Cauter E, et al. Stimulation of the growth hormone (GH)-insulin-like growth factor-I axis by daily oral administration of a GH secretagogue (MK-0677) in healthy elderly subjects. J Clin Endocrinol Metab 1996; 81:4249–4257.

30. Huhn WC, Hartman ML, Pezzoli SS, Thorner MO. 24-hour growth hormone (GH)-releasing peptide (GHRP) infusion enhances pulsatile GH secretion and specifically attenuates the response to a subsequent GHRP-6 bolus. J Clin Endocrinol Metab 1993; 76:1202–1208.

31. Dickson SL. Evidence for a central site and mechanism of action of growth hormone releasing peptide (GHRP-6), In: Bercu BB, Walker RF, eds. Growth Hormone Secretagogues. New York: Springer-Verlag, 1996:237–251.

32. Dickson SL, Doutrelant-Viltart O, Dyball REJ, Leng G. Retrogradely labelled neurosecretory neurones of the rat hypothalamic arcuate nucleus express Fos protein following systemic injection of growth hormone (GH)-releasing peptide (GHRP-6). J Endocrinol 1997; 151:323–331.

33. Dickson SL, Doutrelant-Viltart O, McKenzie DN, Dyball REJ. Somatostatin inhibits arcuate neurons excited by GH-releasing peptide (GHRP-6) in rat hypothalamic slices. J Physiol (Lond) 1996; 495:109–110.

34. Dickson SL, Leng G, Dyball REJ, Smith RG. Central actions of peptide and nonpeptide growth hormone secretagogues in the rat. Neuroendocrinology 1995; 61:36–43.

35. Dickson SL, Leng G, Robinson ICAF. Systemic administration of growth hormone-releasing peptide (GHRP-6) activates hypothalamic arcuate neurones. Neuroscience 1993; 53:303–306.

36. Dickson SL, Luckman SD. Induction of c-*fos* messenger ribonucleic acid in neuropeptide Y and growth hormone (GH)-releasing factor neurones in the rat arcuate nucleus following systemic injection of the GH secretagogue, GH-releasing peptide-6. Endocrinology 1997; 138:771–777.

37. Doutrelant-Viltart O, Dickson SL, Dyball REJ, Leng G. Expression of Fos protein following growth hormone-releasing peptide (GHRP-6) injection in rat arcuate neurones retrogradely labelled by systemic flurogold administration. J Physiol 1995; 483:49–50.

38. Fairhall KM, Mynett A, Robinson ICAF. Central effects of growth hormone-releasing hexapeptide (GHRP-6) on growth hormone release are inhibited by central somatostatin action. J Endocrinol 1995; 144:555–560.

39. Fairhall KM, Mynett A, Thomas GB, Robinson ICAF. Central and peripheral effects of peptide and nonpeptide GH secretagogues of GH release in vivo. In: Bercu BB, Walker RF, eds. Growth Hormone Secretagogues. New York: Springer-Verlag, 1996:219–236.

40. Kamegai J, Hasegawa O, Minami S, Sugihara H, Wakabayashi I. The growth hormone-releasing peptide KP-102 induces c-*fos* expression in the arcuate nucleus. Mol Brain Res 1996; 39:153–159.

41. Fairhall KM, Mynett A, Smith RG, Robinson ICAF. Consistent GH responses to repeated injections of GH-releasing hexapeptide (GHRP-6) and the nonpeptide GH secretagogue, L-692,585. J Endocrinol 1995; 145:417–426.

42. Hickey GJ, Drisko JE, Faidley TD, et al. Mediation by the central nervous system is critical to the in vivo activity of the GH secretagogue L-692,585. J Endocrinol 1996; 148:371–380.

43. Pombo M, Barreiro J, Penalva A, Peino R, Dieguez C, Casaneuva FF. Absence of growth hormone (GH) secretion after the administration of either GH-releasing hormone (GHRH), GH-releasing peptide (GHRP-6), or GHRH plus GHRP-6 in children with neonatal pituitary stalk transection. J Clin Endocrinol Metab 1995; 80:3180–3184.

44. Popovic V, Damjanovic S, Micic D, Djurovic M, Dieguez C, Casaneuva FF. Blocked growth hormone-releasing peptide (GHRP-6)-induced GH secretion and absence of the synergic action of GHRP-6 plus GH-releasing hormone in patients with hypothalalamopituitary disconnection; evidence that GHRP-6 main action is exerted at the hypothalamic level. J Clin Endocrinol Metab 1995; 80:942–947.

45. Bailey A, Smith RG, Leng G. Activation of hypothalamic arcuate neurons in the urethane-anesthetized rat following systemic injection of growth-hormone secretagogues. J Physiol (Lond) 1996; 495:115–116.

46. Sirinathsinghji DJS, Dunnett SB. Imaging gene expression in neural grafts. In: Sharif NA, ed. Molecular Imaging in Neuroscience. Oxford: Oxford University Press, 1993:43–70.

47. Dickson SL, Doutrelant-Viltart O, Leng G. GH-deficient dw/dw rats and lit/lit mice show increases *fos* expression in the hypothalamic arcuate nucleus following system injection of GH-releasing peptide-6. J Endocrinol 1995; 146:519–526.

48. Fletcher TP, Thomas GB, Clarke IJ. Growth hormone-release hormone and somatostatin concentrations in the hypophysial portal blood of conscious sheep during the infusion of growth hormone-releasing peptide-6. Domestic Anim Endocrinol 1996; 13:251–258.

49. Guillaume V, Magnan E, Cataldi M, et al. Growth hormone (GH)-releasing hormone secretion is stimulated by a new GH-releasing hexapeptide in sheep. Endocrinology 1994; 135:1073–1076.

50. Locke W, Kirgis HD, Bowers CY, Abdoh AA. Intracerebroventricular growth hormone-releasing peptide-6 stimulates eating without affecting plasma growth hormone responses in rats. Life Sci 1995; 56:1347–1352.

51. Suzuki N, Okada K, Minami S, Wakabayashi I. Inhibitory effect of neuropeptide Y on growth hormone secretion in rats is mediated by both Y1- and Y2-receptor subtypes and abolished after anterolateral deafferentation of the medial basal hypothalamus. Regul Pept 1996; 65:145–151.

52. Guan X-M, Yu H, Palyha OC, et al. Distribution of mRNA encoding the growth hormone secretagogue receptor in brain and peripheral tissues. Mol Brain Res 1997; 48:23–29.

53. Leung KH, Miller RR, Cohn DA, et al. Pharmacokinetics and disposition of MK-0677, a novel growth hormone secretagogue (abstr). Proceedings of the International Society for the Study of Xenobiotics 7th International Meeting. San Diego, CA, Oct 20–24 1996.

54. Hickey GJ, Jacks TM, Schleim K, et al. Repeat administration of the growth hormone secretagogue MK-0677 increases and maintains elevated IGF-1 levels in beagles. J Endocrinol 1997; 152:183–192.

55. Hickey GJ, Jacks TM, Judith F, Taylor J, Clark JN, Smith RG. In vivo efficacy and specificity of L-692,429, a novel nonpeptidyl growth hormone secretagogue in beagles. Endocrinology 1994; 134:695–701.

56. Jacks TM, Hickey GJ, Judith F, et al. Effects of acute and repeated intravenous administration of L-692,585, a novel nonpeptidyl growth hormone secretagogue, on plasma growth hormone, IGF-1, ACTH, cortisol, prolactin, insulin and thyroxine (T_4) levels in beagles. J Endocrinol 1994; 143:399–406.

57. Jacks TM, Smith RG, Judith F, et al. MK-0677, a potent, novel orally-active growth hormone (GH) secretagogue—GH, IGF-1 and other hormonal responses in beagles. Endocrinology 1996; 137:5284–5289.

58. Jansson JO, Eden S, Isaksson O. Sexual dimorphism in the control of growth hormone secretion. Endocr Rev 1985; 6:128–150.

59. Davis SL, Ohlson DL, Klindt J, Anfinson MS. Episodic growth hormone secretory patterns in sheep: relationship to gonadal steroid hormones. Am J Physiol 1977; 233:E519.

60. Steiner RA, Stewart JK, Barber J, et al. Somatostatin: a physiological role in the regulation of growth hormone secretion in the adolescent male baboon. Endocrinology 1978; 102:1587.

61. Miller JD, Tannenbaum GS, Colle E, Guyda HJ. Daytime pulsatile growth hormone secretion during childhood and adolescence. J Clin Endocrinol Metab 1982; 55:989.

62. Chihara K, Minamitani N, Kaji H, Kodama H, Kita T, Fujita T. Human pancreatic growth hormone-releasing factor stimulates release of growth hormone in conscious unrestrained male rabbits. Endocrinology 1983; 113:2081.

63. Gevers EF, Wit JM, Robinson ICAF. Growth, growth hormone (GH)-binding protein, and GH receptors are differentially regulated by peak and trough components of the GH secretory pattern in the rat. Endocrinology 1996; 137:1013–1018.

64. Wehrenberg WB, Brazeau P, Luhen R, Bohlen P, Guillemin R. Inhibition of the pulsatile secretion by monoclonal antibodies to the hypothalamic growth hormone releasing factor (GRF). Endocrinology 1982; 111:2147–2148.

65. Lumpkin MD, McDonald JK. Blockade of growth hormone-releasing factor (GRF) activity in the pituitary and hypothalamus of the conscious rat with a peptidic GRF antagonist. Endocrinology 1989; 124:1522–1531.

66. Tannenbaum GS, Ling N. The interrelationship of growth hormone (GH)-releasing factor and somatostatin in the generation of the ultradian rhythm of GH secretion. Endocrinology 1984; 115:1952–1957.

67. Plotsky PM, Vale W. Patterns of growth hormone-releasing factor and somatostatin into the hypophysial postal circulation of the rat. Science 1985; 230:461–463.

68. Chihara K, Minamitani N, Kaji H, Arimura A, Fujita T. Intraventricularly injected growth hormone stimulates somatostatin release into rat hypophysial portal blood. Endocrinology 1981; 109:2279–2281.

69. Vance ML, Kaiser DL, Evans WS, et al. Pulsatile growth hormone secretion in normal man during a continuous 24-hour infusion of human growth hormone releasing factor (1–49): evidence for intermittent somatostatin secretion. J Clin Invest 1985; 75:1584–1590.

70. Carlsson LMS, Jansson JO. Endogenous growth hormone (GH) secretion in male rats is synchronized to pulsatile GH infusions given at 3-hour intervals. Endocrinology 1990; 126:6–10.

71. Abe H, Molitch ME, van Wyk JJ, Underwood LE. Human growth hormone and somatomedin C suppress the spontaneous release of growth hormone in unanesthetized rats. Endocrinology 1983; 113:1319–1324.

72. Clark RG, Carlsson LMS, Robinson ICAF. Growth hormone (GH) secretion in the conscious rat: negative feedback of GH on its own release. J Endocrinol 1988; 119:201–209.

73. Clark RG, Robinson ICAF. Growth hormone (GH) responses to multiple injections of a fragment of human GH-releasing factor in conscious male and female rats. J Endocrinol 1985; 106:281-289.

74. Richman RA, Weiss JP, Hochberg Z, Florini JR. Regulation of growth hormone release: evidence against feedback in rat pituitary cells. Endocrinology 1981; 108:2287–2292.

75. Tannenbaum GS. Evidence for autoregulation of growth hormone secretion via the central nervous system. Endocrinology 1980; 107:2117–2120.

76. Oliver C, Mical RS, Porter JC. Hypothalamic-pituitary vasculature: evidence for retrograde blood flow in the pituitary stalk. Endocrinology 1977; 101:598–604.

77. Bergland RM, Page RB. Can the pituitary secrete directly to the brain? (affirmative anatomical evidence). Endocrinology 1978; 102:1325–1338.

78. Fletcher TP, Thomas GB, Willoughby JO, Clarke IJ. Constitutive growth hormone secretion in sheep after hypothalamopituitary disconnection and the direct in vivo pituitary effect of growth hormone releasing peptide 6. Neuroendocrinology 1994; 60:76–86.

79. Clark RG, Carlsson LMS, Rafferty B, Robinson ICAF. The rebound release of growth hormone (GH) following somatostatin infusion in rats involves hypothalamic GH-releasing factor release. J Endocrinol 1988; 119:397–404.

80. Hoffman DL, Baker BL. Effect of treatment with growth hormone on somatostatin

in the median eminence of hypophysectomized rats. Proc Soc Exp Biol Med 1977; 156:265-271.

81. Katakami H, Arimura A, Frohman LA. Growth hormone (GH)-releasing factor stimulations hypothalamic somatostatin release: an inhibitory feedback effect on GH secretion. Endocrinology 1986; 118:1872-1877.

82. Stephens EA, Brown D, Leng G, Smith RG. Computer simulation of a mathematical model of pituitary growth hormone release (abstr). J Physiol 1996; 495:7P.

4

Molecular Cloning and Characterization of Human, Swine, and Rat Growth Hormone Secretagogue Receptors

Lex H. T. Van der Ploeg, Andrew D. Howard, Roy G. Smith, and Scott D. Feighner
Merck Research Laboratories, Merck and Co., Inc., Rahway, New Jersey

I. INTRODUCTION

Growth hormone-releasing peptide-6 (GHRP-6) is believed to stimulate pulsatile GH release through the activation of a receptor distinct from the somatostatin and growth hormone-releasing hormone receptor (GHRH-R; 1–3). Following the discovery of GHRP-6 and its enhancement of GH release, several nonpeptidyl growth hormone secretagogues (GHSs) were developed, as exemplified by L-163,191 (MK-0677; 4,5). Nonpeptidyl GHSs, similar to GHRP-6, stimulate GH release through inositol triphosphate (PI_3)-coupled mechanisms that mediate intracellular Ca^{2+} release and depolarization, leading to exocytosis of GH-containing secretory vesicles (6–10).

Given the potential clinical benefit of GH replacement therapy (11,12), understanding the mechanisms through which nonpeptidyl GHSs amplify pulsatile GH release is important in our efforts to establish safe GH replacement therapy. The identification of a distinct GHS-R (13) expressed in the hypothalamus and pituitary (13–15), provided evidence for the presence of a third neuroendocrine mechanism involved in the control of pulsatile GH release.

II. GHS LIGANDS FOR RECEPTOR IDENTIFICATION

Identification of the GHS-R required knowledge of its cellular localization and an initial biochemical characterization of the receptor protein. The GHS-R characterization was initiated with ^{35}S-radiolabeled MK-0677 with a specific activity of 1000 Ci/mmol (16; EC_{50} for GH release ~ 1.3 nM) and biotinylated MK-0677 (biotin at the end of a long carbon tether; L-164,683), which retained an EC_{50} of 2.5 nM in the GH-release assay (Fig. 1).

A high-affinity ($K_d \sim 200$ pM), low-abundance–binding site (B_{max}, 7 fmol/mg protein) for ^{35}S]MK-0677 could be identified on swine pituitary membranes (17). Binding in rat pituitary membranes revealed a significantly lower B_{max} (~ 3 fmol/mg protein). Appropriate binding pharmacology was observed in which the K_d and EC_{50} for GH release with structurally diverse GHSs exhibited the predicted rank-order of potency.

A nonhydrolyzable GTP analogue, GTPγS, can confer the high-affinity–binding state of GPC-Rs to their low-affinity–binding state. Because addition of GTPγS to the binding assay significantly reduced the K_d for ^{35}S]MK-0677, Pong et al. (17) concluded that the GHS-R is most likely a G–protein-coupled receptor (GPC-R) coupled to phosphoinositol (PI) metabolism and Ca^{2+} mobilization. Experiments were also performed to determine whether the putative GHS-R represented a membrane-binding protein of the outer membrane surface of somatotrophs (S. Feighner, D.

35S-MK-677   L-164,683

(~ 1000 Ci/mmol)   (Biotinylated)

IC_{50} = 0.3 nM IC_{50} = 0.2 nM
(radioreceptor assay)   (radioreceptor assay)

EC_{50} = 1.3 nM EC_{50} = 2.5 nM
(GH release)   (GH release)

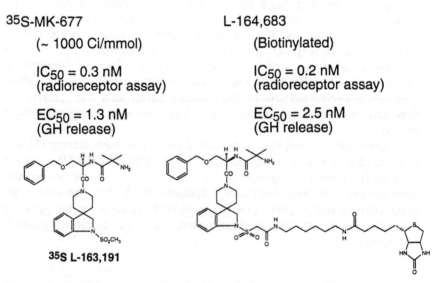

35S L-163,191

Figure 1 GHSs for receptor identification: [35]S]MK-677 and L-164,683 were used for the characterization of the GHS-R.

Hreniuk, and L.H.T. Van der Ploeg, unpublished observations) (18). First, biotinylated MK-0677 (L-164,683) conjugated to avidin, mediated GH release, as expected for a ligand that activates its receptor without having to pass the lipid bilayer. The avidin–biotin–GHS complex had about a 20-fold reduced potency, attributed to steric hindrance. Second, we determined that biotinylated MK-0677 acted uniquely at somatotrophs: L-164,683 was incubated with primary cultures of rat pituitary cells for 3 min at 37°C. Following this incubation, cells were fixed and reacted with streptavidin conjugated to rhodamine. Following membrane permeabilization these same cells were treated with a rabbit anti-rat GH antibody and a fluorescein-conjugated goat-derived anti-rat antibody. Biotinylated MK-0677 bound to only a subpopulation ($\sim 50\%$) of somatotrophs. The GHS-binding sites could not be detected on other cells (S. Feighner, D. Hreniuk, and L.H.T. Van der Ploeg, unpublished observations) (18). From this data we concluded that the GHSs act specifically on a GPC-R of somatotrophs.

III. EXPRESSION CLONING STRATEGIES

We designed an expression-cloning strategy aimed at the isolation of a cDNA clone encoding the GHS-R. Given the low B_{max} obtained for the GHS-R in the pituitary, we expected GHS-R mRNA to be rare. To identify a GHS-R cDNA we employed several molecular-cloning strategies in parallel: (1) Our attempts to establish a whole-cell–binding assay using [35S]MK-0677 to detect cells expressing the GHS-R by cell autoradiography failed. Even [35S]MK-0677 whole-cell–binding experiments with rat somatotrophs, followed by autoradiography, failed to provide a somatotroph-specific signal; therefore, we abandoned this approach. (2) A search for GPC-R orphans was initiated to select for GPC-Rs that were specifically expressed in the pituitary and not in control tissues (liver, which did not exhibit [35S]MK-0677 binding); an exhaustive search for cell lines that expressed an MK-0677-binding site had failed earlier, closing this avenue for the use of cell line-derived poly(A+)RNA on which to base cloning strategies). Candidate cDNAs were evaluated by transient transfection into COS-7 cells, followed by [35S]MK-0677 membrane binding to determine whether they represented a GHS-R.

The GPC-R-cloning effort was based on polymerase chain reaction (PCR) cloning with oligonucleotides designed against conserved transmembrane domains for related GPC-Rs, including members of the angiotensin, neurokinin, and opioid GPC-Rs. The bias toward these families of GPC-Rs was based on the notion that the benzolactam series of GHSs were derived from angiotensin-II-R lead compounds, whereas the design of GHRP-

6 was enkephalin-based. The cDNA libraries were placed on grids and counterselected for irrelevant clones by hybridization with ^{32}P-labeled cRNA probes from tissues that did not express a GHS-R, as determined by [^{35}S]MK-0677 binding. This effort led to identification of more than 30 new GPC-Rs (P. Liberator, M. Hamelin, and L.H.T. Van der Ploeg, unpublished observations), including new chemokine-, angiotensin-, and somatostatin-related GPC-Rs. None of these GPC-Rs represented the GHS-R, possibly because we did not target the appropriate GPC-R families and narrowly focused on opioid and angiotensin receptor families.

(3) Several assumptions were made in setting up expression-cloning strategies. We considered that the human genome encodes approximately 100,000 genes. Because in any tissue, only a fraction of these genes is expressed (assume 50,000) and given the low B_{max} for [^{35}S]MK-0677 binding in the pituitary (7 fmol/mg of protein) we assumed that GHS-R mRNA is rare, requiring a sensitivity of detection exceeding 1:50,000 cDNAs.

Our approach to ensure sensitivity of detection involved generation of pituitary rat cDNA libraries in more than 1000 pools, with a complexity of approximately 1000 cDNAs each (total library complexity $\sim 10^6$ individual cDNAs). To circumvent selection against particular cDNAs during library amplification, we expanded each pool in solid state and recovered DNA from these fractions (library quality was controlled by the determination of average insert size and the isolation of control gonadotropin-releasing hormone (GnRH) receptor cDNAs; we estimated that for mRNAs up to 3 kb most of the cDNAs should be full length).

A third effort aimed at identification of the GHS-R, using these cDNA pools involved transient transfection of rat pituitary cDNA pools into COS-7 cells. Membranes were generated from each transiently transfected cell population, after which these were tested for high-affinity [^{35}S]MK-0677 binding. Approximately 300 pools were evaluated in triplicate. A pool expressing a high-affinity–binding site for [^{35}S]MK-0677 could not be identified. Sensitivity controls using the GnRH receptor (^{125}I-GnRH for receptor detection on COS-7 cell membranes) expressed in a pool of 1000 irrelevant cDNAs indicated that more complex pools could not be used because these were unlikely to support receptor detection.

(4) In parallel with these efforts, we evaluated an expression-cloning strategy using xenopus oocytes. Swine pituitary Poly(A$^+$)-RNA was injected into xenopus oocytes. Following MK-0677 administration we searched for Ca^{2+}-activated chloride currents. These were readily detectable for control genes, such as GnRH and TRH (Fig. 2), but GHS-R responses were initially absent. Following the evaluation of oocytes from numerous different xenopus frogs we finally identified a calcium-activated chloride current in response to 200 nM MK-0677 administration with oocytes from a single

MK 677 200 nM   GnRH 100 nM   TRH 100 nM

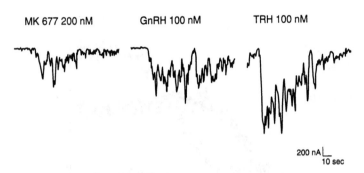

200 nA |_
10 sec

Figure 2 MK-0677 responsiveness of pituitary poly(A⁺)-injected xenopus oocytes: Xenopus oocytes were injected with 50 ng of swine pituitary poly(A⁺)RNA after which calcium-activated chloride currents were reported following the administration of MK-0677 (200 nM). The bar indicates the duration of ligand administration. GnRH (100 nM) and TRH 100 nM represent positive controls.

frog. Overall, the magnitude of the signal varied, and robust signals up to 200 nA could be detected only occasionally (see Fig. 2). Unfortunately, the identification of calcium-activated chloride currents in response to MK-0677 appeared an isolated incident and only about 6% of the frogs had oocytes that could support an MK-0677-induced chloride current. In most instances, the currents were small (10–20 nA).

We assumed that synthetic RNA from cDNA (cRNA) in pools with a complexity of only 1000 cDNAs each, could lead to an enhanced signal; therefore, we initiated a GHS-R-cloning effort in xenopus oocytes. To enhance sample throughput we modified the xenopus oocyte assay, and we switched to the identification of calcium-stimulated bioluminescence, rather than rely on electrophysiology for the identification of a GHS-R-derived signal. Therefore, we injected aequorin protein into the xenopus oocytes, and following administration of the cofactor coelenterazine, we searched for Ca^{2+}-induced chemiluminescence. With GnRH as a control, the sensitivity of detection was estimated at roughly 1:500 cDNAs to 1:1000 unrelated cDNAs (Fig. 3; bioluminescence traces for GnRH-R cRNA at a complexity of 1:1000 expressed in five different oocytes). In the aequorin expression-cloning assay we injected cRNA into oocytes from xenopus frogs on day 1 of the assay. On day 2 we injected aequorin protein, and on days 3–7 we determined whether ligand-dependent Ca^{2+} release could be observed (kinetics of expression for different GPC-R is unpredictable, requiring determination of potential responses at numerous days following injection). To decrease pool complexity and the frequency of GHS-R detection, we also generated cDNA pools from libraries made with sucrose gradient size-fractionated swine pituitary poly(A⁺)-RNA. From each sucrose gradient

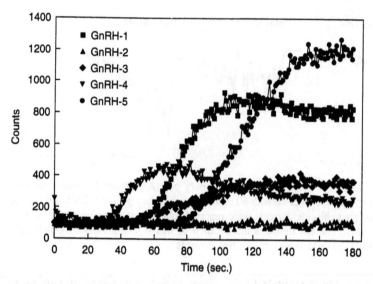

Figure 3 Expression cloning in xenopus oocytes using aequorin to report calcium: Xenopus oocytes were injected with 2 ng of GnRH cRNA (day 1). On day 2 the jellyfish-derived photoprotein aequorin was injected into the oocytes. On days 3-5, just before ligand administration, the oocytes were charged with the cofactor coelantrazine. The kinetics of light readout over time is shown in the graph for five individual oocytes.

fraction, 25 ng was injected into xenopus oocytes and evaluated for its ability to induce a Ca^{2+}-activated inward chloride current as a result of MK-0677 administration. A fraction was identified with mRNA of approximately 1.5-2.0 kb that specifically yielded MK-0677 responses (identified electrophysiologically).

We screened 200 pools of 1000 cDNAs each from an unfractionated swine pituitary cDNA library and evaluated 700 pools with a complexity of 1000 cDNAs each from the sucrose gradient size-fractionated cDNA library. The fractionated mRNA was at least ten-fold enriched (judged by ethidium bromide staining); therefore, the screening effort represented the analysis of approximately 7×10^6 cDNAs from an unfractionated cDNA library. None of these libraries resulted in the identification of a GHS-R cDNA.

As a result, we were left with the assumption that the quality of the xenopus oocytes in which we performed our assays was insufficient because we had also failed to obtain a positive GHS-R signal in most of the batches of xenopus oocytes tested, following pituitary poly(A^+) injection. We concluded that even at a low cRNA complexity, the oocytes were overall not of sufficient quality to allow GHS-R identification.

IV. RESCUE OF RESPONSIVENESS TO MK-0677 IN XENOPUS OOCYTES: G PROTEIN ADDITION

We had earlier considered that specific oocyte populations might be lacking a G protein cofactor, decreasing the efficiency of GHS-R coupling. Therefore, we generated cRNAs for the major classes of G protein subunits, including $G_{\alpha 11}$, G_q, G_o, $G_{\alpha i1}$, $G_{\alpha i3}$, and $G_{\alpha 16}$. cRNAs for these individual G protein subunits were coinjected with swine pituitary poly(A$^+$)-RNA (2 and 25 ng/oocyte, respectively). Challenging the oocytes with 2 µM MK-0677 failed to give bioluminescence responses in the presence of most of the G protein subunits (Fig. 4). However, following administration of swine pituitary poly(A$^+$)-mRNA with $G_{\alpha 11}$, bioluminescence responses from expression of a GHS-R could be obtained ($G_{\alpha 11}$ injection alone did not give a response). This experiment indicated that coexpression of $G_{\alpha 11}$ cRNA and swine pituitary poly(A$^+$)-mRNA rescued GHS-R responsiveness to MK-0677, leading to intracellular Ca^{2+} release (see Fig. 4, middle panel).

A preliminary characterization of the bioluminescence response indicated that concentrations of MK-0677 down to 1 nM resulted in GHS-R-mediated Ca^{2+} release, indicating that receptor activation resulted from interaction with a high-affinity MK-0677-binding site. To increase our ability to screen large numbers of cDNA pools, we further modified the GHS-R assay. Instead of injecting aequorin protein, we coinjected cRNA from the pituitary cDNA pools (50 ng), with 2 ng $G_{\alpha 11}$ cRNA and 2 ng of aequorin cRNA. The coinjection of aequorin cRNA eliminated the need for a second aequorin protein injection of xenopus oocytes on day 2 and increased assay throughput and xenopus oocyte viability. We measured MK-0677 responsiveness at days 3–5, searching for MK-0677-mediated Ca^{2+} fluxes.

With this improved assay, GHS-R expression could be identified with poly(A$^+$)-mRNA from human pituitary, rat pituitary, swine pituitary, and rat hypothalamus (Fig. 5). GHS-R expression in oocytes injected with human pituitary poly(A$^+$)-RNA helped explain the GHS responsiveness of humans, and GHS-R identification with rat hypothalamus poly(A$^+$)-RNA provided firm evidence for the presence of GHS-R expression in the central nervous system (CNS). These findings were consistent with earlier observations in support of GHS-R expression in the CNS (P. Liberator, M. Hamelin, and L.H.T. Van der Ploeg, unpublished observations) (20,21).

V. EXPRESSION CLONING OF A SWINE GHS-R

Rescue of MK-0677 responsiveness by coinjection of $G_{\alpha 11}$ cRNA followed by assay simplification by aequorin–cRNA coinjection enhanced the sen-

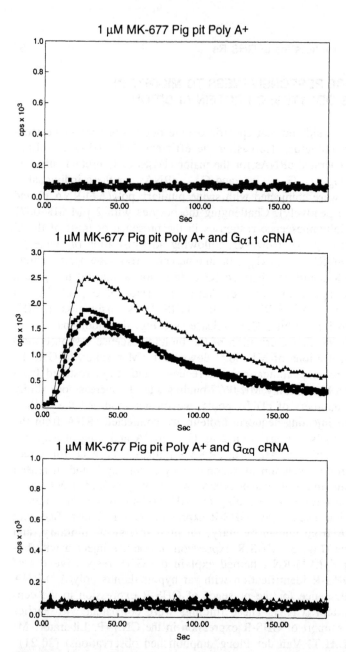

Figure 4 Rescue of GHS-R signal by coinjection with the $G_{\alpha 11}$: (*Top panel*) Injection of xenopus oocytes with 50 ng of swine pituitary poly(A^+)-RNA (application of 1 μM MK-0677); (*middle panel*) injection of xenopus oocytes with 50 ng of swine pituitary poly(A^+)-RNA and 2 ng $G_{\alpha 11}$ RNA results in the identification of calcium-activated chloride currents following the administration of 1 μM MK-0677; (*bottom panel*) injection of xenopus oocytes with 50 ng of swine pituitary poly(A^+)-RNA and 2 ng $G_{\alpha q}$ cRNA, followed by application of 1 αM MK-0677.

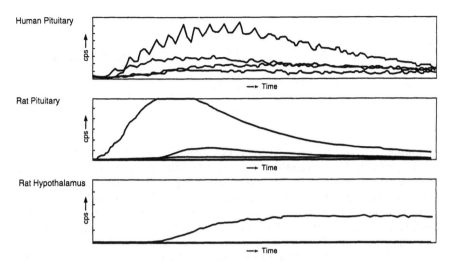

Figure 5 Bioluminescence responses in xenopus oocytes injected with human pituitary poly(A^+)-mRNA or rat hypothalamic mRNA following 1 μM MK-0677 administration: Bioluminescence responses were detected following administration of 17 ng human pituitary poly(A^+)-RNA, rat pituitary poly(A^+)-RNA, or poly(A^+)-RNA from hypothalamus, coinjected with 2 ng of $G_{\alpha 11}$ cRNA and 2 ng of aequorin cRNA.

sitivity, reliability, and throughput of our expression-cloning strategy. We next screened for a GHS-R cDNA in swine pituitary cDNA pools with a complexity of 10,000 (sensitivity based on GnRH and TRH cDNA titrations). Following evaluation of approximately 2×10^6 individual recombinants, a single pool was identified (Fig. 6; S10–20) that gave effective calcium-activated bioluminescence in response to MK-0677 administration. This pool was broken down and a single cDNA clone was identified (clone 7-3), which encoded a putative swine GHS-R that conferred responsiveness of xenopus oocytes to MK-0677. The cRNA of this clone injected into xenopus oocytes also resulted in a robust Ca^{2+}-activated chloride current (see Fig. 6, right panel). Interestingly, once pool complexity dropped below about 50, $G_{\alpha 11}$ administration was no longer required, possibly because even inefficient G protein interactions were now able to rescue MK-0677 responsiveness.

Nucleotide sequence analysis of cDNA clone 7-3 indicated that it represented a G–protein-coupled receptor with seven transmembrane regions (Fig. 7). Clone 7-3 appeared truncated at its NH_2-terminus, and isolation of a full-length cDNA revealed a 366-amino acid-predicted, open-reading frame. Subsequently GHS-R homologues were identified from human and rat pituitary and rat hypothalamus. The human GHS-R cDNA also encoded

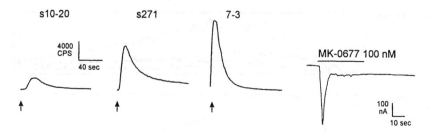

Figure 6 Cloning of a swine GHS-R cDNA from pituitary poly(A⁺)-cDNA libraries: Cloning of GHS-R from swine pituitary poly(A⁺)-cDNA libraries. cDNA pools with a complexity of approximately 10,000 cDNAs per pool were injected into xenopus oocytes with 2 ng of $G_{\alpha 11}$ cRNA and 2 ng of aequorin cRNA. A single pool (S1020) was identified that exhibited a signal over background. This pool was subsequently broken down (pool of 500; S271) leading to the identification of a single clone (7-3) that represented an NH₂ terminally truncated version of the swine GHS-R. Injection of clone 7-3 into xenopus oocytes for the detection of calcium-activated chloride channels exhibited robust MK-0677 responsiveness (*right panel*). Arrows outline the time of MK-0677 administration (1 µM).

a predicted amino acid sequence of 366 amino acids, whereas the two rat cDNAs from pituitary and hypothalamus were identical and encoded a 364-amino acid protein. The GHS-Rs from human, rat, and swine are about 94% identical at the amino acid level. The GHS-R shares several features with members of the family of GPC-Rs, including an ERY GPC-R signature sequence (at the beginning of the second intracellular loop), *N*-linked glycosylation sites, and several serine and threonine phosphorylation sites.

In addition to the full-length seven-transmembrane GPC-R (type 1a cDNA), a shorter truncated version of the GHS-R cDNA (type 1b) was identified that diverges in its amino acid sequence at leucine-265 (the second predicted amino acid of TM 6) and extends for another 24 amino acids (Fig. 8). This short-reading frame is conserved when compared between human and swine type 1b reading frames (13). A type 1b rat cDNA was also identified that differs at this same amino acid, although it encodes a predicted reading frame of 80 amino acids, distinct from the human and swine type 1b-specific reading frames. Analysis of the human GHS-R gene revealed the presence of a single intron, located at the position of type 1a and 1b cDNA sequence divergence. Given the conservation of the intron-derived coding sequence compared between human and swine type 1b cDNAs we anticipate that the type 1b GHS-R cDNA may have functional significance.

```
Rat GHS-R1a   1  MWNATPSEEPEPNVTL-DLDWDASPGNDSLPDEL    33
Hu  GHS-R1a   1  MWNATPSEEPGFNLTLADLDWDASPGNDSLGDEL    34
Pig GHS-R1a   1  MWNATPSEEPGPNLTLPDLGWDAPPENDSLVEEL    34
                                                      TM-1
Rat GHS-R1a  34  LPLFPAPLLAGVTATCVALFVVGISGNLLTMLVV    67
Hu  GHS-R1a  35  LQLFPAPLLAGVTATCVALFVVGIAGNLLTMLVV    68
Pig GHS-R1a  35  LPLFPTPLLAGVTATCVALFVVGIAGNLLTMLVV    68
                                                      TM-2
Rat GHS-R1a  68  SRFRELRTTTNLYLSSMAFSDLLIFLCMPLDLVR   101
Hu  GHS-R1a  69  SRFRELRTTTNLYLSSMAFSDLLIFLCMPLDLVR   102
Pig GHS-R1a  69  SRFREMRTTTNLYLSSMAFSDLLIFLCMPLDLFR   102
                                                      TM-3
Rat GHS-R1a 102  LWQYRPWNFGDLLCKLFQFVSESCTYATVLTITA   135
Hu  GHS-R1a 103  LWQYRPWNFGDLLCKLFQFVSESCTYATVLTITA   136
Pig GHS-R1a 103  LWQYRPWNLGNLLCKLFQFVSESCTYATVLTITA   136
                                                      TM-4
Rat GHS-R1a 136  LSVERYFAICFPLRAKVVVTKGRVKLVILVIWAV   169
Hu  GHS-R1a 137  LSVERYFAICFPLRAKVVVTKGRVKLVIFVIWAV   170
Pig GHS-R1a 137  LSVERYFAICFPLRAKVVVTKGRVKLVILVIWAV   170

Rat GHS-R1a 170  AFCSAGPIFVLVGVEHENGTDPRDTNECRATEFA   203
Hu  GHS-R1a 171  AFCSAGPIFVLVGVEHENGTDPWDTNECRPTEFA   204
Pig GHS-R1a 171  AFCSAGPIFVLVGVEHDNGTDPRDTNECRATEFA   204
                                                      TM-5
Rat GHS-R1a 204  VRSGLLTVMVWVSSVFFFLPVFCLTVLYSLIGRK   237
Hu  GHS-R1a 205  VRSGLLTVMVWVSSIFFFLPVFCLTVLYSLIGRK   238
Pig GHS-R1a 205  VRSGLLTVMVWVSSVFFFLPVFCLTVLYSLIGRK   238
                                                      TM-6
Rat GHS-R1a 238  LWRRR-GDAAVGASLRDQNHKQTVKMLAVVVFAF   270
Hu  GHS-R1a 239  LWRRRRGDAVVGASLRDQNHKQTVKMLAVVVFAF   272
Pig GHS-R1a 239  LWRRKRGEAAVGSSLRDQNHKQTVKMLAVVVFAF   272

Rat GHS-R1a 271  ILCWLPFHVGRYLFSKSFEPGSLEIAQISQYCNL   304
Hu  GHS-R1a 273  ILCWLPFHVGRYLFSKSFEPGSLEIAQISQYCNL   306
Pig GHS-R1a 273  ILCWLPFHVGRYLFSKSLEPGSVEIAQISQYCNL   306
                    TM-7
Rat GHS-R1a 305  VSFVLFYLSAAINPILYNIMSKKYRVAVFKLLGF   338
Hu  GHS-R1a 307  VSFVLFYLSAAINPILYNIMSKKYRVAVFRLLGF   340
Pig GHS-R1a 307  VSFVLFYLSAAINPILYNIMSKKYRVAVFKLLGF   340

Rat GHS-R1a 339  ESFSQRKLSTLKDESSRAWTKSSINT          364
Hu  GHS-R1a 341  EPFSQRKLSTLKDESSRAWTESSINT          366
Pig GHS-R1a 341  EP SQRKLSTLKDESSRAWTESSINT          366
```

	Identity (%)
rat vs. swine	94.5
rat vs. human	96.1
human vs. swine	93.1

Figure 7 Alignment of rat, human, and swine type 1a GHS-R amino acid sequences: Boxed sequences represent identical amino acid residues. Transmembrane domains TM1 through 7 are overlined.

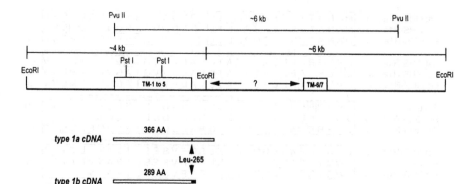

Figure 8 Physical map of human growth hormone secretagogue receptor gene: The human GHS-R gene contains a single intron separating TM1 through 5 from TM6 and 7. The intron is located at the second amino acid of TM6. The predicted structure of the type 1a and 1b cDNA is outlined below the physical map. The open-reading frame of the type 1b cDNA extends into the intron of the GHS-R gene. Preliminary data indicate that an alternative poly(A^+) addition site in the intron generates the 1b mRNA.

VI. EXPRESSION AND PHARMACOLOGICAL CHARACTERIZATION OF GHS-R CDNA

The 1a and 1b GHs-R cDNAs were transiently expressed in COS-7 cells, when searching for a high-affinity [^{35}S]MK-0677-binding site. Only transfection with the 1a cDNA resulted in expression of a single, high-affinity–binding site with a K_d of 400 pM and B_{max} of approximately 800 fmol/mg of cell protein (Fig. 9a). This receptor exhibited the appropriate pharmacology as competition with MK-0677, GHRP-2, and GHRP-6 gave the appropriate rank-order of IC_{50} values compared with their GH-release EC_{50} values (obtained in the rat pituitary GH-release assay; rank-order of potency MK-0677 > GHRP-2 > GHRP-6; see Fig. 9b).

VII. EXPRESSION OF GHS-R MRNA IN RHESUS, HUMAN, AND RAT BRAIN AND RAT PITUITARY

To evaluate a potential CNS function for the GHS-R we performed in situ hybridization in rhesus brain. Two nonoverlapping ^{32}P-labeled GHS-R oligonucleotide probes were used simultaneously in the hybridization. Exposure times averaged 3 weeks, outlining the low level of GHS-R mRNA. Specific signals could be detected in the infundibulum and infundibular

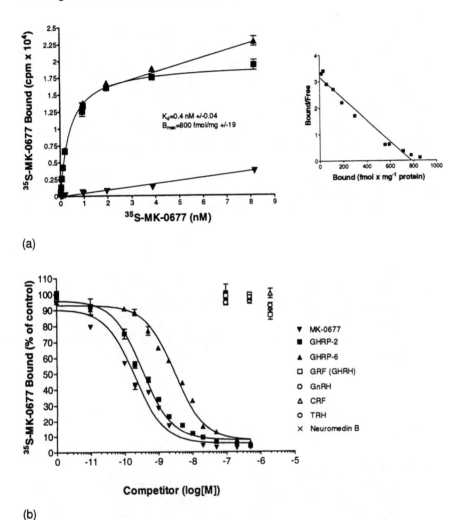

(a)

(b)

Figure 9 High-affinity [³⁵S]K-0677-binding sites on the human GHS-R expressed in COS-7 cells: Expression of the human GHs-R gene into COS-7 cells following lipofectamine transfection of the type 1a GHS-R cDNA (expression of the GHS-R was analyzed after 2 days). (a) A single high-affinity–binding site with a K_d of 0.4 nM and a B_{max} of 800 fmol/mg cell protein was identified. Competition analysis with MK-0677, GHRP-2, and GHRP-6 identifies GHS-R high-affinity–binding sites. (b) Other ligands, including galanin, GHRH, TRH, GnRH, and neurotensin fail to compete for binding of MK-0677 with the GHS-R.

hypothalamus (13). RNase protection was performed to identify GHS-R expression in different human poly(A$^+$)-RNA samples. Detection of the 1a GHS-R mRNA resulted in the generation a 410–nucleotide-protected fragment, whereas precursor mRNA or 1b GHS-R mRNA led to 304– and 110–nucleotide-protected fragments (14,15). Fragments indicative of the expression of type 1a and 1b mRNAs could be detected only when poly(A$^+$)-RNA from the combined thalamic nuclei and the hippocampus, whereas other regions, including cerebellum, caudate nucleus, and thalamic regions did not express GHS-R mRNA.

In situ hybridization on rat brain coronal sections allowed more extensive mapping and resulted in the identification of several additional brain regions in which GHS-R mRNA is expressed (Fig. 10). As predicated, signals

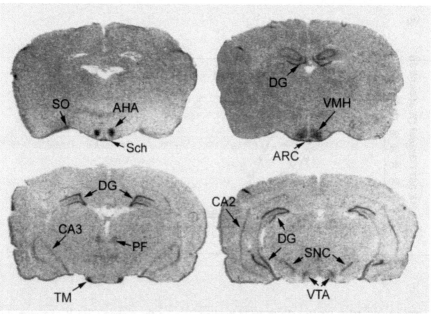

Figure 10 In situ hybridization for rat GHS-R mRNA at coronal rat brain sections: Abbreviations are: AVPO, anteroventral preoptic nucleus; PVN, paraventricular nucleus; AHA, anterior hypothalamic area; Sch, suprachiasmatic nucleus; So, supraoptic nucleus; LA, lateroanterior hypothalamic nucleus; VMH, ventromedial hypothalamic nucleus; ARC, arcuate nucleus; TM, tuberomammillary nucleus; DG, dentate gyrus and CA2, CA3 regions of the hippocampus; PF, parafascicular thalamic nucleus; SNC, pars compacta of the substantia nigra; VTA, ventral tegmental area; EW, Edinger–Westphal nucleus; DR, dorsal raphe nucleus; Mnr; median raphe nucleus; LDTg, laterodorsal tegmental nucleus; Fn, facial nucleus.

could be detected in the arcuate nucleus (ARC), the ventral medial hypothalamus (VMH), and the paraventricular nucleus (15). The suprachiasmatic nuclei (Sch), preoptic nucleus, supraoptic nucleus (S0), anterior hypothalamic area (AHA), lateroanterior hypothalamic area (15), and tuberomammilary nuclei (TM) showed distinct hybridization signals. In addition to hybridization at these sites the dentate gyrus (DG) and CA2 and CA3 regions of the hippocampus and dopaminergic neurons of the pars compacta of the substantia nigra (SNC) and the ventral tegmental area (VTA) hybridized. Several regions in the brain stem, including the Edinger–Westphal, dorsal, and median raphe nuclei gave distinct hybridization signals (15). Finally, the laterodorsal tegmental area and the facial nerve revealed hybridization signals for GHS-R mRNA. In situ hybridization with rat pituitary outlined GHS-R mRNA signals in the anterior pituitary, whereas the posterior pituitary failed to provide any specific signals (15). The distribution pattern of 1a-specific probes or signals for the combined 1a and 1b mRNAs were identical in brain and pituitary. Therefore, we do not have evidence for distinct expression patterns of either cDNA. We conclude that we have isolated a GHS-R cDNA from human, swine, and rat, specifically expressed in hypothalamic and other brain nuclei and in the pituitary. The presence of GHS-R RNA at these sites provides evidence for a third neuroendocrine GH control system.

VIII.   GHS-R FAMILY MEMBERS AND GHS-R GENE CHROMOSOMAL LOCATION

Hybridization of the GHS-R cDNA with restriction enzyme-digested, size-separated nuclear DNA from different species indicated that the GHS-R is encoded by a single, highly conserved gene (Fig. 11, left panel). At a reduced stringency of hybridization additional bands can be detected, indicative of the presence of related genes (see Fig. 11). The human GHS-R is most closely related to the neurotensin receptor ($\sim 35\%$ identity) and the thyrotropin-releasing hormone (TRH) receptor ($\sim 29\%$ identity; Fig. 12). We are in the process of identifying related GHS-R genes to evaluate their potential neuroendocrine functions.

The GHS-R has been mapped to band 3Q26.2 (A. Howard and L.H.T. Van der Ploeg, unpublished observations). This location is close to the proposed map position for the Brachman–de-Lange syndrome a pre- and postnatal growth deficiency (22,23). The determination of the potential significance of this observation is in progress.

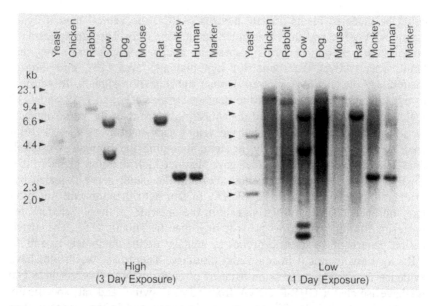

Figure 11 Southern-blotting analysis of the GHS-R: Hybridizations were performed with a human GHS-R cDNA probe to *Eco* RI restriction enzyme-digested DNA from different species (indicated on top of the panel). (*Left panel*) high stringency post-hybridizational washes (65°C and 0.1 × SSC); (*right panel*) low-stringency post-hybridizational washes for the same Southern blot (55°C and 6 × SSC).

IX. CONCLUDING REMARKS

We isolated GHS-R cDNAs encoding a seven-transmembrane GPC-R that confers sensitivity of different vertebrate species to peptide and nonpeptide GHSs. The GHS-R is expressed at several CNS nuclei and in the pituitary; GHS-R expression could not be detected in other tissues. The identification of the GHS-R excludes that MK-0677 or peptide GHSs act on an alternatively spliced form of the GHRH-R or another previously identified GPC-R of somatotrophs or the hypothalamus.

We assume that GHSs mediate GH release through combined action with GHRH, synergistically mediating GH exocytosis in somatotrophs. The essential nature of this synergy is brought out in the loss of GHS sensitivity in hypothalamic–pituitary stalk-sectioned animals, which in the absence of portal blood-derived GHRH, lack GHS responsiveness (24). GH and IGF-1 provide negative regulatory feedback on neurons in the arcuate nucleus, presumably through enhancing somatostatin tone, which antagonizes GH release. The localization of GHS-Rs in the CNS indicates that GHSs may

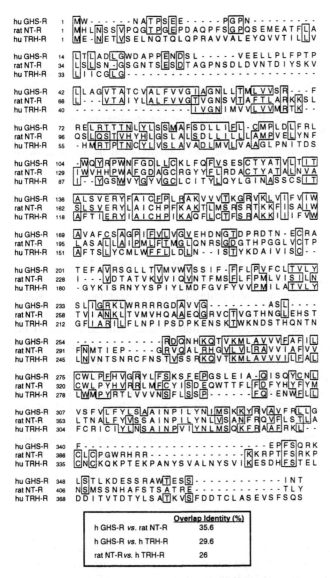

	Overlap Identity (%)
h GHS-R *vs.* rat NT-R	35.6
h GHS-R *vs.* h TRH-R	29.6
rat NT-R *vs.* h TRH-R	26

Figure 12 Alignment of human type 1a GHS-R to rat NT-R and human thyrotropin-releasing hormone receptor (TRH-R) protein sequences: Identical amino acid residues are boxed. Gaps (hyphens) are introduced to optimize alignment.

also mediate GHRH release and possibly reduce somatostatin tone. The combined action of GHSs at hypothalamic neurons and somatotrophs will aid our understanding of the mechanisms by which GHSs restore pulsatile GH release in the elderly. The potential role of GHS-R mRNA on the other neurons in the hypothalamus, hippocampus, and brain stem, including, dopaminergic-, serotonergic-, and NPY-containing neurons (15,25,26), requires further investigation. We conclude that the localization of the GHS-R at several hypothalamic nuclei and the pituitary provides evidence for the presence of a third neuroendocrine pathway involved in the control of GH release, presumably under the control of an as yet undefined hormone.

ACKNOWLEDGMENTS

We thank Jennifer Anderson, Michael Chou, Doris Cully, Carmen Diaz, Mike Dashkevicz, Alex Elbrecht, Michael Hamelin, Donna Hreniuk, Karen Kulju, Paul Liberator, Oksana Palyha, Philip Paress, Charles Rosenblum, Xiao-Ming Guan (Genetics and Molecular Biology); Bret Defranco, Keith Judd, Eve Szekely, Don Thompson (Branchburg Farms); Joe Arena, Lee Chaung, Ken Liu, Anna Pomes, Sheng-Shung Pong, Jim Schaeffer (Cell Biology and Physiology); Dennis Dean, Dave Melillo (Drug Metabolism); Ravi Nargund, Art Patchett (Exploratory Chemistry); Patrick Griffin and Tracy Clark (Analytical Biochemistry); Juli DeMartino (Inflammatory Research); Robert Havens, Mike Riby, Dalip Sirinathsinghji (Terlings Park Neurosciences Center); Kristine Prendergrast, Dennis Underwood (Molecular Design); Molecular Cell Science Inc., Gwen Childs (University of Texas); Keith Elliston (Bioinformatics); and Sunil Gupta (Cellular and Molecular Biology) for their participation and invaluable input into the GHS-R-cloning program.

REFERENCES

1. Bowers CY, Momany FA, Reynolds GA, Hong A. On the in vitro and in vivo activity of a new synthetic hexapeptide that acts on the pituitary to specifically release growth hormone. Endocrinology, 1984; 114:1537–1545.
2. Cheng K, Chan W-S, Barreto A, Convey EM, Smith RG. The synergistic effects of His-D.Trp-Ala-Trp-D.Phe.-Lys-NH$_2$ on GRF stimulated growth hormone release and intracellular cAMP accumulation in rat primary pituitary cell cultures. Endocrinology 1989; 124:2791–2797.
3. Frohman LA. In: Felig P, Baxter JD, Broadus AE, Frohman LA, eds. Endocrinology and Metabolism. New York: McGraw-Hill, 1987.
4. Smith RG, Pong S-S, Hickey G, Jacks T, Cheng K, Leonard R, Cohen CJ, Arena

TP, Chang CH, Drisko J, Wyvratt M, Fisher M, Nargund R, Patchett A. Modulation of pulsatile GH release through a novel receptor in hypothalamus and pituitary gland. Rec Prog Horm Res 1996; 52:261–286.

5. Smith RG, Cheng K, Pong S-S, et al. A nonpeptidyl growth hormone secretagogue. Science 1993; 260:1640–1643.

6. Pong S-S, Chaung L-YP, Leonard RJ. The involvement of ions in the activity of a novel growth hormone secretagogue L-692,429 in rat pituitary cell culture. Proceedings of the 75th Meeting of the Endocrine Society. Bethesda, MD: Endocrine Society, 1993: 172.

7. Bresson-Bepoldin L, Dufy-Barbe L. GHRP-6 induces a biphasic calcium response in rat pituitary somatotrophs. Cell Calcium 1994; 15:247–258.

8. Herrington J, Hille B. Growth hormone-releasing hexapeptide elevates intracellular calcium in rat somatotrophs by two mechanisms. Endocrinology 1994; 135:1100–1108.

9. Lei T, Buchfelder M, Fahlbusch R, Adams EF. Growth hormone releasing peptide (GHRP-6) stimulates phosphatidylinositol (PI) turnover in human pituitary somatotroph cells. J Mol Endocrinol 1995; 14:135.

10. Cheng K, Chan W-S, Butler B, Barreto A, Smith RG. Evidence for a role of protein kinase-C in His-DTrp-Ala-Trp-DPhe-Lys-NH$_2$-induced growth hormone release from rat primary pituitary cells. Endocrinology 1991; 129:3337–3342.

11. Bowers CY. [Editorial] On a peptidomimetic growth hormone-releasing peptide. J Clin Endocrinol Metab 1994; 79:940–942.

12. Rudman D, Feller AG, Nagraj HS, Gergans GA, Llitha PY, Goldberg AF, Schlenker RA, Cohn L, Rudman IW, Mattson DE. Effects of human growth hormone in men over 60 years old. N Engl J Med. 1990; 323:1.

13. Howard AD, Feighner SD, Cully DF, Arena JP, Liberator PA, Rosenblum CI, Hamelin MJ, Hreniuk DL, Palyha OC, Anderson J, Paress PS, Diaz C, Chou M, Liu K, McKee KK, Pong S-S, Chaung L-Y, Elbrecht A, Dashkevicz M, Heavens R, Rigby M, Sirinathsinghji DJS, Dean DC, Melillo DG, Patchett AA, Nargund R, Griffin PR, DeMartino JA, Gupta SK, Schaeffer JM, Smith RG, Van der Ploeg LHT. A new G-protein coupled receptor of the pituitary and hypothalamus involved in pulsatile GH release. Science 1996; 273:974–977.

14. Kulju McKee K, Palyha OC, Feighner SD, Hreniuk DL, Tan C, Phillips M, Smith RG, Van der Ploeg LHT, Howard AD. Molecular analysis of rat pituitary and hypothalamic growth hormone secretagogue receptors. Mol Endocrinol 1997. In press.

15. Guan X-M, Yu H, Palyha OC, Kalju McKee K, Feighner SD, Sirinathsinghji DS, Smith RG, Van der Ploeg LHT, Howard AD. Distribution of mRNA encoding the growth hormone secretagogue receptor in brain and peripheral tissues. Mol Brain Res 1997. In press.

16. Dean DC, Nargund RP, Pong S-S, Chaung L-Y, Griffin P, Melillo DG, Ellsworth RL, Van der Ploeg LHT, Patchett AA, Smith RG. Development of a high specific activity sulfur-35 labelled sulfonamide radioligand that allowed the identification of a new growth hormone secretagogue receptor. J Med Chem 1997; 39:1767–1770.

17. Pong S-S, Chaung L-YP, Dean DC, Nargund RP, Patchett AA, Smith RG. Identification of a new G-protein coupled receptor for growth hormone secretagogues. Mol Endocrinol 1996; 10:57–61.
18. Hreniuk D, Howard A, Rosenblum C, Palyha O, Diaz C, Cully D, Paress P, McKee K, Dashkevicz M, Arena J, Liu K, Schaeffer J, Nargund R, Smith R, Van der Ploeg LHT, Feighner S. Cytochemical identification and cloning of a G-protein coupled growth hormone secretagogue receptor. Soc Neurosci Abstr 1996; 22 (part 2):1300.
20. Sirinathsinghji DJS, Chen HY, Hopkins R, Trumbauer M, Heavens R, Rigby M, Smith RG, Van der Ploeg LHT. Induction of c-*fos* mRNA in the arcuate nucleus of normal and mutant growth hormone-deficient mice by a synthetic non-peptidyl growth hormone secretagogue. Neurol Rep 1995; 6:1989–1992.
21. Dickson SL, Doutrelant-Vilart O, Leng G. GH-deficient dw/dw rats and lit/lit mice show increased Fos expression in the hypothalamic arcuate nucleus following systemic injection of GH-releasing peptide-6. J Endocrinol 1995; 146:519.
22. Jackson LG. de Lange syndrome. Am J Med Genet 1992; 42:377–378.
23. Jackson L, Kline AD, Barr MA, Koch S. de Lange syndrome: a clinical review of 310 individuals. Am J Med Genet 1993; 47:910–946.
24. Smith RG, Van der Ploeg LHT, Cheng K, Hickey G, Wyvratt MW, Fisher M, Nargund R, Patchett A. Peptidomimetic regulation of GH secretion. Endocr Rev 1997. In press.
25. Dickson SL, Doutrelant-Viltart O, Dyball REJ, Leng G. Retrogradely labelled neurosecretory neurones of the rat hypothalamic arcuate nucleus express Fos protein following systemic injection of GH-releasing peptide-6. J Endocrinol 1997; 151:323–331.
26. Dickson SL, Luckman SM. Induction of c-*fos* messenger ribonucleic acid in neuropeptide Y and growth hormone (GH)-releasing factor neurons in the rat arcuate nucleus following systemic injection of the GH secretagogue, GH-releasing peptide-6. Endocrinology 1997; 138:771–777.

5

A Mathematical Model of Growth Hormone Release from the Pituitary

David Brown and Elinor A. Stephens
The Babraham Institute, Cambridge, England

Gareth Leng
University of Edinburgh Medical School, Edinburgh, Scotland

Roy G. Smith
Merck Research Laboratories, Merck and Co., Inc., Rahway, New Jersey

I. INTRODUCTION

The secretion of growth hormone (GH) in the rat is sexually dimorphic. In both sexes secretion is pulsatile, but in males the pulses are larger, less frequent (approximately every 3 h), and arise from a lower interpulse baseline than in females (1). That pulsatile GH secretion fuels faster growth has been demonstrated in animals deficient in growth hormone, and in animals experimentally deprived of hypothalamic GH-releasing factor (GRF; 2–4). Pulses of growth hormone secretion derive from episodic secretion of GRF, but growth hormone secretion is also regulated by the secretion of somatostatin (5–8). However, the response of the pituitary to coordinated release from the hypothalamus of GRF and inhibitory somatostatin somatotropin release-inhibiting factors (SRIF) is complex. First of all, in response to continued application of GRF, the pituitary somatotrophs rapidly desensitize, a process important in determining the pulsatile profile of GH secretion. Secondly, the pituitary amplifies the pulsatile output of GRF from the hypothalamus, the amplification being nonlinear (the dose–response curve of the pituitary to releasing factors is sigmoidal when plotted using a log dose scale). The nonlinearity is partly a consequence of short-term desensitization at the pituitary, which contributes to terminating secretory episodes, resulting in a stereotyping of both pulse amplitude and duration. Furthermore, the pituitary coordinates several inputs from the hypothalamus and elsewhere.

One of the inputs from the hypothalamus is somatostatin, which inhibits the secretion of growth hormone and, while acting at a separate receptor, acts as a functional GRF antagonist, in the sense that in the presence of somatostatin on the growth hormone-releasing ability of GRF is attenuated; both the GRF receptor and the somatostatin receptor are G-protein-linked receptors, and activation of these receptors produce opposing effects on intracellular calcium (Ca^{+2}) and cAMP levels, and on calcium entry, in particular through L-type channels. The sexually dimorphic patterns of growth hormone secretion in the rat appear to derive from sexually dimorphic behavior of the somatostatin neurons, possibly reflecting the sexually dimorphic expression of androgen receptors by these neurons (9). In the male rat, GRF and somatostatin are probably released alternately to produce peaks and troughs of GH release, respectively, whereas in the female, somatostatin is released more continuously. However, inferring the nature of the hypothalamic signals from the pituitary response is not simple, because the pituitary responsiveness to releasing factors is complex, being critically dependent on the previous history of exposure. This complexity arises from the interactions of the hypothalamic factors with each other, from desensitization of the pituitary during sustained exposure to probably either factor, from actions of the hypothalamic factors on the synthesis of growth hormone, and, for somatostatin, a dramatic "off" effect when somatostatin is removed, reflected by an increase of basal release that occurs in the absence of GRF and an increase in the sensitivity of the somatotrophs to GRF (10,11).

Application of artificial growth hormone secretagogues (GHSs) also promotes GH release in a similar dose-dependent manner by different receptors, and there is evidence in vivo of synergism between the two stimulants, in the sense that GH release is in total greater when GRF and secretagogue are applied together than the sum of the releases when they are applied separately.

II. MODELING THE PITUITARY RELEASE OF GROWTH HORMONE

We have based a preliminary mathematical model of pituitary response to GRF and SRIF application (12) on the law of mass action applied to reversible binding of GRF or secretagogue to its receptors (Fig. 1). For the

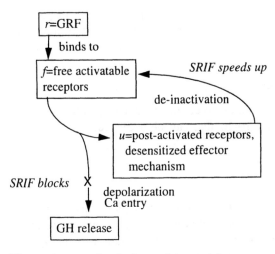

Figure 1 A schematic form of the model.

present, we phrase our general discussion just in terms of GRF. However, in general, analogous arguments hold for the actions of GH-releasing peptide (GHRP) and other artificial secretagogues at the pituitary when these are applied in isolation; accordingly, the following model framework can be extended naturally to encompass these additional factors. How we model the pituitary response to simultaneous application of GRF and artificial secretagogues is beyond the scope of this paper.

The GRF acts by G–protein-coupled receptors, and the transduction pathway involves activation of adenylate cyclase and protein kinase A, elevated levels of cAMP and calcium entry through voltage-gated channels, and calcium–entry-dependent exocytosis. Somatostatin acts at a separate G–protein-coupled receptor to oppose the GRF-induced tranduction pathway, reducing the intracellular concentration of cAMP, and also induces hyperpolarization by increasing potassium conductance. The release signal is probably a function of localized intracellular calcium concentration arising from calcium entry through L-type channels, which are modulated by second-messenger pathways. We could attempt to model all these processes in detail, but we have avoided this for two reasons. First, the necessary quantitative information about at least some of the stages is not currently available. Second, our approach to modeling complex systems is to adopt the simplest model that explains the range of phenomena exhibited by the system under study, and that can be related to experimental data and experimental interventions. In our model, therefore, we assume that the release rate of GH is directly proportional to the rate of binding of GRF, but inversely proportional to the concentration of SRIF, resulting in possibly substantial pulses of growth hormone in the absence of SRIF and much smaller pulses in its presence.

A. The Mathematical Model

The GRF (R, concentration r) binds reversibly to those receptors (F, density f) on the pituitary somatotrophs that can be activated:

$$R + F \underset{k_b}{\overset{k_1}{\rightleftharpoons}} U$$

If the rate of the forward reaction k_1 is sufficiently high compared with the rate of the back reaction k_b, the continued presence of GRF will cause depletion of the activatable receptors and, hence, desensitization of the system. The effects of SRIF are twofold. First of all, it resensitizes the system after densitization. To model this, we assume that the rate of the backward reaction depends on the concentration of SRIF in the extracellular space s, being $k_b = k_2 + k_3\phi(s)$, where $\phi(s) = 1/(1 + \exp[-(s - s_0)/\delta_0])$ is a sigmoidal function of s, rising from zero to unity over a range of concentrations of SRIF of $s_0 \pm 3\delta_0$. The step in the sigmoidal function, $\phi(s)$, can be made more or less steep by reducing or increasing δ_0. k_3 is positive, so the rate of the backward reaction is greater in the presence of somatostatin. We thus obtain the following equations:

$$\frac{df}{dt} = -k_1 rf + [k_2 + k_3\phi(s)]u$$

$$\frac{du}{dt} = k_1 rf - [k_2 + k_3\phi(s)]u$$

where u reflects the number of postactivated, desensitized GRF receptors and r is the concentration of GRF in the extracellular space (see Fig. 1). Some constitutive activation of the secretory pathway (which occurs even in the absence of GRF) probably occurs, which may be equivalent to assuming a nonzero resting level of cAMP. Possibly the simplest way to include this in our current model is to represent it as an equivalent level of GRF, given by r_0. So the equations then become

$$\frac{df}{dt} = -k_1(r + r_0)f + [k_2 + k_3\phi(s)]u \tag{1}$$

$$\frac{du}{dt} = k_1(r + r_0)f - [k_2 + k_3\phi(s)]u \qquad (2)$$

The rate of release of growth hormone is proportional to the rate of binding, but also dependent on the level of SRIF. To model this second action of SRIF, we make the constant of proportionality a function of s. The resulting release rate function is given by

$$\text{Release rate at time } t = \rho(r,s,t) = \{k_4 + k_5[1 - \phi(s)]\}[r + r_0]f \qquad (3)$$

Thus, in the presence of very high levels of SRIF (where $\phi(s) = 1$), the rate of release is $k_4(r + r_0)f$, and in the absence of SRIF it is $\{k_4 + k_5[1 - \phi(0)]\}[r + r_0]f$. If we wish to allow SRIF to dynamically change, we also need a differential equation for s,

$$\frac{ds}{dt} = I_s - k_7 s \qquad (4)$$

where I_s is the rate of release (of infusion) of SRIF, and k_7 its decay rate. The dynamics of SRIF concentration are separate from the other dynamic variables: SRIF affects the dynamics of the interaction between GRF and its receptors, and GH release, but there are no effects of these on SRIF release or decay.

A further differential equation in r could be added

$$\frac{dr}{dt} = I_r - k_1(r + r_0)f - k_6 r$$

where k_6 is the relative rate of decay of GRF in the absence of binding, and I_r is the input of GRF. Frequently, however, other factors have a substantial effect on r, and it is simpler for the purposes of initial modeling to assume that external factors maintain the level of r either as a constant infusion or a pulsatile pattern of stimulation.

An understanding the model may also be easier if we think of SRIF acting as a switch that changes the behavior of the somatotroph when present at suprathreshold concentration (substantially above s_0). In the absence of SRIF, successive pulses of GRF desensitize the somatotroph, as the receptor–effector mechanism is progressively inactivated. Somatostatin resensitizes the somatotroph, by speeding up the recovery of the receptor-effector mechanism to the free, available state. The variable u, being proportional to the number of postactivated, desensitized GRF receptors, can be thought of as an index of the extent of desensitization of the release system. If we make the further assumption that the total number of receptors is constant (i.e., in the short-term, receptors are neither created nor destroyed), and the total concentration of sensitized receptors is f_T. then we can write $u = f_T - f$ and, therefore, f can equally well be considered a complementary *index of sensitivity*: the ability of the somatotrophs to respond to stimulation by GRF.

B. Temporal Profile of GH Release and Its Relation to GRF and SRIF Concentrations

The foregoing formulation of binding is the simplest possible that could be included in this type of dynamic model. We could, for example, have made binding follow Michaelis–Menten kinetics, thereby substantially complicating Eq. (1), and this is an elaboration of the consequences of which we intend studying further. We could also have involved further compartments specifically reflecting desensitized receptors in the free and bound state as has been done by Goldbeter and colleagues (13). An advantage of our current simple model formulation is that, for periods when r and s are constant, the differential equations can be easily solved explicitly to give expressions for the extent of desensitization and for the rate of GH release. Making the substitution $u = f_T - f$ (where f_T is the total concentration of free, activatable receptors in the completely sensitized state (i.e., the maximal concentration of activatable receptors), and $k_b = k_2 + k_3\phi(s)$ in Eq. (1), we obtain

$$\begin{aligned}
\frac{df}{dt} &= -k_1(r + r_0)f + k_b(f_T - f) \\
&= k_b f_T - [k_1(r + r_0) + k_b]f \\
&= k_b f_T - Kf
\end{aligned}$$

where $K = k_1(r + r_0) + k_b$. The equilibrium values of f, f^*, is given by solving $df/dt = 0$, obtaining

$$f = \frac{k_b f_T}{K} = \frac{k_b f_T}{k_1(r + r_0) + k_b} = \frac{[k_2 + k_3\phi(s)]f_T}{k_1(r + r_0) + k_2 + k_3\phi(s)} \tag{5}$$

The f^* is an irreducible minimum sensitivity below which the sensitivity of the system will not fall. This is a *dynamic* equilibrium or minimum, although because it depends on both r and s, it will, therefore, change as they do. Rewriting the differential equation in terms of f^*, we obtain

$$\frac{df}{dt} = -K(f - f^*) \tag{6}$$

with solution

$$f(t) = f^* + (f_0 - f^*)e^{-Kt} \tag{7}$$

assuming that the initial concentration of free receptors at time $t = 0$ is f_0. Thus we see that, in our very simple model, a quantitative index of the current sensitivity of the system is given by f, the density of activatable receptors, which follows the differential Eq. (1) and (2), with solution given by Eq. (7), which can be rewritten

$$f(t) = \frac{k_b f_T}{K} + \left(f_0 - \frac{k_b f_T}{K} \right) e^{-kT}$$

consisting of two components: a basal, equilibrium level, which can be sustained indefinitely, $f^* = k_b f_T/K$, and a more transient response consisting of the excess over this basal level initially equal to $(f_0 - f^*)$, which decays exponentially with relative rate, $K = k_1(r + r_0) + k_b$ (Fig. 2). The equilibrium level f^* (i.e., the sustainable sensitivity of the system) is thus seen to be directly proportional to the rate of the backreaction k_b (and through this to the current level of SRIF, s) and to the total density of free receptors f_T, and thus increases as k_b and f_T increase. On the other hand, it becomes lower as both the rate of the forward reaction k_1, and the current effective density of GRF (i.e. $r + r_0$) increase.

The release rate, as specified by Eq. (3), is given by

$$\rho(r,s,f) = \{k_4 + k_5[1 - \phi(s)]\}(r + r_0)f$$

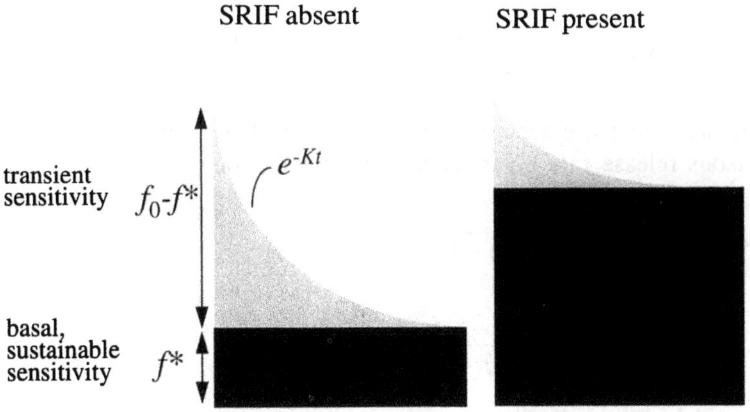

Figure 2 Relation of sensitivity to time: A convenient index of system sensitivity (i.e. the ability of the system to respond to fresh pulses of GRF) is given by f, the density of activatable GRF receptors. This consists of two components, a *basal* or *sustainable component*, which can be sustained indefinitely at the current levels of GRF and SRIF, f^*, and a *transient component* $(f_0 - f^*)e^{-Kt}$ illustrated schematically here at low and high SRIF concentrations.

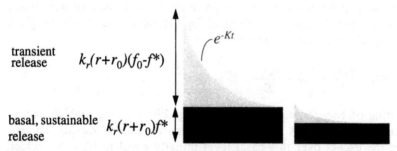

Figure 3 Sustainable and transient components of instantaneous release rate: Corresponding to the sustainable and transient levels of system sensitivity, there are sustainable and transient components of the instantaneous GH release rate, illustrated schematically at low and high SRIF concentrations.

If we write k, as shorthand for $k_4 + k_5[1 - \phi(s)]$, the rate constant for release, and substitute for f using Eq. (7) we get

$$\rho(r,s,t) = k_r(r + r_0)f = k_r(r + r_0) \left[f^* + (f_0 - f^*)e^{-Kt} \right]$$

or more fully,

$$\rho(r,s,t) = k_r(r + r_0) \left[\frac{k_b f_T}{k_1(r + r_0) + k_b} + \left(f_0 - \frac{k_b f_T}{k_1(r + r_0) + k_b} \right) e^{-Kt} \right] \quad (8)$$

Thus, we see that there are transient and sustainable components of the instantaneous release rate given by partitioning Eq. (8)

$$\rho_{sust} = k_r(r + r_0) \frac{k_b f_T}{k_1(r + r_0) + k_b}$$

$$\rho_{trans} = k_r(r + r_0) \left(f_0 - \frac{k_b f_T}{k_1(r + r_0) + k_b} \right) e^{-K_t}$$

as illustrated in Figure 3.

We can then calculate the cumulative amount released in time $(0,t)$ as the integral of Eq. (8), obtaining

$$P(r,s,t) = k_r(r + r_0) \left\{ f^* t + \left[\frac{(f_0 - f^*)(1 - e^{-Kt})}{K} \right] \right\} \quad (9)$$

if again, the initial free receptor concentration is f_0. These equations enable us to obtain theoretical dose–response curves for any concentration of SRIF, s, and for any set of model parameters. We can then fit these mathematical expressions to dose–response curve data, first of all, as a test of the model and, second, to estimate the model parameters.

III. FITTING AND EMPIRICAL TESTING OF THE MODEL

There is a very wide range of experimental data, from in vitro and in vivo experiments, that could be used to test the model and to estimate parameters not directly measurable. We present here the results of such empirical tests for two sets of in vitro data.

A. Infusions of GHRP in the Presence and Absence of SRIF

In a perifusion experiment, somatotroph cells were infused with GHRP-6 at a constant 100-nM concentration, and GH concentrations were measured in the perifusate as displayed in Figure 4. A further experiment was also performed with the same protocol except in the presence of SRIF at 1-nM concentration. We obtain a mathematical expression predicting the GH concentration for both of these cases, as follows. By substituting for f in the expression given for ρ, and assuming the initial concentration of free receptors is f_0, we obtain

$$\rho(r,s,f) = k_r(r + r_0)[f^* + (f_0 - f^*)e^{-Kt}] = A + Be^{-Kt} \qquad (10)$$

Because the rate of change of the concentration of GH in the perifusate h is increased by release at rate ρ and depleted as a result of the perifusion at a rate λh, the rate of change of h is given by the difference of these two quantities

$$dh/dt = \rho - \lambda h \qquad (11)$$

where λ is the perifusion rate. Substituting Eq. (10) for ρ in Eq. (11) and solving the resulting differential equation we obtain

$$h = \frac{A}{\lambda} + \frac{B}{\lambda - K} e^{-Kt} - \left(\frac{A}{\lambda} + \frac{B}{\lambda - K}\right)e^{-\lambda t} \qquad (12)$$

We fitted this nonlinear statistical model by nonlinear least squares to the two data sets, assuming a common value of λ in the two experiments, but different values of A, B, and K. Here and in the next section, the Genstat statistical package (14) was used for fitting nonlinear models. The estimated curves are plotted in Figure 4. Also plotted are the release rate profiles of

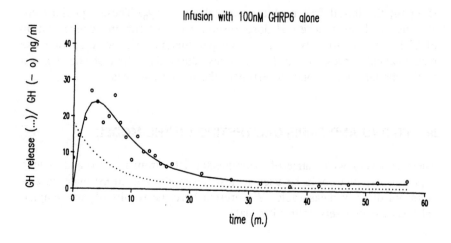

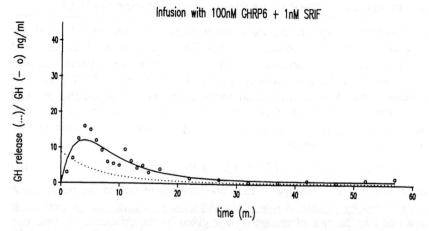

Figure 4 Fitting the model to a GHRP infusion experiment: The open symbols show measured GH release from isolated, dispersed somatotrophs in vitro during an infusion with 100 nM GHRP-6 for 60 min (from time 0) in the top frame, and for a similar experiment but in the presence of SRIF in the lower frame. The model assumes that desensitization is a smooth, continuous process, directly associated with the process of receptor activation that leads to GH release, but this is not readily apparent from the observed data. The measured data at first sight appear consistent with the interpretation that there are successive phases of sensitization and desensitization, or with the interpretation that there is a lag before the onset of desensitization. We have used the model described in the text to generate the best model fits to the two sets of observed data, and the fits are indicated by the thin lines. The dotted lines denote the rate of release of GH that would result in the fitted GH concentration curves, confirming that the observed data are consistent with rapid, smooth desensitization of the GH-release mechanism.

GH from the somatotrophs that would result in these fitted curves. Within the limitations of the data, the model fits adequately. The fitted curves fix A, B, K, and λ. The estimates and their standard errors (se) are in the presence of SRIF: $A = 0.33$ (se 0.40), $B = 8.5$ (se 1.7), $K = 0.16$ (se 0.07); and in the absence of SRIF, $A = 0.90$ (se 0.43), $B = 17.2$ (se 2.3), $K = 0.18$ (se 0.06). The common estimate of λ was 0.40 (se 0.15). A can be interpreted directly as the sustainable GH release rate, B as the initial value of the transient release rate, and K as the relative decay rate of the transient release component. However, we can estimate k_1, k_b, and k_r and equilibrium levels of free receptor (in the presence and absence of SRIF) from these figures, but only if we make certain assumptions—or have prior knowledge— about some parameters. That assumptions or prior knowledge are required can be seen as follows. The relation between A, B, K, and the parameters of the dynamic model are: $A = k_r(r + r_0)f^*$, $B = k_r(r + r_0)(f_0 - f^*)$, $K = k_1(r + r_0) + k_b$ where $f^* = k_b f_T / K$. Thus, for each of the experiments, there are five unknown parameters, k_r, r_0, f_0, k_1, and k_b, and only three constants estimated from the experiment. Only if we know, or at least can make some justifiable assumptions about the extent of desensitization of the receptors at the beginning of the experiment, and if we know the magnitude of the constitutive activation of the GH pathway, can we estimate the dynamic parameters k_r, k_1, and k_b. The inclusion of a preliminary application of somatostatin during which GH release is measured in the protocol of the experiment is one method by which the magnitude of constitutive activation can be assessed and the receptors completely resensitized.

B. GRF Dose–Response Curves in the Presence and Absence of Somatostatin

Stronger inferences can be made from experiments in which several levels of the secretagogue are infused. Consider, as an example, the data of Vale et al. (15), who obtained dose–response curves for range of doses of purified recombinant human GRF (rhGRF) in the presence of (1) zero SRIF, (2) 0.5 nM SRIF-14, or (3) 10 nM SRIF-28 (Fig. 5). Note that data are available for only one time point in this experiment, and so they do not enable us to carry out a direct check on the adequacy of dynamic aspects of the model, only the relation at one time of cumulative release to GRF and SRIF concentration. If we assume that the pituitary cells are initially desensitized to constitutive activation (equivalent to a level of GRF of r_0), we see, by applying Eq. (5), that the initial concentration of free receptors if $f_0^* = k_b f_T / (k_1 r_0 + k_b)$. Equation (9) can then be rewritten

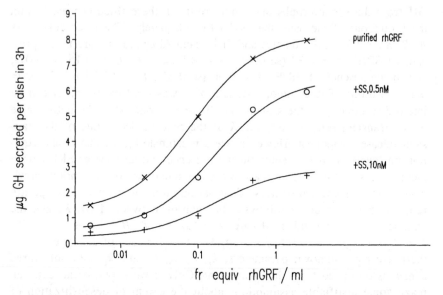

Figure 5 Fitting GRF dose–response curves in the presence and absence of SRIF: To use the model for quantitative, rather than qualitative predictions it is necessary to determine the parameters of the model appropriately, and then to show that the model thus adjusted provides a good predictor of data collected independently in comparable conditions. The symbols indicate measured GRF dose–response curves for GH release from isolated somatotrophs. GH released (in micrograms GH secreted per dish in 3 h) is plotted against log (rhGRF dose) when given alone (X), with 0.5 nM SS (O) and with 10 nM SS (+). The curves show fits of the model, described in the text, to these data; best fits have been found using nonlinear least squares (see text). (Data from Ref. 15.)

$$P(r) = \frac{k_r k_b (r + r_0) f_T}{K_0 + k_1 r} \left[t + \left(\frac{1}{K_0} - \frac{1}{K_0 + k_1 r} \right)(1 - e^{-(K_0 + k_1 r)t}) \right] \tag{13}$$

where $K_0 = k_1 r_0 + k_b$. This was fitted by nonlinear least squares to these three sets of data, allowing use of different SRIF concentration-dependent parameters, k_r, k_b, but all other remaining parameters were the same, resulting in the fitted curves also plotted on Figure 5. The resulting estimated rates (all per minute) were as follows: zero SRIF, k_b – 0.0065, k_r = 0.63; 0.5 nM SRIF, k_b = 0.018, k_r = 0.24; 10 nM SRIF, k_b = 0.017, k_r = 0.11. The common estimates of k_1 and r_0 for the three sets were k_1 = 0.15, r_0 = 0.015. The model predictions fit the data well. The k_r parameters fitted separately for each set show the orderings expected a priori. The k_b estimates are similar for the two doses of SRIF, suggesting that there is no further resensitizing effect for the higher dose of SRIF, although this higher dose inhibits release more substantially.

IV. MODEL PROPERTIES

A. Response to Pulsatile Application of Secretagogue: Desensitization and Resensitization

In vitro and in vivo, the somatotrophs display a striking desensitization to GRF, and to GHRP or other artificial secretagogue and, as discussed earlier, our model also exhibits this property (Fig. 6). Desensitization might occur at many stages between ligand binding to the receptors and exocytosis; however, desensitization to GRF is not accompanied by desensitization to other secretagogues, such as GHRP-6, which act through different G–protein-coupled receptors and different intracellular second-messenger pathways, but that also result in exocytosis by L-type channel-gated calcium entry. Thus desensitization does not reflect depletion of a readily releasable pool of granules, nor inactivation of the calcium channels. In our model the GRF receptor mechanism becomes transiently inactivated during sustained exposure to GRF; hence, the model displays dose- and interval-dependent desensitization of growth hormone release in response to regular pulses of GRF (see Ref. 12).

B. Rebound Hypersecretion After Application of SRIF

In the presence of a sufficiently high concentration of SRIF, both release of growth hormone and desensitization of the GRF receptor mechanism are less (see Fig. 6); it would appear that the bound GRF receptor or the subsequent effector mechanism is transiently inactivated only if it is first functionally activated; this is equivalent to postulating a postactivation latent phase at a stage subsequent to receptor activation. A model incorporating this behavior responds to infusions of GRF with a dose-dependent release of GH and desensitization of the pituitary, whereas coinfusion of SRIF results in inhibition of secretion, followed by a rebound hypersecretion (seen in Fig. 6 at $t = 180$); an "off" effect will occur in the absence of SRIF if some constitutive activation of the secretory pathway is postulated, which may be equivalent to assuming a nonzero resting level of cAMP. Interestingly, if pulses of SRIF are imposed on a background of constant GRF, the pulses result in a paradoxical dose-dependent stimulation of growth hormone release. The manner in which the dependence on SRIF enters in the mathematical formulation of sensitivity is through the dependence of k_b on s; namely, $k_b = k_2 + k_3 \phi(s)$.

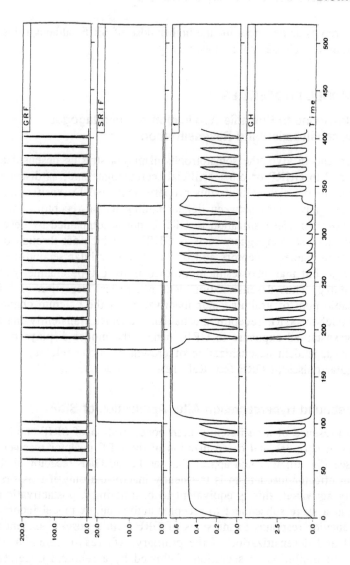

C. Use of the Model for Prediction and Interpolation

To illustrate the implications of these relations for cumulative release, we consider the model using the parameters estimated when fitting the data shown in Figure 5. We can then demonstrate how the extent of desensitization, instantaneous GH release rate, and cumulative GH release change with time (Fig. 7). The infusions in Vale et al.'s experiments were for 180 min, and we show predicted time courses over this time for two doses within the range of doses they used. In the absence of SRIF, desensitization increases as the infusions progress at all GRF doses, but much more markedly at the higher dose. The instantaneous GH release rate is initially much lower than for the higher dose, and hardly falls at all, demonstrating again the low rate of desensitization at this GRF dose. At $r = 0.1$, GH release rate falls to the sustainable level (and the activatable receptors fall to an equilibrium) in just under 3 h, whereas at $r = 0.01$, this process takes longer. Cumulative GH release at the low GRF dose increases approximately linearly, whereas at the higher dose the rate of increase slows down as the infusions progress as a result of increasing desensitization. On the other hand, in the presence of 0.5 nM SRIF, the resensitizing effect of SRIF outweighs the densitization as a result of GRF presence at its low dose. At the higher dose of GRF, in the presence of SRIF, desensitization occurs at approximately the same rate as seen previously in its absence, but the base to which f falls is higher. This low level of SRIF is sufficient to allow a sustainable instantaneous GH release of about 0.3µg per dish/h, which is approximately half of the value in the absence of SRIF. This simple example demonstrates how the model can be used to predict the time course of instantaneous and cumulative GH release and the state of desensitiza-

---

Figure 6 Effects of pulsatile application of secretagogue, and resensitizing and release-inhibiting effects of somatostatin: Model simulation showing pulsatile release of growth hormone as a consequence of pulsatile application of GRF. The labeling of the frames is given on the right: GRF, SRIF = GRF, SRIF concentrations, respectively in the vicinity of the pituitary cells: f = density of free, activatable receptors; GH = GH concentration in the surrounding medium, or in peripheral blood. The simulation shows desensitization to repeated pulses of GRF (at times 60–70, 190–210, 340–360); a complete resensitization as a result of infusion of somatostatin (between $t = 100$ and 180) in the absence of pulses of GRF; almost complete inhibition of release as a result of somatostatin infusion even though the GRF pulses continue (from $t = 250$–330); a partial resensitization as a result of this somatostatin infusion while pulses of GRF continue. Note also the rebound secretion of GH after the end of the somatostatin infusion at about $t = 180$ and 330. Parameters used were $k_1 = 2$, $k_2 = 0.001$, $k_3 = 0.18$, $k_4 = 1.5$, $k_5 = 30$, $k_6 = 5$, $k_7 = 5$, $k_8 = 0.5$, $r_0 = 0.05$, $s_0 = 0$, $\delta_0 = 0.05$.

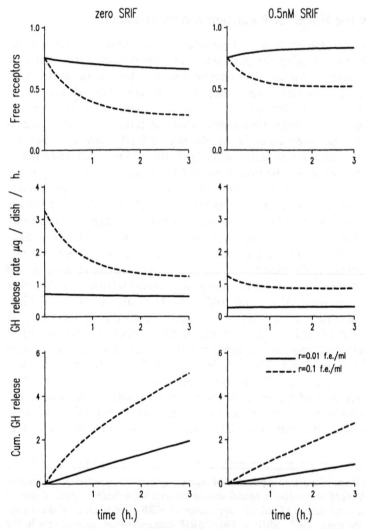

Figure 7 Predicting the time course of infusions of GRF in the absence and presence of SRIF from measurements made at one time: Simulations showing time courses of model variables as follows. *Top row*: the changes in *f*, an index of desensitization of the release mechanism ($f = 1$, indicating no desensitization, and lower value indicating progressively greater desensitization). *Middle row*: the instantaneous GH release rate. *Bottom row*: the cumulative amount of GH released. All are plotted against time for two different GRF doses, $r = 0.01$ fragment equivalents/mL and 0.1 fe/mL (for key see lower right window). These simulations use the parameters derived from fitting to Vale et al.'s data (15) in which GRF alone (left-hand column) and GRF and SRIF in combination (right-hand column) were infused for 180 min.

tion of the system, even though the data to which the model was fitted were obtained at just one time.

REFERENCES

1. Frohman LA, Downs TR, Chomczynski P. Regulation of growth hormone secretion. Front Neuroendocrinol 1992; 13:344–405.
2. Clark RG, Robinson ICAF. Growth induced by pulsatile infusion of an amidated fragment of human growth hormone releasing factor in normal and GHRF-deficient rats. Nature 1985; 314:281–283.
3. Clark RG, Robinson ICAF. Growth hormone responses to multiple injections of a fragment of human growth hormone-releasing factor in conscious male and female rats. J Endocrinol 1985; 106:281–289.
4. Kovacs M, Fancsik A, Hrabovsky E, Mezo I, Teplan I, Flerko B. Effects of continuous and repetitive administration of a potent analog of GH-RH(1-30)-NH$_2$ on the GH release in rats treated with monosodium glutamate. J Neuroendocrinol 1995; 7:703–712.
5. Mason WT, Dickson SL, Leng G. Control of growth hormone at the single cell level. Acta Paeditar Suppl 1993; 388:84–92.
6. Chen C, Zhang J, Vincent JD, Israel JM. Somatostatin increases voltage-dependent potassium currents in rat somatotrophs. Am J Physiol 1990; 259:C854–861.
7. Chen C, Zhang J, Vincent JD, Israel JM. Two types of voltage-dependent calcium current in rat somatotrophs are reduced by somatostatin. J Physiol 1990; 425:29–42.
8. Suzuki M, Kato M. Somatostatin pretreatment facilities GRF-induced GH release and increase in free calcium in pituitary cells. Biochem Biophys Res Commun 1990; 172:276–281.
9. Herbison AE. Sexually dimorphic expression of androgen receptor immunoreactivity by somatostatin neurones in rat hypothalamic periventricularnucleus and bed nucleus of the stria terminalis. J Neuroendocrinol 1995; 7:543–554.
10. Kato M. Withdrawal of somatostatin augments L-type Ca^{2+} current in primary cultured rat somatotrophs. J Neuroendocrinol 1995; 7:855–859.
11. Clark RG, Carlsson LMS, Rafferty YB, Robinson ICAF. The rebound release of growth hormone (GH) following somatostatin infusion in rats involves hypothalamic growth hormone-releasing factor release. Endocrinology 1988; 19:397–404.
12. Stephens E, Brown D, Leng G, Smith RG. A model of the pituitary release of growth hormone. In: Cuthbertson R, Holcombe M, Paton R, eds. Computation in Cellular and Molecular Biological Systems. London: World Scientific, 1996:315–328.
13. Goldbeter A, Li X. Frequency coding in intercellular communication. In: Goldbeter A, ed. Cell to Cell Signalling: From Experiments to Theoretical Models, London: Academic Press, 1989:415–434.
14. Payne RW, Lane PW, Digby PGN, Harding SA, Leech PK, Morgan GW, Todd

AD, Thompson R, Tunnicliffe Wilson G, Welham SJ, White RP. Genstat 5 Release 3 Reference Manual. Oxford: Clarendon Press, 1993.

15. Vale W, Vaughan J, Yamamoto G, Spiess J, Rivier J. Effects of synthetic human pancreatic (tumor) GH releasing factor and somatostatin, triiodothyronine and dexamethasone on GH secretion in vitro. Endocrinology 1983; 112:1553–1555.

6

Evidence for a New Subtype of GHRP Receptors in Ovine Pituitary Gland

Chen Chen, Danxing Wu, and Iain J. Clarke
Prince Henry's Institute of Medical Research, Melbourne, Victoria, Australia

I. INTRODUCTION

Since the discovery of the first potent hexapeptide growth hormone-releasing peptide (GHRP-6) about 12 years ago (1), a range of compounds have been developed with varying potencies, including GHRP-1 and GHRP-2. The chemical structure and relative potency, in terms of GH, for these three peptides are summarized in Table 1. Most recently, Howard et al. (2) identified a novel receptor in rat and swine pituitary glands that binds GHRP and nonpeptidergic GH secretagogues (GHS) with varying levels of affinity. Our work indicates that GHRP-2 acts on a receptor that is different from that employed by GHRP-6 or GHS (3,4). Thus, various subtypes of the GHRP–GHS receptor may exist, in ovine somatotrophs at least, and this chapter reviews our experimental evidence for this.

II. DIFFERENCE IN RECEPTOR CHARACTERS

There is strong evidence that GHRPs do not act through endogenous GH-releasing hormone (GHRH) receptors on somatotrophs as follows. First, a GHRH receptor antagonist inhibits GHRH-stimulated GH release but not GHRP-6- nor GHRP-1-stimulated GH release (3,5). A putative GHRP receptor antagonist does not affect GH release in response to GHRH (5,6). Second, GHRP-6 does not compete with GHRH in a radioreceptor assay for GHRH-binding sites (5). Third, there is an additive effect on GH re-

Table 1 Chemical Structure and the Relative Potency of GHRPs in Terms of Ability to Stimulate GH Release

Name	Chemical structure	In vitro potency (relative to GHRH)[a]
GHRP-6	His-D-Trp-Ala-Trp-D-Phe-Lys-NH$_2$	0.01–0.1
GHRP-1 (KP 101)	Ala-His-D-βNal-Ala-Trp-D-Phe-Lys-NH$_2$	0.1–1
GHRP-2 (KP 102)	D-Ala-D-βNal-Ala-Trp-D-Phe-Lys-NH$_2$	1

[a]Potency is presented in relation to the potency of GHRH (unity) in rat and ovine pituitary cells in vitro.

lease when both GHRH and the GHRP are coadministered to cultured somatotrophs at maximal dosages (3,4). Fourth, there is no cross-desensitization between GHRH and GHRP in terms of GH release, whereas homologous desensitization does occur (3,4).

The GHRPs are able to stimulate GH release in cultured cells in the absence of somatostatin, making it unlikely that they act as antagonists to the somatostatin receptor to stimulate GH secretion. Thus, before the cloning of the receptor there was strong prima facie evidence for a novel GHRP receptor (GHRP-R), and the recent work of Howard et al. (2) indicates this point of view.

We have focused our attention on the action of GHRP-2 in cultured ovine somatotrophs; our first surprise was to find that GH release, stimulated by this secretagogue, could be blocked by a GHRH receptor antagonist (4). The action of GRHP-1 and GHRP-6 was not blocked by the antagonist, raising the possibility that GHRP-2 may act on another receptor, in addition to the recently identified GHRP receptor (2). The result also raised the interesting possibility that GHRP-2 could act on GHRH receptors, but subsequent experiments excluded this possibility. Because GHRP receptors had not been identified at that time, we performed a series of functional experiments to clarify the issue. Somatotroph-enriched ovine pituitary cell cultures were established by Percoll-gradient separation (7), and GH-releasing experiments were performed using a continuous flow perfusion of the cells in petri dishes (10^6 cells per dish) at 37°C (4). Samples of perfusion medium were collected at 2-min intervals, with a flow rate of 1 mL/min, and GH levels were measured by radioimmunoassay (RIA; 3,7). There was an additive effect on GH release when both GHRH and GHRP-2 were coadministered at a maximal dosage (4). In addition, there was no cross-desensitization between GHRH and GHRP-2 sequentially applied within an interval of 1 h, whereas homologous desensitization did occur (4). GHRP-2 also acts on a site different from that used by GHRP-6, as indicated by an additive effect observed on GH release when these two secretagogues were coadministered at a maximal dosage (4). The results of these experiments are summarized in Table 2, and it is clear that GHRP-2 acts on a receptor that is different from the GHRH receptor and also different from the GHRP-6 receptor in ovine somatotrophs.

It has been reported that GHRP-6 interacts with a novel low-affinity GHRH-binding site in rat anterior pituitary cells (8) and Sethumadhavan et al. (9) identified two classes of GHRP-binding sites, using ^{125}I-Tye-Ala-GHRP as a ligand. Therefore, it seems possible that a receptor exists that has a low affinity for GHRH and a high affinity for GHRP-2. Whether or not there is some homology between receptors for GHRH and GHRP-2 is unknown. Another possibility we considered was that GHRP-2 could bind to a site on the GHRH receptor different from that employed by GHRH, and for this we employed a GC cell line with overexpression of the human GHRH receptor (10). In these cells, GHRH increased cAMP levels but GHRP-2 or GHRP-6 did not do so, and the GHRH effect was blocked by a GHRH receptor antagonist. These data strongly suggest that GHRP-2 does not act through GHRH receptors. Nevertheless, it remains possible that there is more than one GHRP receptor, type 1 (GHRP-R1, identified recently by Howard et al.; 2) that is not blocked by the GHRH antagonist ([Ac-Tyr^1,D-Arg^2] GHRH$_{1-29}$) and type 2 (GHRP-R2) that may be blocked by the antagonist. GHRP-R1 binds GHRP-6, GHRP-1, and GHRP-2, whereas GHRP-R2 would be expected to have a higher affinity for GHRP-2 than for GHRP-6 or GHRP-1. GHRP-R2 may also bind GHRH with low affinity, and the antagonist [Ac-Tyr^1,D-Arg^2] GHRH$_{1-29}$, which would explain the ability of the antagonist to block the action of GHRP-2.

Another approach to this issue is to consider the signaling pathways used by different GH secretagogues. Activation of different signaling path-

Table 2 Characterization of the Possibility of Two Different Receptors for GHRP to Induce GH Secretion

	GHRP-R1 (identified by Howard et al. in Ref. 2)	GHRP-R2 (unidentified)
Selectivity	No selectivity for GHRP-6, -1, -2; or nonpeptidergic GHS	GHRP-2 > > GHRP-6, -1, or nonpeptidergic GHS
Sensitive to GHRH-antagonist	No	Yes
Synergistic with GHRH	Yes	No, only additive
Species	Rat, swine, human	Sheep, human?

ways in somatotrophs by the different secretagogues (GHRP-6, GHRP-2, and GHRH) would indicate possible activation by different receptors. We have approached this in two ways: by investigating the membrane ion channels that are activated by the different secretagogues; and by measuring changes in the intracellular second-messager systems that are involved in GH release.

III. INTRACELLULAR FREE CALCIUM CONCENTRATION AND MEMBRANE ION CHANNELS

It is well documented that GH secretion is directly related to intracellular free Ca^{2+} concentration ($[Ca^{2+}]i$) (11). Calcium influx through voltage-dependent Ca^{2+} channels is stimulated by GHRH and reduced by somatostatin (12–17). There is no clear evidence of mobilization of intracellular Ca^{2+} from storage sites by GHRH or somatostatin. On the other hand, the recently identified GHRP receptor (2) is coupled to phospholipase C (PLC) and activation leads to the production of inositol triphosphate (IP_3) which releases Ca^{2+} from IP_3-sensitive Ca^{2+} pools in oocytes expressing this GHRP receptor. In isolated rat somatotrophs, GHRP-6 induces dual-phase increases in $[Ca^{2+}]i$; an initial transient increase owing to Ca^{2+} release from IP_3-sensitive pool, which is not prevented by Ca^{2+} channel blockade, and a second sustained increase caused by Ca^{2+} influx through membrane Ca^{2+} channels (18,19). Indeed, it has been reported that GHRP and nonpetidergic GHSs increase phosphatidylinositol (PI) turnover by an activation of PLC in human acromegalic tumor cells (20,21). Activation of PLC will produce both IP_3 which leads to release Ca^{2+} from intracellular Ca^{2+} storage sites, and diacylglycerol (DAG), which activates PKC. Whether PLC is activated (or PI turnover is increased) in ovine or rat somatotrophs by any GHRP is still an open question. In fact, GHRP-1 causes a subtle and transient increase in $[Ca^{2+}]i$ in ovine pituitary cells, even when extracellular Ca^{2+} is chelated to zero (22). This probably involves the generation of IP_3, as shown in human GH tumor cells in response to GHRP-6 (20,21). In spite of the mobilization of Ca^{2+} from intracellular stores, the major contribution to the elevation of $[Ca^{2+}]i$ is caused by Ca^{2+} influx through Ca^{2+} channels. It would appear that this is an integral factor in the release of GH in response to GHRP, because Ca^{2+} channel blockade abolishes secretion (3, 4, 23). In addition, GHRP-2 causes an increase in $[Ca^{2+}]i$ in ovine pituitary cell suspensions, mainly through the influx of Ca^{2+}, without detectable Ca^{2+} release from intracellular storage pools (24).

In somatotrophs, the major Ca^{2+} channels are of the voltage-gated T- and L-types (14,16). Studies of rat and sheep cells have shown that GHRP-

6 and GHRP-2 depolarize the cell membrane, leading to the opening of these channels (18,24). Because this depolarization can be recorded only with the nystatin-perforated patch clamp, this implies that an intact intracellular second-messenger–signaling system is required for the response. Further definitive evidence indicates that GHRP-2 acts on voltage-gated Ca^{2+} channels through second-messenger systems, leading to an increase in transmembrane L- and T-type Ca^{2+} currents (24). GHRP-6 does not modify Ca^{2+} channel function (18), offering further support to the notion that GHRP-2 and GHRP-6 employ different-signaling pathways. The effects of GHRP-2 on the electrophysiological properties of the membrane of ovine somatotrophs resemble the effects of GHRH, but the kinetic changes of Ca^{2+} currents that are caused by GHRP-2 (24) are different from those caused by GHRH (16), which is consistent with the view that GHRP-2 and GHRH act on two different receptors.

The ion channels that are involved in the depolarization of somatotroph cell membranes are not defined. Studies to date show that GHRP does not activate Na^+ channels in a major way (18,24), and it is thought that the voltage-gated Ca^{2+} channel activation is partly responsible for the depolarization caused by GHRP-2 (24). This action of GHRP on Ca^{2+} channels is in addition to the depolarization-induced opening of Ca^{2+} channels by GHRP because the modification of Ca^{2+} channels is obtained in a voltage-clamp condition, without any slow depolarization (24). Potassium channels may also be involved, for a reduction in K^+ currents in response to GHRP has been reported in rat and ovine somatotrophs, but these have not been fully characterized (24,25).

IV. cAMP STIMULATION

It is well established that GHRH activates the adenylate cyclase in somatotrophs to increase intracellular cAMP levels, and this is fundamental to the release of GH (26, 27). In contrast, GHRP-6 has no direct effect on intracellular cAMP levels in either rat or ovine somatotrophs (6,28), but may synergize with GHRH to elevate cAMP levels (6). L-692,429, a nonpeptidergic GHS, also has a synergistic effect with GHRH on cAMP accumulation in rat pituitary cells (29). That no synergy could be demonstrated with a coadministration of GHRP-1 and GHRH (23), however, suggests that GHRP-6 and GHRP-1 do not stimulate GH release through the cAMP–adenylate cyclase pathway.

The GHRP-2 stimulation of cultured ovine somatotrophs increases intracellular cAMP levels, but this was not observed in rat pituitary cells (28). The GHRP-2-stimulated GH secretion was also blocked in the ovine cells by an inhibitor of adenylate cyclase and a cAMP-binding antagonist, Rp-cAMP. Blockade of adenylate cyclase prevented cAMP increase in response to GHRP-2. Thus, in sheep cells, GHRP-2 activates adenylate cyclase leading to an increase in cAMP levels, which subsequently activates cAMP-dependent protein kinase A (PKA). PKA could phosphorylate transmembrane Ca^{2+} channels to modify their properties in the manner observed by electrophysiological studies (see foregoing; 24) and further work is needed to clarify this point. The significant species difference (sheep vs. rat) in the response to GHRP-2 suggests that there may be different subtypes of GHRP receptors that are variably expressed in the two species. That increase in adenylate cyclase activity is obtained in ovine pituitary cells with GHRP-2, but not within GHRP-6 (28), indicates that a predominant receptor subtype in sheep may have a selective affinity for GHRP-2.

The mechanism of activation of adenylate cyclase by GHRP-2 is unclear. Although some subtypes of this enzyme can be activated by Ca^{2+} (30), the one that is involved in the response to GHRP-2 does not appear to be of this type because blockade of Ca^{2+} influx did not affect the increase in cAMP levels in response to GHRP-2 (28). Although it is clear that GHRH elevates cAMP levels in ovine somatotrophs, it may act through an adenylate cyclase that is different from that used by GHRP-2, because GHRP-2 and GHRH have an additive effect on both cAMP accumulation and GH secretion when both secretagogues are applied at maximal doses (4,28).

In summary, GHRP-2 promotes cAMP accumulation by activation of adenylate cyclase in ovine somatotrophs, but this does not occur in those of the rat. In sheep cells, this appears to be the signaling mechanism for the release of GH by GHRP-2. GHRP-6 and GHRP-1 do not elevate cAMP levels in either ovine, rat, or human somatotrophs, but do amplify the cAMP response to GHRH in rat pituitary cells.

V. PROTEIN KINASE C ACTIVATION

It has been suggested that the action of GHRP-6 and the nonpeptidergic analog L-692,429 to stimulate GH release from rat pituitary cells is mediated by protein kinase C (PKC; 29, 31). The synergistic effect of GHRP-6 and GHRH on cAMP accumulation and GH secretion in rat pituitary cells may also be mediated by PKC (31). However, the specificity of the inhibitor (phloretin) used in the latter study has not been widely tested, and its effect on other kinase systems is not defined. In particular, over the same dose range (10–200 µM), phloretin increased the opening probability of

Ca^{2+}-activated K^+ channels (32), which can hyperpolarize the membrane potential of the cell and prevent stimulation of GH secretion by GHRP-6. Down-regulation of PKC with long-term treatment by phorbol 12-myristate 13-acetate (PMA; 1 μM) partially blocked the effect of GHRP-6 on GH secretion (31), suggesting some involvement of PKC. However, GHRP-1 causes GH release following maximal stimulation of cells with PMA (23). In ovine pituitary cells, GHRP-6 does not cause a detectable translocation of PKC to cell membranes (33). Down-regulation of PKC with phobol 12,13-dibutyrate (PDBu), however, reduces GH release in response to GHRP-6, whereas PMA-stimulated GH release is totally abolished by the same treatment (33). These data suggest that GHRP-6 is a weak stimulator of PKC activation in sheep pituitary cells.

In contrast, GHRP-2 stimulates PKC translocation from cytosol to membrane in cultured ovine somatotrophs (33) over the same ranges of dose and time that are effective in causing GH release (4). The PKC inhibitors (calphostin C, chelerythrine, staurosporine) and down-regulation of PKC by PDBu cause partial attenuation of GH release induced by GHRP-2; therefore, it seems likely that PKC is at least partly involved in the action of GHRP-2 in sheep cells. Interestingly, GHRH also causes PKC translocation in ovine somatotrophs (33). The PKA inhibitor, H_{89}, abolished GH secretion in response to GHRP-2, but did not modify the stimulation of PKC translocation, which is consistent with a predominant role for the cAMP-PKA pathway in response to GHRP-2 and GHRH. Intracellular-signaling systems involved in a particular response to a particular regulatory molecule may vary between species and from tissue to tissue. Furthermore, it is now apparent that more than one-signaling pathway may be employed by a single regulatory molecule. There are several examples of crosstalk between the cAMP-PKA and PKC pathways (34–36). In a bidirectional control model, the two signal transduction systems appear to counteract each other, but in a monodirectional control system, one may potentiate the other (37). In platelets, neutrophils (34), and lymphocytes (35), the PKC pathway activates cellular functions, such as exocytosis and proliferation, but the cAMP-signaling system may antagonize these activations. Conversely, in avian erythrocytes (38), Leydig cells (36), and glioma cells (39), PKC inhibits and desensitizes the adenylate cyclase system. In pinealocytes (40), pituitary cells (41), and lymphoma S49 cells (42), PKC activation greatly potentiates cAMP production. To our knowledge, there is as yet no obvious example of a tissue in which cAMP potentiates PKC activation. In many endocrine cells, such as pancreatic islets (43), these two signal transduction pathways frequently act in concert to induce hormone secretion. As such, it is reasonable to assume that various combinations of the two sec-

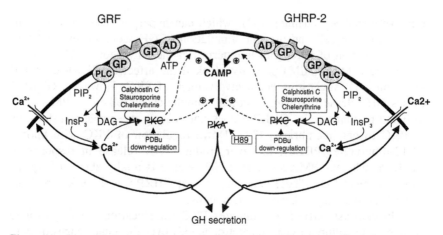

GH secretion

Figure 1 A model for crosstalk between PKC and cAMP–PKA pathways in response to GRF or GHRH and GHRP-2 in ovine somatotrophs: The receptors for GRF or GHRH and GHRP-2 are coupled by G proteins (GP) to both adenylate cyclase (AD) and phospholipase C (PLC), which activate PKA and PKC, respectively. The main-signaling pathway is through PKA (heavy line) to increase Ca^{2+} influx through Ca^{2+} channels and GH secretion. Activation of PKC could potentiate the PKA pathway (dashed line) by increasing the activity of AD or by enhancing the ability of cAMP to activate PKA. PIP_2, phosphatidylinositol (4,5)-biphosphate; IP_3, inositol 1,4,5-triphosphate; DAG, diacylglycerol.

ond-messenger systems may cooperate and intensify responses in a physiological process. Our studies on GHRP provide support for this hypothesis, with evidence that GHRP-2 and GHRH activate both the PKA and PKC pathways, and PKC activation potentiates the PKA-signaling system to cause GH secretion. The crosstalk between PKC and cAMP pathways in the regulation of GH secretion is illustrated in Figure 1.

VI. CONCLUSION

Different forms of GHRP appear to act in different ways in different species and we have described some specific examples of this. These differences in signaling systems may reflect the extent to which two subtypes of receptor for GHRP are expressed. We propose that at least two subtypes exist as GHRP-R1 (identified recently by Howard et al.; 2) that is without selectivity for different versions of GHRP, and GHRP-R2 (as yet unidentified) that is selective for GHRP-2.

REFERENCES

1. Bowers CY, Monany FA, Reynolds GA, Chang D, Hong A, Chang K. Structure-activity relationships of a synthetic pentapeptide that specifically releases growth hormone in vitro. Endocrinology 1984; 106:663–667.
2. Howard AD, Feighner SD, Cully DF, Arena JP, Liberator PA, Rosenblum CI, Hamelin M, Hreniuk DL, Palyha OC, Anderson J, Sparess PS, Diaz C, Chou M, Liu KK, McKee KK, Pong SS, Chaung LY, Elbrecht A, Dashkevicz M, Heavens R, Tigby M, Sirinathsinghji DJS, Dean DC, Melillo DG, Patchett AA, Nargund R, Griffin PR, DeMartino JA, Gupta SK, Schaeffer JM, Smith RG, Van der Ploeg LHT. A receptor in pituitary and hypothalamus that functions in growth hormone release. Science 1996; 273:974–977.
3. Wu D, Chen C, Zhang J, Katoh K, Clarke IJ. Effects in vitro new growth hormone releasing peptide (GHRP-1) on growth hormone secretion from ovine pituitary cells in primary culture. J Neuroendocrinol 1994; 6:185–190.
4. Wu D, Chen C, Katoh K, Zhang J, Clarke IJ. The effect of GH-releasing peptide-2 (GHRP-2 or KP102) on GH secretion from primary cultured ovine pituitary cells can be abolished by a specific GH-releasing factor (GRF) receptor antagonist J Endocrinol 1994; 140:R9–R13.
5. Thorner MO, Hartman ML, Gaylinn BD, et al. Current status of therapy with growth hormone-releasing neuropeptides. In: Savage MO, Bourguignon J, Grossman AB, eds. Frontiers in Paediatric Neuroendocrinology. Oxford: Blackwell Scientific, 1994:161–167.
6. Cheng K, Chan WW, Barreto A Jr, Convey EM, Smith RG. The synergistic effects of His-D-Trp-Ala-Trp-D-Phe-Lys-NH$_2$ on growth hormone (GH)-releasing factor stimulated GH release and intracellular adenosine 3', 5'-monophosphate acid accumulation in rat pituitary cell culture. Endocrinology 1989; 124:2791–2798.
7. Chen C, Heyward P, Zhang J, Wu D, Clarke IJ. Voltage-dependent potassium current sin ovine somatotrophs and their function in growth hormone secretion. Neuroendocrinology 1994; 59:1–9.
8. Lau YS, Camoratto AM, White LM, Moriarty CM. Effect of lead on TRH and GRF binding in rat anterior pituitary membranes. Toxicology 1991; 68:169–179.
9. Sethumadhaven K, Veeraragavan K, Bowers CY. Demonstration and characterization of the specific binding of growth hormone-releasing peptide to rat anterior pituitary and hypothalamic membranes. Biochem Biophys Res Commun 1991; 178:31–37.
10. Chen C, Farnworth P, Musgrave I, Petersenn S, Melmed S, Clarke IJ. Cyclic AMP levels, [Ca^{2+}]i and growth hormone (GH) secretion in GC cell lines overexpressing human GH-releasing factor (hGRF) receptors in response to GRF or GH-releasing peptide-2 (GHRP-2). 4th International Pituitary Congress. San Diego, CA, 1996.
11. Chen C, Vincent JD, Clarke IJ. Ion channels and the signal transduction pathways in the regulation of growth hormone secretion. Trends Endocrinol Metab 1994; 5:227–233.

12. Chen C, Israel JM, Vincent JD. Electrophysiological responses to somatostatin of rat hypophysial cells in somatotroph-enriched primary cultures. J Physiol 1989; 408:493–510.

13. Chen C, Israel JM, Vincent JD. Electrophysiological responses of rat pituitary cells in somatotroph-enriched primary culture to human growth hormone-releasing factor. Neuroendocrinology 1989; 50:679–587.

14. Chen C, Zhang J, Vincent JD, Isreal JM. Two types of voltage-dependent calcium currents in rat somatotrophs are reduced by somatostatin. J Physiol 1990; 425:29–42.

15. Chen C, Zhang J, McNeill P, Pullar M, Cummins J, Clarke IJ. Growth hormone releasing factor modulates calcium currents in human growth hormone secreting adenoma cells. Brain Res 1992; 604:345–348.

16. Chen C, Clarke IJ. Modulation of Ca^{2+} influx in the ovine somatotroph by growth hormone-releasing factor. Am J Physiol 1995; 268:E204–E212.

17. Lussier BT, French MB, Moor BC, Kraicer J. Free intracellular Ca^{2+} concentration ($[Ca^{2+}]i$) and GH release from purified rat somatotrophs. III. Mechanism of action of GH-releasing factor and somatostatin. Endocrinology 1991; 128:592–603.

18. Herrington J, Hille B. Growth hormone-releasing hexapeptide elevates intracellular calcium in rat somatotropes by two mechanisms. Endocrinology 1994; 135:1100–1108.

19. Bresson-Bepoldin L, Dufy-Barbe L. GHRP-6 induces a biphasic calcium response in rat pituitary somatotrophs. Cell Calcium 1994; 15:247–258.

20. Lei T, Buchfelder M, Fahlbusch R, Adams EF. Growth hormone releasing peptide (GHRP-6) stimulates phosphatidylinositol (PI) turnover in human pituitary somatotroph cells. J Mol Endocrinol 1995; 14:135–138.

21. Adams EF, Petersen B, Lei T, Buchfelder M, Fahlbusch R. The growth hormone secretagogue, L-692,429, induces phosphatidylinositol hydrolysis and hormone secretion by human pituitary tumors. Biochem Biophy Res Commun 1995; 208:555–561.

22. Katoh K, Chen C, Engler D. Effects of GHRP on GH release and intracellular calcium concentration in primary cultured ovine anterior pituitary cells. J Reprod Dev 1996; 42(suppl):106–108.

23. Akman MS, Gieard M, O'Brien LF, Ho AK, Chik CL. Mechanisms of action of a second generation growth hormone-releasing peptide (Ala-His-D-beta-Nal-Ala-Trp-D-Phe-Lys-NH$_2$) in rat anterior pituitary cells. Endocrinology 199; 132:1286–1291.

24. Chen C, Clarke IJ. Effect of growth hormone-releasing peptide-2 (GHRP-2) on membrane Ca^{2+} permeability in cultured ovine somatotrophs. J Neuroendocrinol 1995; 7:179–186.

25. Pong SS, Chaung LY, Smith RG. GHRP-6 (His-D-Trp-Ala-Trp-D-Phe-Lys-NH$_2$) stimulates growth hormone secretion by depolarization in rat pituitary cell cultures. 73th Annual Meeting of the Endocrine Society. Washington, DC, 1991. Abstr 230.

26. Frohman LA, Downs TR, Chomczynski P. Regulation of growth hormone secretion. Front Neuroendocrinol 1992; 13:344-405.
27. Harwood JP, Grew C, Aguilera G. Action of growth hormone-releasing factor and somatostatin on adenylate cyclase and growth hormone release in rat anterior pituitary. Mol Cell Endocrinol 1984; 37:277-284.
28. Wu D, Chen C, Clarke IJ. The effect of GH-releasing peptide-6 (GHRP-6) and GHRP-2 on intracellular adenosine 3',5'-monophosphate (cAMP) levels and GH secretion in ovine and rat somatotrophs. J Endocrinol 1996; 14:197-205.
29. Cheng K, Chan WW, Butler B, Wei L, Schoen WR, Wyvratt MJ Jr, Fisher MH, Smith RG. Stimulation of growth hormone release from rat primary pituitary cells by L-692,429, a novel non-peptidyl GH secretagogue. Endocrinology 1993; 132:2729-2731.
30. Cooper EMF, Mons N, Karpen JW. Adenylyl cyclases and the interaction between calcium and cAMP signalling. Nature 1995; 374:421-424.
31. Cheng K, Chan WWS, Barreto A, Butler B, Smith RG. Evidence for a role of protein kinase-C in His-D-Trp-Ala-Trp-D-Phe-Lys-NH$_2$-induced growth hormone release from rat primary pituitary cells. Endocrinology 1991; 129:3337-3342.
32. Koh DS, Reid G, Vogel W. Effect of the flavoid phloretin on Ca^{2+}-activated K$^+$ channels in myelinated nerve fibres of Xenopus laevis. Neurosci Lett 1994; 165:167-170.
33. Wu D, Chen C, Clarke IJ. The functional relationship of protein kinase C (PKC) activity translocation and growth hormone (GH) release by GH-releasing factor (GRF) and GH-releasing peptides (GHRP-2 and GHRP-6). 10th International Congress of Endocrinology. San Francisco, CA, 1996.
34. Takai Y, Kaibuchi K, Sano K, Nishizuka Y. Counteraction of calcium-activated phospholipid-dependent protein kinase activation by adenosine 3',5'-monophosphate and guanosine 3',5'-monophosphate in platelets. J Biochem 1982; 91:403-406.
35. Kaibuchi K, Takai Y, Ogawa Y, Kimura S, Nishizuka Y, Nakamura T, Tomomura A, Ichihara A. Inhibitory action of adenosine 3',5'-monophosphate on phosphatidylinositol turnover: difference in tissue response. Biochem Biophys Res Commun 1982; 104:105-112.
36. Mukhopadhyay AK, Schumacher M. Inhibition of hCG-stimulated adenylate cyclase in purified mouse Leydig cells by the porbol ester PMA. FEBS Lett 1985; 187:56-60.
37. Houslay MD. "Cross-talk": a pivotal role for protein kinase C in modulating relationships between signal transduction pathways. Eur J Biochem 1991; 195:9-27.
38. Kelleher DJ, Pessin JE, Ruoho AE, Johnson GL. Phorbol ester induces desensitization of adenylate cyclase and phosphorylation of the β-adrenergic receptor in turkey erythrocytes. Proc Natl Acad Sci USA 1984; 81:4316-4320.
39. Kassis S, Zaremba T, Patel J, Fishman PH. Phorbol esters and beta-adrenergic agonists mediate desensitization of adenylate cyclase in rat glioma C6 cells by distinct mechanisms. J Biol Chem 1985; 260:8911-8917.

40. Sugden D, Vanecek J, Klein DC, Thomas TP, Anderson WB. Activation of protein kinase C potentiates isoprenaline-induced cyclic AMP accumulation in rat pinealocytes. Nature 1985; 314:359–361.
41. Cronin MJ, Canonico PL. Tumor promoters enhance basal and growth hormone releasing factor stimulated cyclic AMP levels in anterior pituitary cells. Biochem Biophys Res Commun 1985; 129:404–410.
42. Bell DJ, Buxton IL, Brunton LL. Enhancement of adenylate cyclase activity in S49 lymphoma cells by phorbol esters. Putative effect of C kinase on alpha s-GTP-catalytic subunit interaction. J Biol Chem 1985; 260:2625–2628.
43. Tamagawa T, Niki H, Niki A. Insulin release independent of a rise in cytosolic free Ca^{2+} by forskolin and phorbol ester. FEBS Lett 1985; 183:430–432.

7

The Design, Synthesis, and Biological Activities of a Series of Tetrahydroquinoline- and Isoindoline-Based Growth Hormone Secretagogues

Bruce A. Lefker, Philip A. Carpino, Lydia C. Pan, Kristin L. Chidsey-Frink, Paul DaSilva Jardine, Frank M. DiCapua, David A. Griffith, William A. Hada, Jr., John K. Inthavongsay, Sharon K. Lewis, F. Michael Mangano, Marianne C. Murray, OiCheng Ng, Jane G. Owens, Christine M. Pirie, Colin R. Rose, Nancy I. Ryan, John R. Schafer, John L. Tucker, Ann S. Wright, Michael P. Zawistoski, and David D. Thompson
Pfizer Central Research, Groton, Connecticut

I. INTRODUCTION

Growth hormone (GH) is an important endocrine regulator of growth, metabolism, body composition, and physiology. GH deficiency in children causes dwarfism, whereas the clinical profile of GH-deficient adults includes visceral adiposity, reduced lean body mass and muscle/fat ratio, osteopenia, fatigue and muscle weakness, decreased cardiac function, defective thermoregulation, and reduced extracellular fluid volume (1). Current therapy for GH deficiency is replacement with recombinant growth hormone (rGH). Although largely effective in promoting growth in children and normalizing the metabolic and functional deficits associated with adult GH deficiency, rGH therapy is not ideal owing to dose-limiting side effects, parenteral delivery, and cost. Recombinant GH is a peptide drug that is currently approved only in injectable form and costs

Numbers in boldface in this chapter refer to compound numbers as described in figures and tables.

$10,000–$15,000/year and higher. Many adults discontinue therapy after experiencing adverse effects, such as arthralgia and carpal tunnel syndrome, which may be due to excess sodium and fluid retention (2). Side effects have been linked to the nonphysiological pattern of hormone replacement produced by rGH administration, which fails to mimic the pulsatility of normal secretion.

Endogenous pulsatility is established principally by the interplay of two hypothalamic regulators, growth hormone-releasing hormone (GHRH) and somatostatin, acting on the pituitary gland to control GH secretion (3,4). A normal profile of GH secretion shows nyctohemeral rhythmicity, with larger and more frequent secretory bursts during the night. The amplitude of GH pulses can be further modulated by a variety of steroid hormones, neurotransmitter pathways (cholinergic, adrenergic, opioid), feedback loops (GH and IGF-1), nutritional status, sleep, and exercise (5).

Recently, a novel pathway for stimulation of GH secretion was identified based on the activity of GH-releasing peptide (GHRP)-6, which was first described by Bowers in 1984 (6). GHRP-6 is a potent and selective inducer of GH release that does not concomitantly stimulate release of luteinizing hormone (LH), follicule-stimulating hormone (FSH), thyroid-stimulating hormone (TSH), or insulin. GHRP-6 acts directly on the somatotroph through specific receptors and signaling mechanisms distinct from those used by GHRH; furthermore, it can synergize with GHRH to stimulate GH secretion to an even greater degree (7). GHRP-6 increases phosphoinositol (PI) turnover and activates protein kinase C (PKC), without increasing cAMP (8,9). In addition to augmenting GH release at the level of the pituitary, GHRP-6 activates discrete populations of neurons in the arcuate nucleus of the hypothalamus (10).

Most cases of GH deficiency result from secretory defects, rather than loss of pituitary hormone production (11). Reductions in serum GH associated with acquired GH deficiency, obesity, hypothyroidism, and aging are generally due to diminution of pulse amplitude, without change in pulse frequency (12,13). Treatment of the elderly with GH secretagogues augments GH pulse amplitude and produces a more youthful pattern of hormone secretion (14–16). Secretagogues can also accelerate growth velocity in short children who have the capacity to respond with enhanced GH release (17). Most importantly, prolonged administration of secretagogues does not appear to disrupt or override the intrinsic feedback regulation of GH and produces a physiological profile of GH secretion. Therefore, the use of GH secretagogues represents an alternative therapeutic strategy for many patients with relative GH insufficiency. The ability of secretagogues to stimulate GH secretion under a wide variety of experimental conditions has fueled growing interest in potential clinical applications.

II. NONPEPTIDYL GH SECRETAGOGUES

Small-molecule mimetics of GHRP-6 may overcome many of the potential disadvantages associated with a peptide therapeutic agent for increasing the endogenous secretion of GH. Peptide agents suffer from problems such as poor absorption and rapid metabolism. The identification of such nonpeptidyl GHRP-6 mimetics has proved challenging, for GHRP-6 is an agonist at the receptor, and any such small molecule must not only bind to the receptor, but must also activate signal transduction. There are only a few examples reported in the literature of small molecular agonists for endogenous peptides. Researchers at Merck used information about the key structural features of GHRP-6 that were necessary for biological activity to identify a weakly active chemical structure that they were able to modify into L-692,429 (EC_{50} = 60 nM; Fig. 1), the first reported nonpeptidyl molecule to act as a GHRP-6 mimetic and demonstrate efficacy for increasing GH levels in humans (18). This compound was later dropped from clinical development, presumably because of poor oral bioavailability. A second structurally distinct class of small molecule GH secretagogues (GHSs), exemplified by L-162,762 and MK-0677 (EC_{50} = 3 nM), was subsequently identified by combining "privileged structures" for G–protein-coupled receptors with fragments containing key functional groups for GHRP-6 agonist activity (19). MK-0677 is currently under development as an oral agent.

In this report, we disclose a new series of potent GHSs that were derived from a molecular-modeling analysis of known small-molecule mimetics, the benzolactam L-692,429 and the acyclic dipeptides L-162,752, and MK-0677. We sought to replace the COOH-terminal spiropiperidine group in the acyclic series with novel amines that could superimpose a phenyltetrazoyl or phenylurea substituent in the region of the receptor as occupied by these same pharmacophores in the benzolactam series (Fig. 2). A set of compounds containing heterocyclic amino groups linking the phenyl urea moiety with the D-Trp-Aib fragment were analyzed using the modeling software Sybil 6.0 (Tripos Associates, St. Louis, MO). From a satisfactory conformational overlay, a series of compounds containing substituted tetrahydroisoquinoline and isoindoline rings were synthesized (Fig. 3). Investigation of the structure–activity relations (SAR) led to the identification of several compounds that exhibit robust GH-releasing

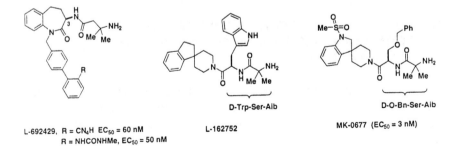

D-Trp-Ser-Aib

D-O-Bn-Ser-Aib

L-692429, R = CN₄H EC_{50} = 60 nM
R = NHCONHMe, EC_{50} = 50 nM

L-162752

MK-0677 (EC_{50} = 3 nM)

Figure 1   Nonpeptidyl GHSs.

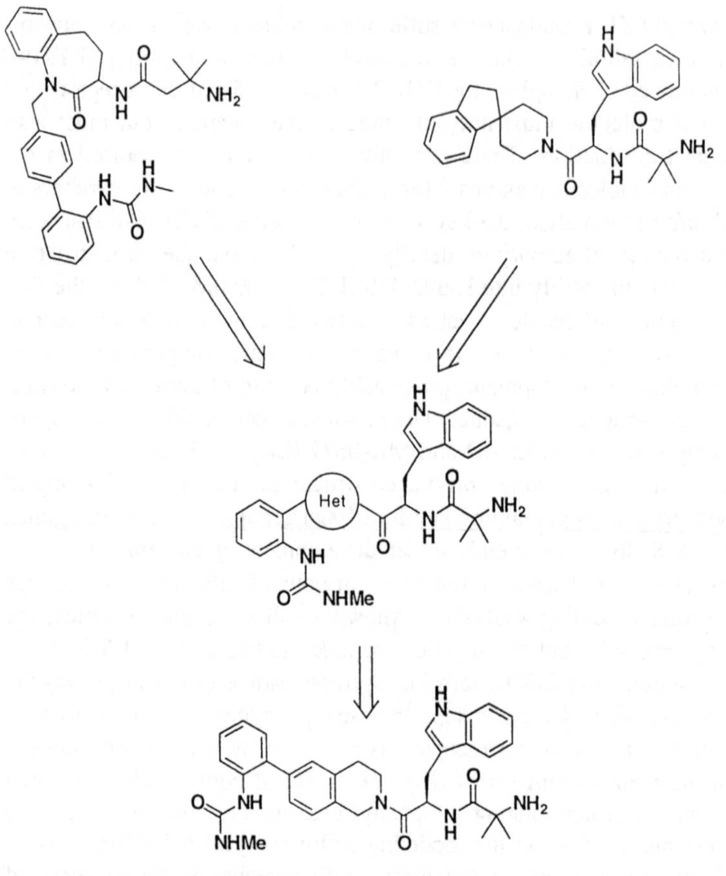

Figure 2 Design rationale for compound **1**.

activity following intravenous injection into rats and weak oral activity in the beagle dog.

III. STRUCTURE–ACTIVITY RELATIONS OF THE TETRAHYDROISOQUINOLINE AND ISOINDOLINE SERIES OF GHSs

The syntheses of the tetrahydroisoquinoline (THQ)- and isoindoline-based GHSs are shown in Figures 4 and 5. The amino- and sulfonyl-THQ and isoindoline heterocycles were prepared by known procedures and then coupled with the dipeptide moiety to provide the desired targets. The 5-amino-isoindoline het-

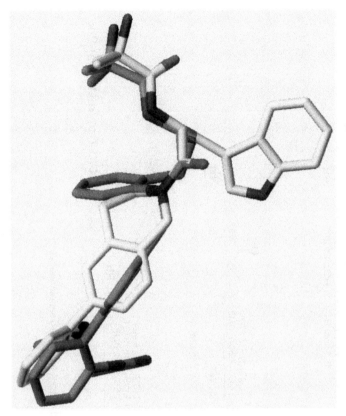

Figure 3 Overlay of CP-320,802 (dark gray) and compound **1** (light gray).

erocycles were prepared by alkylation of benzylamine with 2,3-bis-
(bromomethyl)nitrobenzene (21), reduction of the nitro group, followed by either
acylation or sulfonylation of the amino group, and removal of the *N*-benzyl-
protecting group. The 7-amino-THQ analogs were prepared by nitration of
tetrahydroisoquinoline (22), treatment with the dipeptide moiety, reduction of
the nitro group, and acylation of the amino group with an appropriate acid
chloride. The 6-amino-THQ derivatives were prepared using a similar synthetic
sequence from 7-bromo-tetrahydroisoquinoline (23). The 7-sulfonamido-THQ
heterocycles were prepared by sulfonylation of *N*-trifluoroacetamido-tetrahy-
droisoquinoline with chlorosulfonic acid (24), followed by treatment with an
appropriate amine, and removal of the *N*-trifluoroacetamido-protecting group.
The 5-sulfonamido-THQ heterocycles were prepared from 5-isoquinoline sul-
fonic acid using standard methods.

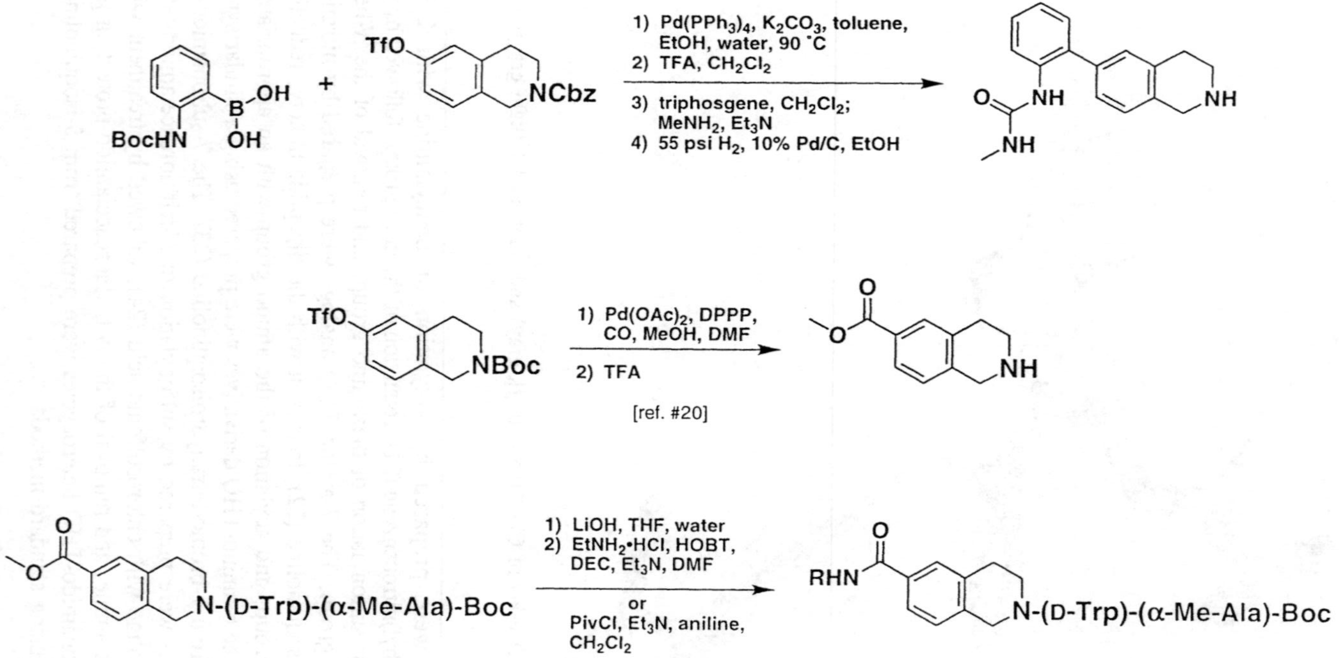

Figure 4 Scheme for synthesis of aryl- and acyl-substituted tetrahydroisoquinolines.

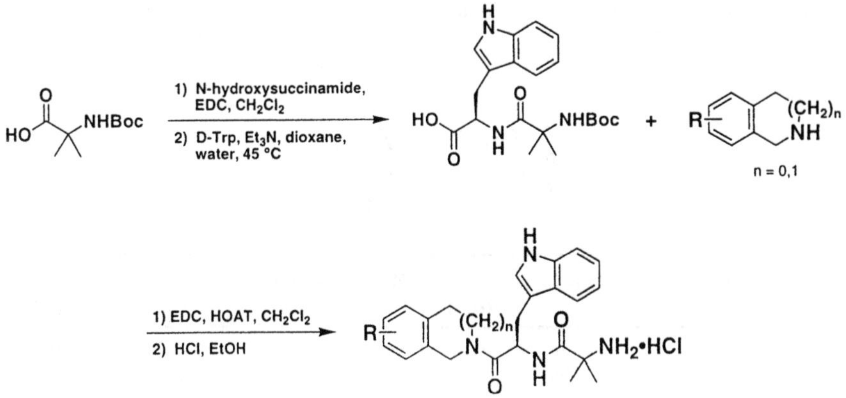

Figure 5 Scheme for synthesis of putative GH secretagogues.

The in vitro activity of the GHSs containing substituted THQ and iso-indoline heterocycles was determined by incubating each test compound for 15 min in rat pituicyte cell cultures and then measuring GH release into the culture medium with a rat-specific GH radioimmunoassay (NIDDK reagents from A. F. Parlow). Active compounds typically produced a two to four fold increase over basal GH release in this assay. The response to each compound over a range of concentrations was determined and reported as the concentration necessary to produce a half-maximal GH response (EC_{50}). As shown in Table I, compound 1, with a THQ group substituted at the C-6 position with a 2-phenylurea moiety was half as potent as the benzolactam L-692,429. Removal of the urea group on the C-6 phenyl substituent resulted in no change in potency. Interestingly, the analog without any substituents on the THQ ring showed a threefold improvement in potency (13, EC_{50} = 30 nM). This SAR is completely different from that observed in the benzolactam series, in which similar changes on the middle phenyl ring in the biaryl moiety resulted in a loss of activity. The potency of this THQ series was further improved by the introduction of polar substituents at the C-6 position, such as a N-ethylaminocarbonyl group (7, EC_{50} = 8 nM) or a N-tolylsulfonylamino group (3, EC_{50} = 3 nM). The potency of THQ analogs substituted at the C-7 position with polar functional groups was comparable with that for the C-6 compounds (4, EC_{50} = 7 nM; 6, EC_{50} = 25 nM; 9, EC_{50} = 20 nM). Methylation of the amino groups on the C-7 aminosulfonyl analog 9 did not affect in vitro potency. The C-5 aminosulfonyl analog 12, was significantly less active (EC_{50} = >1000 nM). The potencies of isoindoline analogs 14 and 15 were comparable with the potencies of analogous THQ compounds.

Table 1   SAR of THQ and Isoindoline GHSs for In Vitro GH Secretion

No.	Position	R	EC$_{50}$ (nM)	No.	Position	R	EC$_{50}$ (nM)
1	6-		100	8	6-		40
2	6-		100	9	7-		20
3	6-		3	10	7-		12
4	7-		7	11	7-		20
5	6-		100	12	5-		>1000
6	7-		25	13	-	H	30
7	6-		6				

Compound 14
(EC$_{50}$ = 10 nM)

Compound 15
(EC$_{50}$ = 20 nM)

The THQ GHSs were further evaluated for their ability to stimulate GH secretion after intravenous injection into pentobarbital-anesthetized rats (Table 2), a model for the evaluation of in vivo potency (25,26). The serum GH response was measured 10 min after dosing. At doses up to 10 mg/kg, compounds

Table 2 SAR of THQ GHSs in Pentobarbital-Anesthetized Rats

No.	Position	R	ED$_{50}$ (mg/kg)
1	6-		>30
3	6-		>10
7	6-		>10
9 CP-361039	7-		0.2
10 CP-458316	7-		0.8
11 CP-406767	7-		3
12	5-		>3

3, 7, and 12 exhibited weak activity (< 200 ng/mL GH), whereas high doses of compounds 9 (CP-361,039), 10 (CP-458,316), and 11 (CP-406,767) produced robust GH responses in the range of 800–1000 ng/mL GH. The most active secretagogues in this model contained small, polar functional groups on the THQ ring. A correlation between lipophilicity and biological activity was shown for the series of C-7 aminosulfonyl analogs, in which the addition of methyl groups on the aminosulfonyl substituent resulted in a decrease in the immediate GH response, despite the absence of change in the in vitro pituicyte activity (26).

To ascertain whether these GHSs have oral activity, two THQ compounds, having high in vitro potency and robust in vivo activity in the rat model, were chosen for administration to beagle dogs by oral gavage (Fig. 6). Blood samples were taken by direct venipuncture of the jugular vein for determination of plasma GH levels and concentrations of the test compound. Plasma GH was measured by a canine-specific radioimmunoassay (NIDDK reagents from A. F. Parlow). Plasma concentrations of test compounds were analyzed by liquid chromatography–mass spectroscopy. Although CP-458,316 [10] was more potent than CP-406,767 [11] in rats, the opposite pattern was observed in the oral dog model. Plasma levels of CP-458,316 were 8 ng/mL or less, at all time points after dosing. Therefore, CP-458,316 failed to elevate plasma GH, most likely because the drug failed to achieve efficacious concentrations in the blood and target tissues. In contrast, CP-406,767 produced a transient elevation of plasma GH levels at 30 min after dosing, at which time plasma drug concentrations were approximately 50 ng/mL.

IV. MECHANISM OF ACTION STUDIES

CP-458,316 was selected as a representative of this THQ series of GHS compounds for mechanistic studies at the cellular level. By using rat pituicyte cultures, we demonstrated that CP-458,316 stimulated GH release in a manner that was similar to GHRP-6 and distinct from GHRH (Fig. 7). The maximal amounts of GH released by escalating concentrations of CP-458,316 or GHRP-6 were comparable (data not shown). Similar to GHRP-6, CP-458,316 synergized with

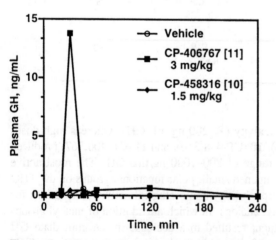

Figure 6   Effect of selected compounds on GH secretion after oral dosing in the dog.

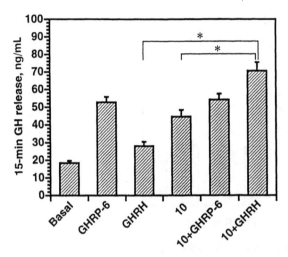

Figure 7 Ability of CP-458,316 to increase GH release stimulated by GHRH, but not by GHRP-6: Rat pituicyte cultures were treated with 100 nM GHRP-6, 10 nM GHRH, 1000 nM CP-458,316 [**10**] or the indicated combinations in the 15-min GH-release assay. *Indicates $p < 0.05$ by Student's t-test.

GHRH to elicit a significantly greater GH release than either CP-458,316 or GHRH alone. However, the GH response to a combination of CP-458,316 and GHRP-6 was not distinguishable from that elicited by either single agent. These results argue that the signaling pathway used by CP-458,316 to stimulate GH secretion is the same as that used by GHRP-6, but different from that used by GHRH.

Another distinguishing feature of the isolated pituitary cell response to GHRP-6 is rapid homologus desensitization. Rat pituitary cultures preincubated with CP-458,316 for 1 h were completely unable to respond to a subsequent challenge with fresh GHRP-6 or with CP-458,316 (Fig. 8). Pretreatment of cultures with GHRP-6 similarly desensitized the cultures to CP-458,316. This pattern of homologous and reciprocal desensitization at the level of the pituitary cell further supports the notion that the THQ GH secretagogues share a receptor-mediated mechanism with GHRP-6.

V. SUMMARY

A new series of GHSs containing tetrahydroisoquinoline (THQ) and isoindoline heterocyles has been identified. Investigation of the SAR in this structural class led to the discovery of potent compounds containing polar functional groups at

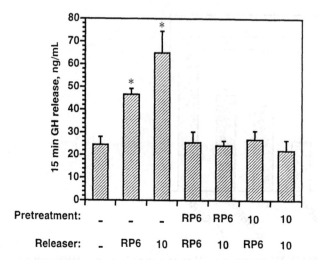

Figure 8 Homologous and reciprocal desensitization of somatotrophs by CP-458,316 [10] and GHRP-6. Rat pituitary cultures were preincubated for 1 h with the indicated compounds, then the culture medium was replaced with fresh medium containing the releaser for 15 min and assayed for GH. RP6 = 100 nM GHRP-6; [10] = 1000 nM CP-458,316; *$p < 0.05$ versus untreated control cultures by Student's t-test.

the C-6 and the C-7 positions for the tetrahydroisoquinoline compounds and at the C-5 position for the isoindoline analogs. CP-458,316 [10] stimulates GH secretion in a GHRP-like manner, in that it potentiates the effect of GHRH on GH secretion and exhibits densensitization of the in vitro secretory response to itself and to GHRP-6. This compound stimulated GH secretion in an anesthesized rat model with ED_{50} = 0.8 mg/kg. The N-methylsulfonyl analog, CP-406,767 [11], exhibited modest oral activity in a beagle dog model at a dose of 3 mg/ kg. These nonpeptidyl mimetics of GHRP-6 may be useful as pharmacological tools to investigate the usefulness and mechanism of action of this class of therapeutic agents.

REFERENCES

1. Jørgensen JOL, Müller J, Møller J, Wolthers T, Vahl N, Juul A, Skakkebæk NE, Christiansen JS. Adult growth hormone deficiency. Horm Res 1994; 42:235–241.
2. Corpas E, Harman SM, Blackman MR. Human growth hormone and human aging. Endocr Rev 1993; 14:20–39.
3. Tannenbaum GS, Ling N. The interrelationship of growth hormone (GH)-releasing factor and somatostatin in the generation of the ultradian rhythm of GH secretion. Endocrinology 1984; 115:1952–1957.

4. Clark RG, Robinson ICAF. Paradoxical growth-promoting effects induced by patterned infusions of somatostatin in female rats. Endocrinology 1988; 122:2675–2682.

5. Strobl JS, Thomas MJ. Human growth hormone. Pharmacol Rev 1994; 46:1–34.

6. Bowers CY, Momany FA, Reynolds GA, Hong A. On the in vitro and in vivo activity of a new synthetic hexapeptide that acts on the pituitary to specifically release growth hormone. Endocrinology 1984; 114:1537–1545.

7. Cheng K, Chan WW, Barreto A Jr, Convey EM, Smith RG. The synergistic effects of His-D-Trp-Ala-Trp-D-Phe-Lys-NH$_2$ on growth hormone (GH)-releasing factor-stimulated GH release and intracellular adenosine-3',5'-monophosphate accumulation in rat primary pituitary cell culture. Endocrinology 1989; 124:2791–2798.

8. Cheng K, Chan WW, Butler B, Barreto A Jr, Smith RG. Evidence for a role of protein kinase-C in His-D-Trp-Ala-Trp-D-Phe-Lys-NH$_2$-induced growth hormone release from rat primary pituitary cells. Endocrinology 1991; 129:3337–3342.

9. Lei T, Buchfelder M, Fahlbusch R, Adams EF. Growth hormone releasing peptide (GHRP-6) stimulates phosphatidylinositol (PI) turnover in human pituitary somatotroph cells. J Mol Endocrinol 1995; 14:135–138.

10. Dickson SL, Leng G, Dyball RE, Smith RG. Central actions of peptide and nonpeptide growth hormone secretagogues in the rat. Neuroendocrinology 1995; 61:36–43.

11. Pintor C, Loche S, Puggioni R, Cella SG, Locatelli V, Lampis A, Muller EE. Growth hormone deficiency states: approach by CNS-acting compounds. In: Muller EE, Cocchi D, Locatelli V, eds. Advances in Growth Hormone and Growth Factor Research. Heidelberg: Springer-Verlag, 1989: 375–388.

12. Vermeulen A. Nyctohemeral growth hormone profiles in young and aged men: correlation with somatomedin C levels. J Clin Endocrinol Metab 1987; 64:884–888.

13. Iranmanesh A, Grisso B, Veldhuis JD. Low basal and persistent pulsatile growth hormone secretion are revealed in normal and hyposomatotropic men studied with a new ultrasensitive chemiluminescence assay. J Clin Endocrinol Metab 1994; 78:526–535.

14. Huhn WC, Hartman ML, Pezzoli SS, Thorner MO. Twenty-four-hour growth hormone (GH)-releasing peptide (GHRP) infusion enhances pulsatile GH secretion and specifically attenuates the response to a subsequent GHRP bolus. J Clin Endocrinol Metab 1993; 76:1202–1208.

15. Aloi JA, Gertz BJ, Hartman ML, Huhn WC, Pezzoli SS, Wittreich JM, Krupa DA, Thorner MO. Neuroendocrine responses to a novel growth hormone secretagogue, L-692,429, in healthy older subjects. J Clin Endocrinol Metab 1994; 79:943–949.

16. Bach MA, Chapman I, Farmer M, Schilling L, Van Cauter E, Taylor AM, Bolognese J, Krupa D, Gormley G, Thorner MO. An orally active growth hormone secretagogue (MK-0677) increases pulsatile GH secretion and circulating IGF-1 in the healthy elderly. 10th International Congress of Endocrinology. San Francisco, CA, June 12–15, 1996.

17. Laron Z, Frenkel J, Deghenghi R, Anin S, Klinger B, Silbergeld A. Intranasal administration of the GHRP hexarelin accelerates growth in short children. Clin Endocrinol (Oxf) 1995; 43:631–635.
18. Smith RG, Cheng K, Schoen WR, et al. A nonpeptidyl growth hormone secretagogue. Science 1993; 260:1640–1643.
19. Patchett AA, Nargund RP, Tata JR, et al. Design and biological activities of L-163,191 (MK-0677): a potent, orally active growth hormone secretagogue. Proc Natl Acad Sci USA 1995; 92:7001–7005.
20. Dolle RE, Schmidt SJ, Kruse LI. Palladium catalysed alkoxycarbonylation of phenols to benzoate esters. J Chem Soc Chem Commun 1987; 12:904–905.
21. Yatsunami T, Yazaki A, Inoue S, Yamamoto H, Yokomoto M, Nomiyama J, Noda S. Preparation of 7-(2-isoindolinyl)-1,4-dihydro-4-oxoquinoline- and 1,8-naphthyridine-3-carboxylic acid derivatives as antibacterials. EP 343560 A2 891129. Eur. Patent Application: EP: 89-109165 890522. CAN 112:235326..
22. Ochiai, E Nakagome T. Bemerkung zur Nitrierung des 1,2,3,4-Tetrahydro-isochinolins und seiner Derivate. Chem Pharm Bull 1958; 6:497–500.
23. Carpino PA, DaSilva-Jardine PA, Lefker BA, Ragan JA. Preparation of heterocyclic dipeptide derivatives which promote release of growth hormone. PCT Int. Appl. WO 9638471 A1 961205. Application WO 95-IB410 950529. CAN 126:104431.
24. Blank B, Krog AJ, Weiner G, Pendleton RG. Inhibitors of phenylethanolamine N-methyltransferase and epinephrine biosynthesis. 2. 1,2,3,4-Tetrahydroisoquinoline-7-sulfonanilides. J Med Chem 1980; 23:837–840.
25. Sartor O, Bowers CY, Reynolds GA, Momany FA. Variables determining the growth hormone response of His-D-Trp-Ala-Trp-D-Phe-Lys-NH$_2$ in the rat. Endocrinology 1985; 117:1441–1447.
26. McDowell RS, Elias KA, Stanley MS, Burdick DJ, Burnier JP, Chan KS, Fairbrother WJ, Hammonds RG, Ingle GS, Jacobsen NE. Growth hormone secretagogues: characterization, efficacy, and minimal bioactive conformation. Proc Natl Acad Sci USA 1995; 92:11165–11169.

8

Development of a New and Simple Screening System for the Evaluation of Growth Hormone Secretagogues

Vojislav Pejović
Institute for Chemistry, Technology, and Metallurgy, Belgrade, Yugoslavia

Jelena Joksimović and Vukić Šoškić
Institute for Biological Research, Belgrade, Yugoslavia

I. INTRODUCTION

During the past several years much attention has been focused on peptidyl and nonpeptidyl growth hormone (GH) secretagogues (GHSs). Such an interest can be primarily ascribed to their promising therapeutic application both in animals and humans, but also to their use in fundamental research to better understand their mechanism of action, especially their interaction with specific receptors. In 1984, a synthetic hexapeptide (His-D-Trp-Ala-Trp-D-Phe-Lys-NH$_2$) acting as a potent GHS termed growth hormone-releasing peptide (GHRP)-6 was developed that bound to a receptor different from a specific GH receptor (1).

Reduced GH secretion in children leads to delayed physical development, whereas in adults, deficiency of this hormone is manifested as diminished subjective well-being and exercise capacity, general and continuing fatigue, obesity, reduction of muscle mass and bone density; these states represent indications for GH administration. Although recombinant GH (rGH) has been available for some time, its usefulness is limited because of its poor oral bioavailability, as well as undesirable side effects, such as carpal tunnel syndrome (2) and increase in cortisol and prolactin levels (3,4). Clinical efficacy and specificity of GHRPs in elevating GH level have already been demonstrated (5). The application of hexarelin*, a GHRP-6 analog, results in GH release in humans (6). On the other hand, L-692,429 (MK-0751), a nonpeptidic GHS elevates GH in young men, with only a small transient increase in cortisol and prolactin levels (7). Recently, the benzolactam L-692,585, a 2-hydroxypropyl derivative of L-692,429 that expressed a potent GHS activity on oral administration to dogs was synthesized (8,9).

However, although much effort has been devoted to the design, synthesis, and evaluation of both peptidyl and nonpeptidyl GHS, suitable screening systems for a rapid evaluation of a large number of novel compounds representing potential candidates as GHSs are still lacking. This prompted us to develop a simple screening system based on polyclonal antibodies. For this purpose a pentapeptide (Gly-Ala-Asn-Ala-Gly) was bound to the COOH- or NH_2-terminus of GHRP-6 and the resulting undekapeptides termed P1 and P2, respectively, further conjugated with bovine serum albumin were used as the antigens for the immunization of rabbits and preparation of the antisera.

II. MATERIALS AND METHODS

A. Chemical Syntheses

Compound 1a (3-amino-3-methyl-N-[2,3,4,5-tetrahydro-2-oxo-1-(2'-carboxy-methyl)-1,1'-biphenyl-4-yl)-1H-1-benzazepine-3(R,S)-yl]-butanamide was synthesized as described by Schoen et al. (9).

For synthesis of compound 1b (3-succinylamido-3-methyl-N-[2,3,4,5-tetrahydro-2-oxo-1-(2'-carboxymethyl)-1,1'-biphenyl-4-yl)-1H-1-benzazepine-3(R,S)-yl]-butanamide), 1a (25 mg, 0.042 mmol) was dissolved in 2 mL of dry N,N-dimethylformamide, and succinic anhydride (4.5 mg, 0.046 mmol), pyridine (3.6 mg, 0.046 mmol), and 4-dimethyl aminopyridine (0.56 mg, 0.046 mmol; DMAP) were added. After stirring under anhydrous conditions at 60°C for 8 h, the same amounts of succinic anhydride, pyridine, and DMAP were again introduced and the stirring continued for the next 8 h. After that, the excess of succinic anhydride was quenched with water, and the resulting succinylated benzolactam ester 1b was used to prepare the conjugates.

B. Preparation of the Antigens

Undekapeptides, herein referred to as P1 and P2, were coupled to bovine serum albumin (BSA) and chicken egg albumin as protein carriers, yielding antigens 1, 2, 3, and 1', 2', and 3', respectively, as follows:

1. After blocking free amino groups of P1 and P2 with citraconic anhydride (10), antigens 1, 1', 2 and 2' were prepared by linking the COOH-

---

*Hexarelin is a trademark for examorelin (I.N.N.).

termini to protein carriers using 1-ethyl-3-(3-dimethylaminopropyl)-carbodiimide hydrochloride (EDC) as a coupling agent. To block free amino groups of P1 and P2, the peptides (20 mg each) were dissolved in 2 mL of 20 mM phosphate buffer and the pH was adjusted to 8.50. Then, 2 mL of citraconic anhydride solution (10 mg/mL in 20 mM phosphate buffer, pH 8.0; buffer A) was slowly added to either peptide solution, keeping the pH between 8 and 9 with 5% NaOH. After that, the reaction mixtures were incubated at ambient temperature for 60 min. The peptides with blocked free amino groups were separated from the remaining anhydride by column chromatography (Sephadex G-10, 1.8 × 74 cm, buffer A), concentrated in vacuo to 5.0 mL, and further diluted to 1.0 mg peptide per milliliter. EDC was added to a final concentration of 10 mg/mL, pH adjusted to 8.0, and the mixtures were incubated for 6 min at ambient temperature. After that, equal volumes of carrier protein solutions in buffer A were added (1.0 mg and 1.6 mg of BSA and ovalbumin per milligram peptide, respectively). After a 4-h incubation, the reaction was terminated by sodium acetate (pH 4.2; 100 mM final concentration), and the incubation was continued for 1 h at room temperature. The peptide–protein conjugates were separated from the peptides by column chromatography (Sephadex G-25, 1.8 × 62 cm, buffer A) and concentrated by solid PEG 20,000.

2. Antigens 3 and 3′ were prepared by linking the NH_2-terminus of P1 to protein carriers using glutaraldehyde as a coupling agent. Carrier proteins were added to the peptide solutions (5 mg/mL, buffer A). To the stirring mixtures an equal volume of 0.2% glutaraldehyde solution in buffer A was introduced dropwise. After 60 min of incubation at room temperature, the mixture was supplemented with 1 M glycine in buffer A (pH adjusted to 7.0) to a final concentration of 200 mM and incubation continued for 60 min with constant stirring. The peptide-protein conjugates were separated from the peptide and concentrated as described under paragraph 1.

3. Benzolactam secretagogue analogs were coupled to ovalbumin by the methods described in the foregoing (1a by glutaraldehyde, 1b by EDC), providing the same molar ratio as in the peptide–protein reaction mixture.

C. Immunizations

Three male rabbits were each sc injected with the corresponding antigen (1, 2, or 3; 200 mg antigen in 1.5 mL) in Freund's complete adjuvans (Sigma) for the first injection and in Freund's incomplete adjuvans (Sigma) for the following four booster injections, spaced 3 weeks apart. Test blood was taken from the marginal ear vein of each animal and 3 weeks after the fifth immunization the animals were humanely killed by heart exsanguination.

D. Enzyme-Linked Immunosorbent Assays

The enzyme-linked immunosorbent assays (ELISAs) were performed as already described (11), omitting Tween 20 from the washing solutions. Ovalbumin and ovalbumin conjugates (2 μg) (i.e., BSA and BSA conjugates; 1.6 μg) were introduced into the plate wells: monoclonal antirabbit IgG coupled with alkaline phosphatase was a Sigma product and p-nitrophenyl phosphate was supplied by Boehring Mannheim. Polyclonal antisera dilutions ranged from 1:5000 to 1:100.

III. RESULTS

Avidity changes of the polyclonal antisera developed against conjugates 1, 2, and 3 are presented in Figure 1. As seen, the immunization schedule of the rabbits applied in the present study resulted in a sufficiently high rise of the antibody titers. The antisera were tested against the two protein carriers, BSA and ovalbumin, as blanks and against the corresponding conjugates. Antisera 1 and 2 recognized both antigens 1 and 2, as well as ovalbumin analogs 1′ and 2′, whereas antiserum 3 interacted with all three antigens and ovalbumin analogs (Fig. 2). However, because the titer of antibodies against BSA was high,

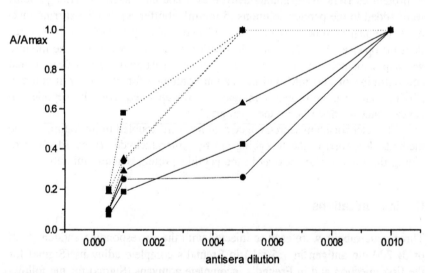

Figure 1 Relative avidity changes of the rabbit antisera raised against the three antigens. Solid lines, antibody titers 3 weeks after the first injection; dotted lines, antibody titers after the fourth booster injection. Squares, triangles, and circles, antiserum vs. antigen 1 vs. 1, 2 vs. 2, and 3 vs. 3, respectively.

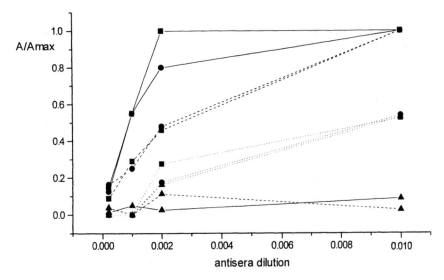

Figure 2   Recognition of the antigens by the three antisera: squares, antigen 1; circles, antigen 2; triangles, antigen 3; solid lines, antiserum 1; dashed lines, antiserum 2; dotted lines, antiserum 3.

as shown in Figure 3, whereas that against ovalbumin was negligible (data not shown), only the ovalbumin conjugates were used for further tests. To evaluate the quality of the antisera as reliable tools for detecting GHRP-6 mimotopes, compounds 1a and 1b were coupled to ovalbumin by glutaraldehyde and EDC, respectively, affording the corresponding conjugates. The latter conjugate was recognized by all three antisera (Fig. 4), whereas the former interacted conspicuously with only antiserum 3, weakly with antiserum 1, and was not recognized by antiserum 2 (Fig. 5). The results described earlier for the benzolactam secretagogue analogs 1a and 1b were obtained using racemic compounds. The data on the tests performed with enantiomerically enriched 1b conjugate (Fig. 6) clearly demonstrate a stronger interaction than observed when the racemic 1b conjugate was employed.

IV. DISCUSSION

The aim of the present study was to develop a reliable, easy-to-use system for the screening of new compounds, potential GHS. For this purpose, rabbit polyclonal antisera against GHRP-6 were prepared using the conjugates of BSA and undekapeptides P1 and P2 which include nonamide GHRP-6 sequences, with palindromic pentapeptide Gly-Ala-Asn-Ala-Gly linked to COOH- and NH_2-ter-

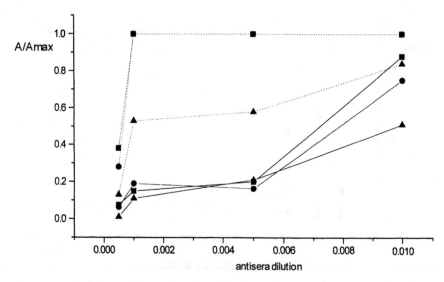

Figure 3 Relative avidity changes of the rabbit antisera raised against bovine serum albumin: squares, antiserum 1; triangles, antiserum 2; circles, antiserum 3; solid lines, antibody titers 3 weeks after the first injection; dotted lines, antibody titers after the fourth booster injection.

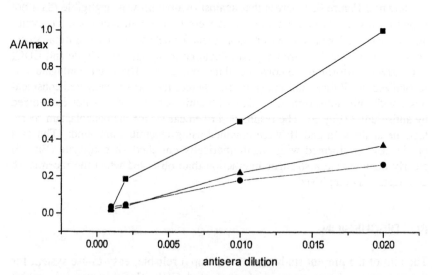

Figure 4 Relative avidity of the antisera raised against the racemic 1b conjugate: squares, antiserum 1; triangles, antiserum 2; circles, antiserum 3.

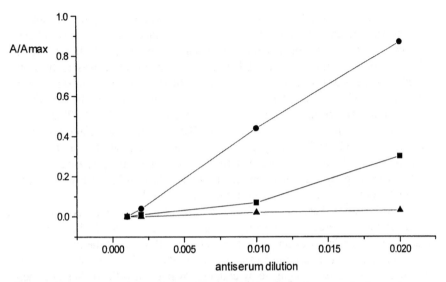

Figure 5 Relative avidity of the antisera raised against the racemic 1a conjugate: squares, antiserum 1; triangles, antiserum 2; circles, antiserum 3.

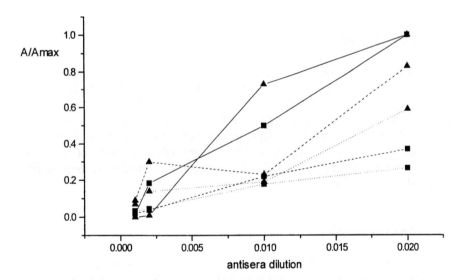

Figure 6 Relative avidity of the antisera against $R(+)$-form-enriched and racemic 1b conjugates: squares, racemic 1b conjugate; triangles, $R(+)$-form-enriched 1b conjugate; solid lines, antiserum 1; dashed lines, antiserum 2; dotted lines, antiserum 3.

mini of GHRP-6, respectively, as the antigens. This particular "spacer" sequence was chosen because Gly, Asn, and Asp residues frequently occur in α-helices. Because it has been demonstrated by molecular modelling that GHRP-6 and L-692,429 share common structural features (9), our intention was to find out whether these features could also be recognized by polyclonal antisera. After achieving a high antibody titer, the antisera were evaluated using the conjugates of ovalbumin and compounds 1a and 1b, analogous to L-692,429. The results depicted in Figures 4 and 5 indicate that EDC-coupled conjugates are more promising candidates for this type of investigations, because 1b conjugate was recognized by all three antisera, whereas the 1a conjugate, prepared by glut-araldehyde coupling, interacted markedly with antiserum 3, weakly with anti-serum 1, and showed no interaction with antiserum 2.

Furthermore, because L-692,429 is an *R*-enantiomer and agonist of the GHS receptor, whereas the *S*-enantiomer acts as an antagonist (9,12), we were interested in learning whether the antisera obtained throughout the present study could differentiate between the *R*- and *S*-configurations of the benzolactam. With this aim we prepared a conjugate by coupling ovalbumin with an *R*-enantiomer-enriched fraction (70% *R*(+)- vs. 30% *S*(+)-form) of compound 1b, which expressed a much stronger interaction with all three antisera than the correspond-ing conjugate of the racemate (see Fig. 6).

V. CONCLUSION

A simple screening system for the evaluation of new compounds, potential GHSs has been developed. This system is based on polyclonal rabbit antibodies raised against the conjugates of bovine serum albumin and undekapeptides produced by binding Gly-Ala-Asn-Ala-Gly to COOH- or NH$_2$-termini of the sequence of GHRP-6, a known peptidyl secretagogue. Given the results obtained through-out this study, it can be expected that this system will be valuable, not only for the screening of peptidyl, but also for nonpeptidyl compounds, as potential candidates for GHSs.

ACKNOWLEDGMENTS

The authors are very grateful to Dr. Jasminka Godovac-Zimmermann of the Institute for Molecular Biotechnology (IMB), Jena, Germany, for kindly pro-viding the undekapeptides. The results presented here were obtained within the scope of the research projects 02E24 (VP and VŠ) and 03E20(VŠ and JJ) sup-ported by Ministry for Science and Technology of Serbia.

REFERENCES

1. Bowers CY, Momany FA, Reynolds GA, Hong A. On the in vitro and in vivo activity of a new synthetic hexapeptide that acts on the pituitary to specifically release growth hormone. Endocrinology 1984; 114:1537–1545.
2. Hintz RL. Untoward events in patients treated with growth hormone in the USA. Horm Res 1992; 38(suppl 1):44–49.
3. Bowers CY, Reynolds GA, Durham D, Barrera CM, Pezzoli SS, Thorner MO. Growth hormone (GH)-releasing peptide stimulates GH release in normal men and acts synergistically with GH-releasing hormone. J Clin Endocrinol Metab 1990; 70:975–982.
4. Hayashi S, Okimura Y, Yagi H, Uchiyama T, Takeshima Y, Shakutsui S, Oohashi S, Bowers CY, Chiara K. Intranasal administration of His-D-Trp-Ala-Trp-D-Phe-Lys-NH$_2$ (growth hormone releasing peptide) increased plasma growth hormone and insulin-like growth factor-I levels in normal men. Endocrinol Jpn 1991; 38:15–21.
5. Bowers CY. GH releasing peptides: structure and kinetics. J Pediatr Endocrinol 1993; 6:21–31.
6. Ghigo E, Awat E, Gianotti L, Imbimbo BP, Lenaerts V, Deghenght R, Camanni F. GH-releasing activity of hexarelin, a new synthetic hexapeptide, after intravenous, subcutaneous, intranasal and oral administration in men. J Clin Endocrinol Metab 1994; 78:693–698.
7. Gertz B, Barret JS, Eisenhandler R, Krupa DA, Wittreich JM, Seibold JR, Schneider SH. Growth hormone response in man to L-692,429, a novel nonpeptide mimic of growth hormone-releasing peptide-6. J Clin Endocrinol Metab 1993; 77:1393–1397.
8. Smith RG, Cheng K, Schoen WR, Pong SS, Hickey H, Jacks T, Butler B, Chan WW-S, Chaung L-YP, Judith F, Taylor J, Wyvratt MJ, Fisher MH. A nonpeptidyl growth hormone secretagogue. Science 1993; 260:1640–1643.
9. Schoen WR, Pisano JM, Prendergast K, Wyvratt MJ, Fisher MH, Cheng K, Chan WW-S, Butler B, Smith RG, Ball RG. A novel 3-substituted benazepinone growth hormone secretagogue (L-692,429). J Med Chem 1994; 37:897–906.
10. Atassi MZ, Habeeb ASSA. Reaction of proteins with citraconic anhydride. Methods Enzymol 1972; 25:546–553.
11. Johnston A, Thorpe R. Enzyme linked assays. In: Johnston A, Thorpe R, eds. Immunochemistry in Practice. Oxford: Blackwell Scientific, 1982:252–254.
12. Howard AD, Feighner SD, Cully DF, Arena JP, Liberator PA, Rosenblum CI, Hamelin M, Hreniuk DL, Palyha OC, Anderson J, Paress PS, Diaz C, Chou M, Liui KK, McKee KK, Pong S-S, Chaung L-Y, Elbrecht A, Dashkevicz M, Heavens R, Rigby M, Sirinathsinghji DJS, Dean DC, Melillo DG, Patchett AA, Nargund R, Griffin PR, DeMartino JA, Gupta SK, Schaeffer JM, Smith RG, Van der Ploeg LHT. A receptor in pituitary and hypothalamus that functions in growth hormone release. Science 1996; 273:974–976.

9

Agonistic Analogs of Growth Hormone-Releasing Hormone: Endocrine and Growth Studies

Andrew V. Schally and Ana Maria Comaru-Schally
Tulane University School of Medicine and Veterans Affairs Medical Center, New Orleans, Louisiana

I. INTRODUCTION

Various studies carried out in the 1960s and 1970s established that the secretion of growth hormone (GH) from the anterior pituitary gland is regulated by a dual system of hypothalamic control, and materials that could stimulate (1–3) or inhibit (3) the release of GH, were demonstrated in hypothalamic extracts. In 1970s the tetradecapeptide, which inhibits the release of GH and was named somatostatin, was isolated from ovine (4) and, subsequently, porcine hypothalami (5) and synthesized. The breakthrough on the identification of growth hormone-releasing hormone (GHRH) was provided by Frohman (6), who demonstrated ectopic production of GHRH by carcinoid and pancreatic cell tumors. In 1982, two groups isolated GHRH from human pancreatic tumors that caused acromegaly (7,8). Guillemin et al. reported a 44-amino acid amidated peptide [GHRH $(1–44)NH_2$] (8) and Rivier et al. (7) a 40-amino acid peptide [GHRH (1–40)OH and GHRH(1–29)NH_2]. Subsequently, GHRH(1–44)NH_2 was isolated from human and animal hypothalami (9,10). Virtually full intrinsic biological activity is present in the 29 NH_2-terminal amino acid residues [GHRH(1–29)NH_2] (7).

Clinical studies with GHRH(1–40) and GHRH(1–29)NH_2 were carried out by Thorner et al. (11,12) and Ross et al. (13), and with GHRH(1–44)NH_2 by Laron et al. (14). These studies will be reviewed later.

Many agonistic analogs of GHRH(1–29)NH$_2$, intended for potential clinical and veterinary applications have been synthesized by various groups (15–31). Analogs with enhanced biological activities in vivo and in vitro and, therefore, with a greater potential therapeutic usefulness, have resulted from changes in the molecule, such as derivatization of functional groups or exchange of amino acids (19,31). Replacement of tyrosine-1 by desNH$_2$-Tyr[1] (Dat) led to analogs with increased biological potency as a result of enhanced resistance of the NH$_2$-terminus to enzymatic degradation (16,17,23,25,26,30,31), and D-amino acid substitutions in positions 1, 2, or 3 resulted in analogs with increased GH-releasing activity both in vitro and in vivo (24,27,29). Studies with deletion of amino acids at the COOH-terminus show that GHRH(1–29)NH$_2$ is the shortest fragment endowed with the full intrinsic activity (27,28). We have previously synthesized GHRH analogs with agmatine (4-guanidinobutylamine), an amine formed by decarboxylation of arginine with the structure H$_2$N-(CH$_2$)$_4$-NH-C(NH$_2$)=NH in position 29 (22,23,25,26,30,31). In the nomenclature used initially, these agonists were listed as analogs of GHRH(1–28)Agm, but in the terminology subsequently adopted, the name [Agm[29]]GHRH(1–29) is preferred. We showed that some analogs of GHRH(1–28)Agm are very potent in vitro and in vivo (22,23,25,26,30,31). To avoid oxidative inactivation of the analogs, Met[27] was replaced by Nle in some peptides. Most of our analogs that displayed enhanced GH-releasing activity contained Nle[27] (25,26,30).

The aim of this chapter is to review various endocrine studies performed with the latest series of superactive agonists of Agm[29] GHRH(1–29) containing ornithine instead of lysine in positions 12 and 21 and other modifications (19). Effects of long-term administration of these analogs on growth of young rats will be also summarized. Clinical studies already performed with GHRH will be briefly discussed as well as potential therapeutic applications of superactive GHRH agonists.

II. SYNTHESIS AND IN VITRO AND IN VIVO EVALUATION OF NEW ANALOGS OF GHRH CONTAINING ORNITHINE IN POSITIONS 12 AND 21

A. Synthesis

To produce GHRH agonists with greater metabolic stability and increased activity in vivo, nine new analogs of human GHRH(1-29)NH$_2$ were synthesized by solid-phase methods and studied in vitro and in vivo (19). In all these peptides, some substitutions previously used, were incorporated to increase the resistance to degradation. The NH$_2$-terminal tyrosine was replaced by Dat (22,30,31) and COOH-terminal Arg-NH$_2$ by Agm (25,26,30). The Met[27] was also replaced by Nle to prevent chemical oxidation (19). The Gly[15] was substituted by Abu to increase the resistance to degradation in blood by chymotrypsin-like enzymes (19) and to enhance the affinity to the receptors. The Asn in position 8 was left unchanged or replaced by Thr, Gln, or Ser. Three analogs contained Asp instead of Ser in position 28. In two analogs, one Lys was replaced by Orn, and in six analogs, two Lys residues were substituted by Orn in position 12 and 21. Trypsin exhibits a relatively low activity toward the ornithyl substrates, and it was expected that this modification would increase the stability of the bonds between residues 12 and 13, as well as 21 and 22 (19). After the synthesis, the peptides were purified by high-performance liquid chromatography (HPLC; 19). Substitution of the Lys in positions 12 and 21 by Orn, introduced in this series of analogs, together with the other substitutions just cited, produced highly potent and long-acting compounds. The analogs [Dat[1], Orn[12,21], Abu[15], Nle[27], Agm[29]]hGHRH(1–29) (JI-22), [Dat[1], Orn[12,21], Abu[15], Nle[27], Asp[28], Agm[29]]hGHRH(1–29) (JI-34), [Dat[1], Thr[8], Orn[12,21], Abu[15], Nle[27], Asp[28], Agm[29][hGHRH (JI-36), and [Dat[1], Gln[8], Orn[12,21], Abu[15], Nle[27], Asp[28], Agm[29]]hGHRH(1–29) (JI-38) appeared to be the most potent in various tests (19).

B. In Vitro Tests

In vitro GH-releasing activity of the analogs was studied in the superfused rat pituitary cell system (26). All peptides showed much greater activity than hGHRH(1–29)NH$_2$. The binding affinity of analogs to GHRH membrane receptors on rat anterior pituitary cells was determined by using ^{125}I-labeled [His[1],Nle[27]]hGHRH(1–32)NH$_2$. The affinities of analogs were compared with hGHRH(1–29)NH$_2$ as a standard (IC$_{50}$ = 4.54 ± 0.09 nM), and they were much higher (19). The relative binding affinity of analogs JI-34, JI-36, and JI-38 to GHRH receptors was 29.3, 25.3, and 16.8 times higher, respectively, than that of h(GHRH)(1–29)NH$_2$ (19).

C. In Vivo Tests

For the determination of GH-release in vivo, male Sprague–Dawley rats were anesthetized with pentobarbital (Nembutal) and 20 min later the analogs and hGHRH(1–29)NH$_2$, dissolved in saline, were injected intravenously (iv) or subcutaneously (sc); control groups were injected with only saline (26). Blood samples were drawn before and 5 and 15 min after iv administration and 15, 30, and occasionally, 60 min after sc administration. After iv administration, all analogs were more potent than hGHRH(1–29)NH$_2$. The relative potencies of JI-36, JI-38, and JI-22 were 3.2–3.8 times higher compared with hGHRH(1–29)NH$_2$ at 5 min and 6.1–8.5 times higher at 15 min after the injection (19).

Following sc administration, all the analogs displayed much greater activities than after iv injection (19). The relative potencies of analogs JI-34, JI-36, JI-38, and JI-22 were 80.9, 95.8, 71.4, and 44.6 times higher, respectively, when compared with hGHRH(1–29)NH$_2$ at 15 min, and 89.7, 87.9, 116.8, and 217.7 times higher, respectively, at 30 min after the sc injection. The potencies of the other analogs were also higher, compared with the standard. The GH-releasing activity could be also expressed on the basis of the area under the 0–30 min GH-response curve for each analog (19). The GH-release produced by JI-34, JI-36, and JI-38 was much greater than that resulting from the administration of hGHRH(1–29)NH$_2$ at doses 50 times higher (19). These results indicate that peptides JI-22, JI-34, JI-36, and JI-38 were more potent after sc administration (19) than other analogs studied in our laboratory (22,23,26, 30,31). The most active analogs JI-34, JI-36, and JI-38 contained aspartic acid in position 28 instead of serine. The increase in potency of these analogs is also likely caused by higher affinity to the receptors (19).

A comparison of the results obtained after sc and iv administration suggests that very high potency of analogs after sc injection, compared with hGHRH(1–29)NH$_2$, is partly due to their increased resistance to enzymatic degradation in sc tissue (19). The substitution of the asparagine in position 8 by threonine or glutamine in analogs JI-36 and JI-38, respectively, was intended to avoid an inactivation through chemical isomerization (19). Peptide JI-34 had native asparagine in this position. All three analogs displayed high activities in vivo and increased binding affinities to the receptors (19). Intense activities of the analogs [Dat[1],Thr[8],Orn[12,21],Abu[15],Nle[27],Agm[29]]hGHRH(1–29) (JI-36) and [Dat[1],Gln[8],Orn[12,21],Abu[15],Nle[27],Agm[29]]hGHRH(1–29) (JI-38) after sc administration suggest that these agonists could be used clinically for therapy of patients with growth hormone deficiency (19).

D. Other Studies with Agm[29] GHRH Analogs

It is of interest that Agm[29] GHRH analogs are active in rats not only when administered iv or sc, but also intranasally and by pulmonary inhalation (32). Our results indicate a high bioavailability of a GHRH Agm[29] analog MZ-3-149 administered by a convenient pulmonary inhalation route (32).

Serum levels of various Agm[29] GHRH analogs can be measured by radioimmunoassays, using antisera developed against GHRH agonist Dat[1], Ala[15], Nle[27] Agm[29]GHRH(1–29)(MZ-2-51) and its COOH-terminal Nle[27] Agm[29] GHRH(17–29) fragment (33).

III. STUDIES ON LONG-TERM ADMINISTRATION OF GHRH AGONISTS IN RATS

A. Effect of Long-Term Administration of GHRH Agonist JI-36 on Linear Growth and GH Responsiveness in Male Rats

The results of a repeated or continuous administration of a potent agonistic analog of (GHRH), [Dat[1], Thr[8], Orn[12,21], Abu[15], Nle[27], Asp[28], Agm[29]]hGHRH-(1–29) (JI-36), on the linear growth and the GH responses to GHRH(1–29)NH$_2$ were investigated in normal male rats 7 weeks old (34). The GHRH agonist JI-36 was administered by continuous release from dorsally implanted Alzet osmotic minipumps at the rate of 0.2 μg/h or given by twice daily sc injections of 0.5 and 5 μg/rat for 4 weeks. Body weight and tail length were monitored (34). Basal serum GH and insulin-like growth factor I (IGF-I) concentrations and GH responses to GHRH(1–29)NH$_2$ were measured by radioimmunoassay (RIA). Prolonged administration of JI-36 significantly sped up the growth of rats, as measured by the tail length (34). Tail growth in animals treated with JI-36 in doses of 0.5 and 5 μg/rat bid was significantly increased to 17.0 ± 0.1 and 16.7 ± 0.2 cm, respectively, within 1 week from the start of the experiment, compared with the controls, which measured 16.0 ± 0.1 cm. Similarly, tail length in the group receiving JI-36 released from osmotic minipumps was also significantly increased to 16.7 ± 0.2 cm after 1 week of treatment. Tail length remained significantly increased in all groups receiving JI-36 during the whole experimental period (34). At the end of the experiment, the increase in tail length was 8.1% in the osmotic minipump group, 4.3% in animals receiving 0.5 μg bid of JI-36, and 7.0% in the group treated with 5 μg/day bid of the agonist, compared with the control group (34). The acceleration of growth was associated with stimulation of IGF-I secretion. Prolonged administration of the GHRH analog JI-36 induced a significant elevation in the basal serum IGF-I concentration after 2 weeks of treatment, increasing it by 17–23% in the three groups that received the treatment, compared with controls. The different doses and modes of administration of JI-36 resulted in a similar elevation of IGF-I levels. Treatment with JI-36 released from osmotic minipumps or administered by daily injections at a dose of 0.5 μg/rat bid did not cause significant changes in the mean basal serum GH concentration 2 weeks after initiation of the experiment (34). However, basal serum levels of GH were significantly elevated in rats injected with 5 μg/rat bid of JI-36. Thus, it appears that serum IGF-I levels may be related to the magnitude of GHRH-stimulated GH pulses, rather than to basal GH levels between pulses (34). The GH response to iv bolus injection of GHRH(1–29)NH$_2$ was preserved in all groups. No attenuation of the GH-response occurred in rats treated with agonist JI-36, when compared with the control group. After 14 days of treatment with JI-36, the GH response to intravenous bolus injection of hGHRH(1–29)NH$_2$ in all three groups that received the analog was similar to that in the control group. No decreases in GH responses to hGHRH(1–29)NH$_2$ occurred in the three groups of rats treated for 28 days with the GHRH agonist according to different regimens (34). These results indicate that long-term administration of GHRH agonist JI-36 significantly increases the growth rate in rats, without affecting somatotroph responsiveness (34).

This study suggests that long-term administration of agonistic GHRH analogs, such as JI-36, can increase growth rate in rats without evidence of somatotroph depletion or desensitization. Consequently, this class of GHRH agonists might be useful clinically for the treatment of GH deficiencies, caused by a hypothalamic dysfunction, even in regimens involving long-term administration (34).

B. Induction of Compensatory Linear Growth in MSG-Lesioned, GH-Deficient Rats by Long-Term Administration of GHRH Agonist JI-38

We investigated the effects of prolonged administration of GHRH agonist [Dat[1], Gln[8], Orn[12,21], Abu[15], Nle[27], Asp[28], Agm[29]]hGHRH-(1–29) (JI-38), on growth responses in monosodium glutamate (MSG)-lesioned and normal young rats (35). Body weight, body length, tibia length, and tail length were monitored. Basal serum GH concentrations, GH responses to bolus injections of GHRH(1–29)NH$_2$, pituitary GH, and serum IGF-I concentrations were measured by RIA. The levels of pituitary GHRH receptors and binding affinity were also evaluated after the treatment (35). Neonatal treatment with MSG resulted in blunted growth, with a decrease in serum and pituitary GH concentration, decreased serum IGF-I levels, and a reduction in GHRH receptor concentration (35). Long-term administration of GHRH agonist JI-38 in doses of 2 µg bid for 2 weeks significantly stimulated linear growth in the MSG-treated animals, by 11% in males and 13% in females at 6 weeks of age, with MSG-treated animals achieving the growth rate of normal controls. Administration of the GHRH analog for 2 weeks caused an increase of 27 and 28% in body weights of the MSG-treated male and female rats, respectively, at 6 weeks of age (35). Treatment with JI-38 for 2 weeks evoked a compensatory growth of tibia in MSG-lesioned rats: tibia length increasing by 9% in the males and 15% in the females (35). Treatment with the GHRH analog induced more distinct changes in the tail length of the MSG-lesioned female than of male rats (14 and 8% increase, respectively). Acceleration of growth by JI-38 was associated with stimulated pituitary GH synthesis, although basal serum GH levels did not change (35). Twice-daily injections of the GHRH agonist JI-38 resulted in a 39% elevation of IGF-I levels in both male and female rats after 2 weeks. Pituitary GHRH receptor concentration in MSG rats was slightly, but not significantly, increased by the treatment with JI-38 (35). Repeated injections of JI-38 increased the GH responses to bolus injections of GHRH(1–29)NH$_2$ by 112% in MSG-lesioned males and by 81% in females, as compared with respective MSG-lesioned untreated controls (35). The mean GH response was 484.7 $\pm$ 92.8 ng/mL in the MSG-lesioned male rats treated with JI-38 and 228.4 $\pm$ 68.2 ng/mL in the untreated MSG-lesioned controls. In MSG-lesioned female rats treated with JI-38, the GH response was 307.2 $\pm$ 64.4 ng/mL, whereas in the MSG-lesioned untreated controls it was only 169.3 $\pm$ 25.2 ng/mL (35). Treatment of normal young growing rats with agonist JI-38 did not further increase the normal growth acceleration in these rats, but it did stimulate the GH synthesis and augment the GH secretory responsiveness. In normal young rats, treatment with agonist JI-38 also augmented GH responses to bolus injections of GHRH by 59% in males and by 66% in females, compared with respective untreated controls (35). The treatment of MSG-lesioned rats with GHRH agonist was generally more effective in female than in male animals. These findings provide evidence that the blunted growth rate of the MSG-lesioned rats is associated with a decreased pituitary GHRH receptor concentration (35). This work provides proof for the therapeutic efficacy of our new GHRH analog in MSG-lesioned rat model, which clinically might reflect the condition of human hypothalamic GHRH deficiency. The demonstration that administration of GHRH agonist JI-38 is able to restore the normal growth rate of the GH-deficient rats by stimulating GH synthesis and IGF-I secretion may have clinical importance. Collectively, our studies suggest that agonistic GHRH analogs, such as JI-38, might be possibly useful clinically for the treatment of GH deficiencies caused by hypothalamic dysfunction.

IV. CLINICAL STUDIES WITH GHRH

The role of GHRH in clinical medicine was recently reviewed by Gelato (36). It is well established that most children with GH deficiency have hypothalamic dysfunction and intact anterior pituitary function (11–13,36,37). Various studies demonstrated that most children and adults with documented GH deficiency will release GH in response to brief or prolonged administration of GHRH (11–14,36–40). In addition, some patients, initially unresponsive to GHRH, show an improvement in their GH response after repeated administrations of GHRH (36). Several groups have clearly demonstrated that hGHRH(1–44)NH$_2$, hGHRH(1–40), and the shortened fragment, hGHRH(1–29)NH$_2$, promote endogenous GH secretion, increase serum IGF-I levels and stimulate linear growth (height velocity) in GH-deficient children (11–13,36–38). In initial studies, GHRH(1–40) was given in discrete 1-min pulses every 3 h by an infusion pump (11). In subsequent trials, GHRH(1–29)NH$_2$ was administered sc twice daily (13). Even once-daily sc therapy with GHRH(1–29)NH$_2$ could accelerate growth in GH-deficient children (12). These studies indicated that long-term treatment with GHRH in GH-deficient children is well tolerated, and no adverse effects were observed in general biochemical or hormonal analyses. Although the growth-promoting effects of prolonged administration of GHRH appear equivalent to GH therapy, the influence of this new modality of treatment on final height is still unknown (12,36).

It has been also proposed that reduced availability of GH in late adulthood may contribute to the decline in several physiological functions associated with aging (38,41). Evidence accumulated during the last decade indicates that GH deficiency in adults is associated with adverse changes in body composition, lipid profile, insulin status, physical performance, bone mineral density, and quality of life (38,41–45). The benefits of the treatment with hGH in adults have been demonstrated in several clinical trials (41–46). Initial studies using about 5 IU/day of hGH were associated with a high incidence of side effects, mainly salt and water retention (44,46). In later studies using half this dose, the incidence of side effects was decreased, and the supranormal levels of serum IGF-I could be avoided (45,46). Chronic elevations of GH and IGF-I levels, as seen in untreated acromegalics, are associated with an increase in morbidity and mortality, mainly from cardiovascular disorders (42,46). Thus, it is important to avoid supraphysiological levels of serum GH and IGF-I by using smaller doses of hGH. It was also demonstrated that short-term sc administration of GHRH(1–29)NH$_2$ to healthy men over 60 years old reverses age-related decreases in GH and IGF-I (38). This suggests that prolonged treatment with GHRH agonists could improve alteration in the body composition, as well as the health, strength, and functional capacity in an aging population, as demonstrated with GH replacement therapy (41–46). Treatment with GHRH(1–29)NH$_2$ did not cause elevations in the systolic blood pressure, fasting plasma glucose levels, or salt and water retention, which have been observed during treatment with GH, when excessive doses were used (38,41,44,46).

Prolonged administration of hGHRH does not cause somatotroph desensitization or depletion in normal men and a GH-deficient boy (39). Continuous administration of GHRH to normal men results in continued GH release and augmentation of naturally occurring GH pulses (39). Moreover, in a recent study, administration of GHRH(1–29)NH$_2$ to GH-deficient children for a period of 12 months normalized the height velocity and induced catch-up growth in the majority of patients (12). In our studies with Agm[29] GH–RH(1–29) analogs in rats, we obtained stimulation of linear growth, without somatotroph desensitization (34,35). Overall results indicate that a therapy based on hGHRH could be used in most children and adults with GH deficiency.

Few clinical studies have been performed with GHRH analogs. The activity of an early agonistic analog D-Ala^{2}GH-RH(1–29)NH$_2$ was similar to that of GHRH(1–29)NH$_2$ after iv administration (40,47). When another agonistic analog of GHRH [Ac-D-Tyr^1,DAla^2]-GHRH(1–29)NH$_2$, which showed greatly enhanced activity in rats, was administered sc to normal men, the peak GH responses and the area under the GH curve were similar to those for GHRH(1–29)NH$_2$ (48). Thus, there is a great need for analogs that would be more stable and superactive in humans. The availability of such analogs would allow a reduction in the doses and frequency of administration, compared with GHRH(1–29)NH$_2$. The GHRH analogs would have advantages over hGH by virtue of maintaining GH secretion by an action on the pituitary gland. Because GHRH does not cause a supraphysiological elevation in endogenous GH levels, its clinically active analogs should produce similar release patterns. Thus, this class of compounds might have a better safety profile in elderly patients than the recombinant GH administration.

V. POTENTIAL CLINICAL APPLICATIONS OF GHRH ANALOGS

The GHRH analogs could find clinical therapeutic applications in children with growth retardation caused by hypothalamic GHRH deficiency, constitutional growth delay, Turner's syndrome (gonadal dysgenesis), familial short stature, prepubertal children with chronic renal insufficiency and severe growth retardation, and children on long-term treatment with glucocorticoids and growing at subnormal rate.

Possible clinical applications of agonistic analogs of GHRH in adult population would include uses in geriatric patients; adults with hypopituitarism, catabolic states, wound healing, delayed healing of fractures, osteoporosis, obesity; as an adjunct to total parenteral nutrition in malnourished patients with chronic obstructive pulmonary disease, to increase respiratory muscle strength; and improve pulmonary function in respiratory muscle dysfunction; cardiac failure, including dilated cardiomyopathy and AIDS wasting syndrome.

After the agonistic analogs of GHRH are evaluated in clinical trials, it will be possible to determine if these compounds can be used for the treatment of these multiple conditions, in addition to growth retardation and for the elderly population. It will be necessary to determine the optimal dose, optimal frequency of administration, long-term benefits, and long-term complications of GHRH agonists.

Considerable evidence exists to suggest that the novel synthetic hexapeptide GH-releasing peptide (GHRP) stimulates GH release in normal men synergistically with GHRH (49,50). The GHRH agonists could be given alone or together with GHRP (49,50). Because these peptides act on different receptors, the use of the combination of both types of analogs produces a more powerful stimulation of secretion of GH from the pituitary. Thus, a combination of GHRP with GHRH agonists could result in greater therapeutic responses in patients with hypothalamic GHRH deficiency.

ACKNOWLEDGMENTS

Some experimental work described in this paper was supported by the Medical Research Service of the Veterans Affairs Department.

REFERENCES

1. Deuben RR, Meites J. Stimulation of pituitary growth hormone release by a hypothalamic extract in vitro. Endocrinology 1964; 74:408–414.
2. Schally AV, Steelman S, Bowers CY. Effect of hypothalamic extracts on the release of growth hormone in vitro. Proc Soc Exp Biol Med 1965; 119:208–212.
3. Krulich L, Dhariwal APS, McCann SM. Stimulatory and inhibitory effects of purified hypothalamic extracts on growth hormone release from rat pituitary in vitro. Endocrinology 1968; 83:783–790.
4. Brazeau P, Vale W, Burgus R, Ling N, Butcher M, Rivier J, Guillemin R. Hypothalamic polypeptide that inhibits the secretion of immunoreactive pituitary growth hormone. Science 1973; 179:77–79.
5. Schally AV, Dupont A, Arimura A, Redding TW, Nishi N, Linthicum GL, Schlesinger DH. Isolation and structure of growth hormone-release inhibiting hormone (somatostatin) from porcine hypothalami. Biochemistry 1976; 15:509–514, 1976.
6. Frohman LA, Szabo M. Ectopic production of growth hormone-releasing factor by carcinoid and pancreatic islet tumors associated with acromegaly. Prog Clin Biol Res 1981; 74:259–271.
7. Rivier, J, Spiess J, Thorner M, Vale W. Characterization of a growth hormone-releasing factor from a human pancreatic islet tumour. Nature 1982; 300:276–278.
8. Guillemin R, Brazeau P, Bohlen P, Esch F, Ling N, Wehrenberg WB. Growth hormone-releasing factor from a human pancreatic tumor that caused acromegaly. Science 1982; 218:585–587.
9. Ling N, Esch F, Bohlen R, Brazeau P, Wehrenberg WB, Guillemin R. Isolation, primary structure, and synthesis of human hypothalamic somatocrinin: growth hormone-releasing factor. Proc Natl Acad Sci USA 1984; 81:4302–4306.

10. Guillemin R, Zeytin F, Ling N, Böhlen P, Esch F, Brazeau P, Bloch B, Wehrenberg WB. Growth hormone-releasing factor: chemistry and physiology. Proc Soc Exp Biol Med 1984; 175:407–413.

11. Thorner MO, Reschke J, Chitwood J, Rogol AD, Furlanetto R, Rivier J, Vale W, Blizzard RM. Acceleration of growth in two children treated with human growth hormone-releasing factor. N Engl J Med 1985; 312:4–9.

12. Thorner M, Rochiccioli P, Colle M, Lanes R, Grunt J, Galazka A, Landy H, Eengrand P, Shah S, on behalf of the Geref International Study Group. Once daily subcutaneous growth hormone-releasing hormone therapy accelerates growth in growth hormone-deficient children during the first year of therapy. J Clin Endocrinol Metab 1996; 81:1189–1196.

13. Ross RJM, Tsagarakis S, Grossman A, Preece MA, Rodda C, Davies PSW, Rees LH, Savage MO, Besser GM. Treatment of growth-hormone deficiency with growth-hormone-releasing hormone. Lancet 1987; 1:5–8.

14. Laron Z, Keret R, Bauman B, Pertzelan A, Ben-Zeev Z, Olsen DB, Comaru-Schally AM, Schally AV. Differential diagnosis between hypothalamic and pituitary hGH deficiency with the aid of synthetic GH-RH 1–44. Clin Endocrinol (Oxf) 1984; 21:9–12.

15. Coy DH, Hocart SJ, Murphy WA. Human growth hormone-releasing hormone analogues with much improved in vitro growth hormone-releasing potencies in rat pituitary cells. Eur J Pharmacol 1991; 204:179–185.

16. Felix A M, Heimer EP, Mowles T F, Eisenbeis H, Leung P, Lambros TJ, Ahmad M, Wang C-T. Synthesis and biological activity of novel growth hormone releasing factor analogs. In: Theodoropoulos D, ed. Peptides. New York: Walter de Gruyter; 1986:481–484.

17. Felix AM, Heimer EP, Wang C-T, Lambros TJ, Fournier A, Mowles TF, Maines S, Campbell RM, Wegrzynski BB, Toome V, Fry D, Madison VS. Synthesis, biological activity and conformation analysis of cyclic GRF analogs. Int J Pept Protein Res 1988; 32:441–454.

18. Gaudreau P, Boulanger L, Abribat T. Affinity of human growth hormone-releasing factor (1–29)NH$_2$ analogues for GRF binding sites in rat adenopituitary. J Med Chem 1992; 35:1864–1869.

19. Izdebski J, Pinski J, Horvath JE, Halmos G, Groot K, Schally AV. Synthesis and biological evaluation of superactive agonists of growth hormone-releasing hormone. Proc Natl Acad Sci USA 1995; 92:4872–4876.

20. Kubiak TM, Kelly CR, Krabill LF. In vitro metabolic degradation of a bovine growth hormone-releasing factor analog Leu27-bGRF(1–29)NH$_2$ in bovine and porcine plasma. Drug Metab Dispos 1989; 17:393–397.

21. Sato K, Hotta M, Kageyama J, Chiang T-C, Hu H-Y, Dong M-H, Ling N. Synthesis and in vitro bioactivity of human growth hormone-releasing factor analogs substituted with a single D-amino acid. Biochem Biophys Res Commun 1987; 49:531–537.

22. Zarandi M, Serfozo P, Zsigo J, Bokser L, Janaky T, Olsen DB, Bajusz S, Schally AV. Potent agonists of growth hormone-releasing hormone. Int J Pept Protein Res 1992; 39:211–217.

23. Bokser L, Zarandi M, Schally AV. Evaluation of in vivo biological activity of new agmatine analogs of growth hormone-releasing hormone (GH-RH). Life Sci 1989; 46:999–1005.
24. Coy DH, Murphy WA, Sueiras-Diaz J, Coy EJ, Lance VA. Structure–activity studies on the N-terminal region of growth hormone-releasing factor. J Med Chem 1985; 28:181–185.
25. Gulyas J, Bajusz S, Kovacs M, Schally AV. New methods for synthesis of potent GH-RH analogs. In: Penke B, Torok A, eds. Peptide Chemistry, Biology. Interactions with Proteins. New York: Walter de Gruyter, 1988:113–116.
26. Kovacs M, Gulyas J, Bajusz S, Schally AV. An evaluation of intravenous, subcutaneous, and in vitro activity of new agmatine analogs of growth hormone-releasing hormone hGH-RH. Life Sci 1988; 42:27–35.
27. Lance VA, Murphy WA, Sueiras-Diaz J, Coy DH. Super-active analogs of growth hormone-releasing factor (1–29)-amide. Biochem Biophys Res Commun 1984; 119:265–272.
28. Ling N, Baird A, Wehrenberg WB, Ueno N, Munegumi T, Brazeau P. Synthesis and in vitro bioactivity of human growth hormone-releasing factor analogs substituted at position-1. Biochem Biophys Res Commun 1984; 123:854–861.
29. Ling N, Baird A, Wehrenberg WB, Ueno N, Munegumi T, Regno TC, Brazeau P. Synthesis and in vitro bioactivity of human growth hormone-releasing factor analogs substituted at position-1. Biochem Biophys Res Commun 1984; 122:304–310.
30. Zarandi M, Csernus V, Bokser L, Bajusz S, Groot K, Schally AV. Synthesis and biological activities of analogs of growth hormone-releasing hormone (GH-RH) with C-terminal agmatine. Int J Pep Protein Res 1990; 36:499–505.
31. Zarandi M, Serfozo P, Zsigo J, Deutch AH, Janaky T, Olsen DB, Bajusz S, Schally AV. Potent agonists of growth hormone-releasing hormone. II. Pept Res 1992; 5:190–193.
32. Pinski J, Yano T, Groot K, Zsigo J, Rekasi Z, Comaru-Schally AM, Schally AV. Comparison of GH-stimulation by GH-RH(1–29)NH$_2$ and an agmatine[29] GH-RH analog, after intravenous, subcutaneous and intranasal administration and after pulmonary inhalation in rats. Int J Pept Protein Res 1993; 41:246–249.
33. Groot K, Csernus VJ, Pinski J, Zsigo J, Rekasi Z, Zarandi M, Schally AV. Development of a radioimmunoassay for some agonists of growth hormone-releasing hormone. Int J Pept Protein Res 1993; 41:162–168.
34. Pinski J, Izdebski J, Jungwirth A, Groot K, Schally AV. Effect of chronic administration of a new potent agonist of GH-RH(1–29)NH$_2$ on linear growth and GH responsiveness in rats. Regul Pept 1996; 65:197–201.
35. Kovacs M, Halmos G, Groot K, Izdebski J, Schally AV. Chronic administration of a new potent agonist of growth hormone releasing hormone induces compensatory linear growth in growth hormone-deficient rats: mechanism of action. Neuroendocrinology 1996; 64:169–176.
36. Gelato MC. Growth hormone-releasing hormone: clinical perspectives. Endocrinologist 1994; 4:64–68.

37. Gelato MC, Ross JL, Malozowski S, Pescovitz OH, Skerda M, Cassorla F, Loriaux DL, Merriam GR. Effects of pulsatile administration of growth hormone (GH)-releasing hormone on short term linear growth in children with GH deficiency. J Clin Endocrinol Metab 1985; 61:444–450.

38. Corpas E, Harman SM, Pineyro MA, Roberson R, Blackman MR. Growth hormone (GH)-releasing hormone-(1–29) twice daily reverses the decreased GH and insulin-like growth factor-I levels in old men. J Clin Endocrinol Metab 1992; 75:530–535.

39. Vance ML, Kaiser DL, Martha PM Jr, Furlanetto R, Rivier J, Vale W, Thorner MO. Lack of in vivo somatotroph desensitization or depletion after 14 days of continuous growth hormone (GH)-releasing hormone administration in normal men and a GH-deficient boy. J Clin Endocrinol Metab 1989; 68:22–28.

40. Grossman A, Savage MO, Lytras N, Preece MA, Sueiras-Diaz J, Coy DH, Rees LH, Besser GM. Responses to analogues of growth hormone-releasing hormone in normal subjects, and in growth-hormone deficient children and young adults. Clin Endocrinol 1984; 21:321–330.

41. Rudman D, Feller AG, Nagraj HS, Gergans GA, Lalitha PY, Goldberg AF, Schlenker RA, Cohn L, Rudman IW, Mattson DE. Effects of human growth hormone in men over 60 years old. N Engl J Med 1990; 323:1–6.

42. Vance ML. Growth hormone for the elderly? N Engl J Med 1990; 323:52–54.

43. Shalet SM. Growth hormone deficiency and replacement in adults. Br J Med 1996; 313:314.

44. Salomon F, Cuneo RC, Hesp R, Sönksen PH. The effects of treatment with recombinant human growth hormone on body composition and metabolism in adults with growth hormone deficiency. N Engl J Med 1989; 321:1797–1803.

45. Mardh G, Lindeberg A, on behalf of the investigators. Growth hormone replacement therapy in adult hypopituitary patients with growth hormone deficiency: combined clinical safety data from clinical trials in 665 patients. Endocrinol Metab 1995; 2:11–16.

46. Janssen YJH, Frölich M, Roelfsema F. A low starting dose of genotropin in growth hormone-deficient adults. J Clin Endocrinol Metab 1997; 82:129–135.

47. Barron JL, Coy DH, Miller RP. Growth hormone responses to growth hormone-releasing hormone (1–29)NH_2 and a D-Ala^2 analog in normal men. Peptides 1985; 6:575–577.

48. Aitman TJ, Rafferty B, Coy D, Lynch SS, Clayton RN. Bioactivity of growth hormone releasing hormone (1–29) analogues after sc injection in man. Peptides 1989; 10:1–4.

49. Bowers CY, Reynolds GA, Durham D, Barrera CM, Pezzoli SS, Thorner MO. Growth hormone (GH)-releasing peptide stimulates GH release in normal men and acts synergistically with GH-releasing hormone. J Clin Endocrinol Metab 1990; 70:975–982.

50. Bowers CY. [Editorial] On a peptidomimetic growth hormone-releasing peptide. J Clin Endocrinol Metab 1994; 7:940–942.

10

Antagonistic Analogs of Growth Hormone-Releasing Hormone: Endocrine and Oncological Studies

Andrew V. Schally, Magdolna Kovacs, Katalin Toth, and Ana Maria Comaru-Schally
Tulane University School of Medicine and Veterans Affairs Medical Center, New Orleans, Louisiana

I. INTRODUCTION

Isolation and structural elucidation of growth hormone-releasing hormone (GHRH) (1–3), made possible the synthesis of various analogs of GHRH. Most of them were agonists intended for potential clinical and veterinary applications (4–7). For a review see Chapter 9 in this volume (8).

There is an even greater clinical need for antagonistic analogs of GHRH (9–11). The GHRH antagonists could be tried in conditions such as acromegaly, diabetic retinopathy, or diabetic nephropathy (glomerulosclerosis). However, the main applications of GHRH antagonists would be in the field of cancer (10; see Sec. VI on potential clinical uses).

The clinical need for GHRH antagonists was first advocated by Pollak et al. (9,12,13) on the grounds that somatostatin analogs do not adequately suppress GH and insulin-like growth factor (IGF)-I levels in patients with neoplasms potentially dependent on IGF-I (9,12,13). The GHRH antagonists could be given alone or in combination with superactive somatostatin analogs, such as octreotide (Sandostatin) or vapreotide (RC-160) (10,12–14). The use of combination of both analogs could achieve a more complete suppression of GH and IGF-I levels (10,14). The GHRH antagonists could be also used for inhibiting growth of tumors that do not express somatostatin receptors, such as osteosarcomas or pancreatic cancers (10,12).

In the course of synthesis of various agonists of hGHRH(1–29)NH$_2$, it was found that the replacement of Ala^2 by D-Arg^2 produced antagonists (15,16). An early GHRH antagonist [Ac-Tyr^1,D-Arg^2]hGHRH(1–29)NH$_2$ is able to inhibit the GHRH-stimulated adenylate cyclase activity in rat pituitary cells (16). This antagonist also suppresses the endogenous pulsatile GH secretion and GHRH-stimulated GH release after intravenous (iv) injection into rats (17,18). Sato et al. (19,20) also reported that various analogs of [D-Arg^2]GHRH(1–29)NH$_2$ inhibited GH release in tissue cultures. Subsequent studies (21,22) confirmed the essential role of D-Arg at position 2 for generating GHRH antagonistic activity.

The aim of this chapter is to summarize our work on the synthesis and biological evaluation of new series of potent GHRH antagonists. We will also cite our studies on the mechanism of action of GHRH antagonists and review oncological investigations performed in various cancer models with two of these antagonists that appeared to be the most effective in inhibiting GH release in vivo in rats. Potential clinical applications of GHRH antagonists for the treatment of IGF-I- or IGF-II-dependent cancers will be discussed in the light of the increasing evidence for the involvement of IGF-I and IGF-II in the progression and metastases of various malignancies.

II. SYNTHESIS AND BIOLOGICAL EVALUATION OF NEW GHRH ANTAGONISTS

A. Synthesis

In an endeavor to prepare GHRH antagonists with increased activity for oncological uses, two series of analogs were synthesized by solid-phase methods and purified by high-performance liquid chromatography (HPLC; 23,24). These analogs were based on the NH$_2$-terminal sequence of 28 or 29 amino acid residues of hGHRH, but contained D-Arg^2 and Nle^{27} modifications. Most analogs had Phe(pCl)^6 and Agm^{29} substituents. In the first series of 22 analogs, all the peptides, except one, were acylated at the NH$_2$-terminus with different hydrophobic acids (e.g. isobutyric acid [Ibu] or 1-naphthylacetic acid [Nac]) to study the effect of NH$_2$-terminal acylation on the antagonistic activity (23).

Acylation with Nac or Ibu of the analogs that contain D-Arg^2 modification, combined with Phe(4-Cl)^6, Abu^{15}, Nle^{27}, and Agm^{29} substitution, causes a great increase in inhibitory activity on GH release in vitro, as well as in binding affinity to the receptors (23). Among peptides acylated with Nac or Ibu, those containing Tyr^1 showed greater antagonistic potencies than the corresponding analogs with His^1 or Glu^1.

In the second series of 20 antagonists, the effects of other substitutions such as Abu^8 (or Abu^{15} and Ala^{15} and various hydrophobic and hydrophilic D- or L-amino acids at position 8) were also investigated (24). Thus, substitutions, such as Abu, Ser, Phe, and D-amino acids (D-Ala, D-Thr, D-Leu), were incorporated into antagonistic GHRH analogs at position 8. Once more, all the peptides were acylated at the NH_2-terminus. Because previously synthesized antagonist $[Nac^0,D\text{-}Arg^2,Phe(pCl)^6,Abu^{15},Nle^{27}]hGHRH(1-28)Agm$ (MZ-4-243), acylated with Nac, showed very powerful antagonistic action in vitro (23), but its inhibitory activity on GH release in vivo was low, different acyl groups were used in this series of new GHRH antagonists, to increase the biological activity. Thus, the new analogs were acylated with Ac-, Ac-Nal, formyl, Ibu, or phenylacetic acid (PhAc) at the NH_2-terminus.

B. In Vitro Tests

1. Determination of GHRH Antagonistic Activity In Vitro

Antagonistic activities of the peptides on GH release were determined by using a superfused rat pituitary system (25,26). The antagonistic peptides were perfused through the cells for 9 mins at various concentrations (1–300 nM). After this preincubation, the cells were exposed to a mixture of the GHRH antagonists and 1 nM of $hGHRH(1-29)NH_2$ for an additional 3 min (23,24). To check the duration of the antagonistic effect of the analog, 1 nM of $hGHRH(1-29)NH_2$ was applied 30, 60, 90, and 120 min later. Net integral values of the GH responses were evaluated with a computer program. The GH responses following administration of antagonists were compared with GH responses induced by 1 nM of $hGHRH(1-29)NH_2$. The inhibitory activities of the antagonists were expressed as percentage inhibition of the original GH response. The inhibitory potencies of the modified antagonists were then compared with that of $[Ac\text{-}Tyr^1,D\text{-}Arg^2]hGHRH(1-29)NH_2$ (standard antagonist). Among the antagonists synthesized in the first series, $[Ibu^0,D\text{-}Arg^2,Phe\ (4\text{-}Cl)^6,$ $Abu^{15},\ Nle^{27}]hGHRH(1-28)Agm$ (MZ-4-71), $[Nac^0,D\text{-}Arg^2,Phe\ (4\text{-}Cl)^6,\ Abu^{15},$ $Nle^{27}]hGHRH(1-28)Agm$ (MZ-4-243), $[Nac^0,D\text{-}Arg^2,Phe\ (4\text{-}Cl)^6,\ Abu^{15},$ $Nle^{27}]hGHRH(1-29)NH_2$ (MZ-4-169), $[Nac^0\text{-}His^1,D\text{-}Arg^2,Phe\ (4\text{-}Cl)^6,Abu^{15},$ $Nle^{27}]hGHRH(1-29)NH_2$ (MZ-4-181), inhibited GH release at 3×10^{-9} M and were also long acting in vitro. Antagonist MZ-4-243 inhibited GH release 100 times more powerfully than the standard antagonist and was the most potent in vitro among the GHRH antagonists synthesized in this series (23).

In the second series of antagonists, most analogs inhibited the GH release induced by GHRH more powerfully than the standard antagonist and some were long acting (24). Among the peptides synthesized, antagonist PhAc-$[D\text{-}Arg^2,$ $Phe(pCl)^6,\ Abu^{15},\ Nle^{27}]hGHRH(1-28)Agm$ (MZ-5-156) appeared to be the most potent and inhibited GH release in vitro 63–200 times more powerfully than the standard antagonist (24).

C. Binding Assay

The binding of GHRH antagonists to membrane receptors of rat anterior pituitary cells was determined by using ^{125}I-labeled [His^1,Nle^{27}]hGHRH(1–32)NH$_2$ as radioactive ligand (27). In competitive binding analysis ^{125}I-labeled [His^1,Nle^{27}]hGHRH(1–32)NH$_2$ (0.2 nM) was displaced by GHRH antagonists at 10^{-6} to 10^{-12} M. Relative binding affinities were compared with that of [Ac-Tyr^1,D-Arg^2]hGHRH(1–29)NH$_2$ (23,24). Analogs with high inhibitory effects in vitro also have high affinities for rat pituitary GHRH receptors. The binding affinities of the first group of analogs—MZ-4-71, MZ-4-243, MZ-4-169, and MZ-4-181—to membrane receptors on rat anterior pituitary cells were 26.2, 85.1, 82.5, and 67.1 times greater than that of the standard GHRH antagonist, respectively (23). These antagonists showed K_i values in the range of 0.04–0.12 nM, compared with standard antagonist that had a K_i of 3.2 nM (23). The binding affinities of the second series of analogs—MZ-5-78, MZ-5-116, MZ-5-156, MZ-5-192, and MZ-5-208—to membrane GHRH receptors on rat anterior pituitary cells were 51.8, 37.0, 48.8, 57.1, and 44.3 times greater, respectively, than the affinity of the standard GHRH antagonist (24). These antagonists showed the highest binding affinities and displayed K_i values in the range of 0.059–0.076 nM (24).

D. In Vivo Tests

To measure inhibition of GH release in vivo, adult male Sprague–Dawley rats anesthetized with pentobarbital were used (23,28). One group of animals received hGHRH(1–29)NH$_2$ as control (3 µg/kg). Other groups of rats were injected with [Ac-Tyr^1,D-Arg^2]hGHRH(1–29)NH$_2$ as standard antagonist (80–100 and 320–400 µg/kg body weight), or with modified antagonists (20 and 80 µg/kg body weight) 0.5–2 min before administration of hGHRH(1–29)NH$_2$. Blood samples were taken from the jugular vein before and 5 and 15 min after the injection of the antagonists for measurement of serum GH levels by radioimmunoassay (RIA; 23,28). In some experiments, the inhibitory effects of selected antagonists, such as MZ-5-156, were also measured after intramuscular, subcutaneous, or intraperitoneal administration (28).

In the first series of antagonists, [Ibu^0,D-Arg^2,Phe(4-Cl)^6,Abu^{15}, Nle^{27}]hGHRH(1–28)Agm (MZ-4-71), [Nac^0,D-Arg^2,Phe(4-Cl)^6,Abu^{15},Nle^{27} hGHRH(1–29)NH$_2$, (MZ-4-169), and [Nac-His^1,D-Arg^2,Phe(4-Cl)^6,Abu^{15}, Nle^{27}]hGHRH(1–29)NH$_2$ (MZ-4-181) induced a significantly greater inhibition of GH release than the standard antagonist. In these in vivo tests, antagonist MZ-4-71 inhibited GH release 18.9 times more powerfully than the standard antagonist. This compound showed the highest inhibitory potency in vivo among the first series of peptides. In spite of its powerful and protracted activity in vitro and high-binding affinity to rat pituitary receptors, antagonist MZ-4-243 acylated with 1-naphthylacetic acid had only a very low activity in vivo (24,28).

The high and protracted antagonistic activity of these analogs was thought to be due to the combination of a substitution of D-Arg at position 2 with a hydrophobic acyl group at the NH_2-terminus, and other substitutions in the molecule (23). Acylation of the NH_2-terminus of the GHRH antagonists with hydrophobic acyl groups resulted in analogs with high-binding affinities for the pituitary GHRH receptors (23). The structure–activity studies on GHRH antagonists showed that substitutions in GHRH(1–29)NH_2 at positions 1, 2, 6, 15, 27, 28, and 29 paired with a hydrophobic, acylated NH_2-terminus produces high antagonistic potency (23).

Biological activity of five antagonists from the second series, containing a formyl or phenylacetyl group at the NH_2-terminus, D-Arg[2], Phe(4-Cl)[6], and Nle[27] modifications, as well as various substitutions in positions 8, 15, or 28, and Agm or Arg in position 29, was also evaluated in vivo in rats (28). All five antagonists, administered intravenously at a 27-fold molar excess, suppressed the GH-releasing effect of exogenous GHRH(1–29)NH_2 by 64–75%. The inhibitory effects lasted for more than 15 min. The most potent analog from this series PhAc-[D-Arg[2], Phe(4-Cl)[6],Abu[15],Nle[27]]hGHRH(1–28)Agm (MZ-5-156), showed an in vivo potency 7–16 times higher than the standard antagonist [Ac-Tyr[1],D-Arg[2]]hGHRH(1–29)NH_2 (28). MZ-5-156 was also capable of inhibiting GHRH-induced GH release after intraperitoneal or intramuscular administration, but the doses required were larger than iv ones (28). It is intriguing that sc administration of MZ-5-156 at a dose 27 times larger on molar basis than that of the exogenous GHRH did not cause inhibition of the GH response. A much slower absorption of the antagonist, compared with GHRH, from the subcutaneous tissues might account for the lack of effect after sc administration in rats. However, the inhibitory effect of sc administration of MZ-5-156 and other GHRH antagonists on serum GH and IGF-I levels and tumor growth has been repeatedly demonstrated in nude mice (29,30). These results show that NH_2-terminal acylation, with phenylacetic acid, of the sequence [D-Arg[2],Phe(4-Cl)[6], Nle[27]]hGHRH(1–29)NH_2, containing modifications in positions 8, 15, 28, or 29 results in antagonists with high and protracted potency in vivo (24,28). It is possible that GHRH antagonists of this class (23,24) might find clinical application in the therapy of GH- and IGF-I- and IGF-II-dependent tumors and in the treatment of disorders characterized by excessive GH secretion.

III. INVESTIGATION OF THE EFFECTS OF BRIEF AND LONG-TERM ADMINISTRATION OF GHRH ANTAGONISTS IN RATS AND THEIR MECHANISMS OF ACTION

A. Effects of Long-Term Administration of GHRH Antagonists

We evaluated the effects of long-term intramuscular (im) treatment with GHRH antagonist [Ibu^0,D-Arg^2,Phe(4-Cl)^6,Abu^{15},Nle^{27}]hGHRH(1–28)Agm (MZ-4-71) on the growth rate, serum GH and IGF-I concentration, GH responsiveness to exogenous GHRH, as well as the pituitary GH content and GHRH receptor concentration, in young female rats (31). Twice daily im injections of 20 μg MZ-4-71 for 2 weeks resulted in 21% decrease in body weights and 36% reduction in body length of these rats. The GH responses to bolus injections of GHRH were reduced by 22%, and serum IGF-I concentrations by 15% at the end of the treatment. After prolonged treatment with antagonist MZ-4-71 (31), the total pituitary GH content was decreased by 15% and GHRH receptor concentration by 48%. These results demonstrate that GHRH antagonist MZ-4-71 can inhibit somatic growth and secretion of GH and IGF-I in rats. Our findings also suggest that GHRH may regulate synthesis of GH and GHRH receptors (31).

B. Effect of Brief Administration of Large Doses of GHRH Antagonists

We studied the effects of brief, high-dose intravenous (iv) application of antagonist MZ-4-71 on the basal GH and IGF-I levels in adult male rats (31). A bolus injection of high doses of MZ-4-71 (400 μg iv) induced a marked and prolonged (6 h) inhibition of serum GH concentration and a parallel inhibition of the serum IGF-I levels. The greatest decrease in serum GH and IGF-I levels was found at 3 h after the injection (31). The marked inhibition of the basal serum GH concentration induced by single high doses of the GHRH antagonist and the prolonged suppression of the episodic GH pulses, even with unchanged basal GH level were both associated with a decrease in the serum IGF-I levels. These results indicate that IGF-I secretion is highly dependent on the pulsatile pattern of GH release (31). This work demonstrates the efficacy of GHRH antagonist MZ-4-71 in blocking GHRH-regulated GH and IGF-I secretion (31).

C. Studies In Vitro

The ability of GHRH antagonist MZ-4-71 to prevent GH release induced by GHRH pulses was also determined in vitro in the superfused rat pituitary cell system (31). To study the effect of the antagonist MZ-4-71 on GH responses to repeatedly administered GHRH pulses, 3-min pulses of 1 nM GHRH were coadministered with the antagonist at 30, 60, and 90 min of this continuous perfusion (31). Continuous perfusion of the pituitary cells with GHRH antagonist MZ-4-71 resulted in a dose-dependent inhibition of the GHRH-induced GH pulses. At 10 nM concentration, MZ-4-71 caused a gradually increasing inhibition of the GH pulses induced by GH-RH at 30-min intervals. At 100 nM concentration, MZ-4-71 entirely nullified the effect of GHRH (31). These in vitro studies showed that MZ-4-71 can dose-dependently inhibit the GH-releasing effect of GHRH pulses.

D. Effect of GHRH Antagonists on GH and CyclicAMP Release in Rat Pituitary Cells

Antagonists MZ-4-71 ([Ibu^0,D-Arg^2,Phe(4-Cl)^6,Abu^{15},Nle^{27}]hGHRH(1–28)Agm) and MZ-4-243 ([Nac^0,D-Arg^2,Phe(4-Cl)^6,Abu^{15},Nle^{27}]hGHRH(1–28)Agm) were evaluated for their long-term effect on the release of GH and cAMP in the superfused rat pituitary cells (32). After a 9-min preincubation, antagonists MZ-4-71 and MZ-4-243 at 1–3 nM concentrations caused an inhibition of GH release, stimulated by 1 nM of GHRH (32). The inhibition caused by MZ-4-71 at 30 nM decreased gradually to 30%, 120 min after the treatment, but the 30-nM dose of MZ-4-243 reduced the GH response by more than 90%, even 270 min after its administration (32). During a 2-h incubation with 1 nM of GHRH in combination with a 30-min infusion of the antagonists MZ-4-71 or MZ-4-243 from the 30th to the 60th min, the decrease in GH discharge preceded the inhibition of cAMP release. After the infusion of the antagonists was stopped, GH release resumed sooner than that of cAMP (32). Simultaneous determinations of cAMP and GH in the samples showed that changes in GH levels were never preceded by a rise or decrease in cAMP release. Because it has been hypothesized that the effects of GHRH are mediated by cAMP, our work suggests that other signal transduction mechanisms may also mediate the effect of GHRH.

E. The Specificity of GHRH Antagonists for GHRH Receptors

It should be mentioned that GHRH is structurally related to vasoactive intestinal peptide (VIP) and secretin family (16); GHRH can recognize the VIP receptors in pancreatic membranes (16). However, using antagonist [N-Ac-Tyr^1-D-Arg^2]GHRH(1–29)NH$_2$ it was possible to demonstrate that GHRH and VIP receptors are distinct in the rat pituitary (16). Even at 10^{-6}-M concentrations, VIP does not inhibit the binding of monoiodinated [His^1, ^{125}I-Tyr^{10},

Nle^{27}]hGHRH(1–32)NH$_2$ to the GHRH receptors on rat pituitaries (27). In the reverse hemolytic plaque assay in rat pituitaries, specific antagonists to GHRH and GH-releasing peptide (His-D-Trp-Ala-Trp-D-Phe-Lys-NH$_2$; [GHRP]) inhibit GH secretion induced by the respective agonist, but not that induced by the other peptide (33). GHRH antagonist (N-Ac-Tyr^1,D-Arg^2)GHRH(1–29)NH$_2$ inhibited only the increases in GH secretion induced by rat GHRH (33). Conversely, the GHRP antagonist His-D-Trp-D-Lys-Trp-D-Phe-Lys-NH$_2$ selectively blocked GHRP-induced increases in GH, but had no significant inhibitory effect on rat GHRH-induced GH release (33). This shows that GHRP and GHRH stimulate GH release through distinct pituitary receptors (33) and that the respective antagonists are specific for these receptors (33).

IV. EFFECTS OF GHRH ANTAGONISTS ON VARIOUS EXPERIMENTAL CANCERS

The GHRH antagonists were investigated in various animal tumor models and in nude mice bearing transplanted human cancer lines on the basis that the blocking of the GH–IGF axis might inhibit the growth of IGF-I- or IGF-II-dependent cancers (10,29,30).

A. Inhibition of Growth of Human Osteosarcomas by Antagonists of GHRH

Osteogenic sarcomas represent the most common type of primary bone tumors in children and young adults (29); IGF-I can stimulate growth of osteosarcomas (12,13). The presence of receptors for IGF-I on osteogenic sarcomas has also been demonstrated (12,13). In view of these findings, we have evaluated the effect of GHRH antagonist MZ-4-71 on the growth of the human osteosarcoma cell lines SK-ES-1 and MNNG/HOS in vivo and in vitro (29).

Nude mice bearing SK-ES-1 and MNNG/HOS osteosarcomas were treated for 3–4 weeks with MZ-4-71 administered from Alzet osmotic minipumps at a dose of 40 µg/animal per day. Tumor volume and weight, and levels of receptors for IGF-I, were determined. The IGF-I levels in serum, tumors, and liver tissue were also measured. In other experiments, tumor-bearing nude mice were treated subcutaneously for 3 weeks with the hGHRH(1–29)NH$_2$ or with MZ-4-71 for 13 days at doses of 50 µg/animal per day. Effects of MZ-4-71 on cell proliferation were also evaluated in SK-ES-1 and MNNG/HOS cells in vitro (29).

Growth of SK-ES-1 and MNNG/HOS tumors in nude mice was significantly inhibited by MZ-4-71 administered from osmotic minipumps, as shown by a reduction in tumor volume and weight. Treatment with MZ-4-71 of animals bearing SK-ES-1 or MNNG/HOS osteosarcomas decreased IGF-I levels in tumor issue. IGF-I levels in serum of tumor-bearing nude mice treated subcutaneously for 13 days with MZ-4-71 were also decreased. High-affinity–binding sites for IGF-I were demonstrated on cell membranes of SK-ES-1 and MNNG/HOS osteosarcomas (29). The growth rate of the two cell lines in vitro was also suppressed by 10^{-6} to 10^{-5} M MZ-4-71. Interestingly, GHRH(1–29)NH$_2$ stimulated growth of MNNG/HOS xenografts.

   These findings demonstrate that the GHRH antagonists such as MZ-4-71 can inhibit the growth of SK-ES-1 and MNNG/HOS osteosarcomas in nude mice (29). Our results suggest that GHRH antagonists might find applications in the treatment of osteogenic sarcomas and other tumors that are dependent on IGF-I.

B. Inhibition of Growth of Human Small-Cell and Non– Small-Cell Lung Carcinomas by Antagonists of GHRH

Lung carcinoma is the leading cause of cancer-related deaths in the western world (10). Small-cell lung carcinoma (SCLC) accounts for 20–25% of all cases of lung cancer and non-SCLC includes the three remaining major histological subtypes of lung cancer: squamous carcinoma, adenocarcinoma, and large cell carcinoma (10). Surgery, radiation, and chemotherapy are of limited effectiveness in the treatment of lung carcinomas, and other therapeutic approaches must be explored (10). The presence of receptors for IGF-I and IGF-II on lung carcinomas has been demonstrated (34); also, human lung cancer cell lines express IGF-I and IGF-II genes (35).

   Because IGF-I and IGF-II appear to be involved in the proliferation of human lung carcinomas, we investigated the effects of two antagonists of GHRH, MZ-4-71 and MZ-5-156, on the growth of the H69 human small-cell lung cancer (SCLC) and H157 non-SCLC lines transplanted into nude mice or cultured in vitro (30). Nude mice bearing SCLC H69 and non-SCLC H157 tumors were treated for 3–5 weeks with MZ-4-71 or MZ-5-156 injected sc twice a day at a dose of 20 μg/animal. Growth of SCLC H69 and non-SCLC H157 tumors in nude mice was significantly inhibited by MZ-4-71 and MZ-5-156, as shown by a reduction in tumor volume and weight (30). Levels of IGF-I in serum and liver tissue of H157 tumor-bearing nude mice treated with MZ-4-71 were decreased (30). In animals bearing H157 non-SCLC, treatment with MZ-4-71 also reduced IGF-I and IGF-II levels in tumor tissue. In cell cultures, the proliferation rate of H69 SCLC cells or the H157 non-SCLC line was inhibited by 10^{-7} to 10^{-5} M antagonist MZ-4-71. These findings demonstrate that the GHRH antagonists, exemplified by MZ-4-71 and MZ-5-156, can inhibit the growth of both SCLC and non-SCLC (30). These results suggest that GHRH antagonists might find applications in the treatment of lung carcinomas and other tumors that are influenced by IGFs.

C. Inhibition of In Vivo Proliferation of Androgen-Independent Prostate Cancers by an Antagonist of GHRH

Carcinoma of the prostate is the most common malignant tumor in men (10). Medical castration based on agonists of luteinizing hormone-releasing hormone (LHRH) is an established therapy for advanced prostate cancer, with a response rate of 70–80% (10). However, all hormonal therapies aimed at androgen deprivation can provide only a remission, and most patients with advanced prostatic carcinoma eventually relapse (10). The treatment of androgen-independent prostate cancer is a major oncological challenge and new approaches must be developed (10).

Both IGF-I and IGF-II can regulate the function and growth of the prostate, especially after malignant transformation. Various cancer cells are able to produce and secrete IGF-I or IGF-II (36), including PC-3 and DU-145 prostate cancer cell lines (37,38). Type I IGF-receptors have been shown on membranes of DU-145 and PC-3 cells (37,38). There is also evidence that IGF might be responsible for the progression of prostate cancer in the advanced stages. Consequently, the inhibition of the GH–IGF axis by GHRH antagonists might provide a new approach for the treatment of advanced, androgen-independent prostate cancer.

Tumor inhibitory effects of GHRH antagonist MZ-4-71 were evaluated in nude mice bearing androgen-independent human prostate cancer cell lines DU-145 and PC-3 and in Copenhagen rats implanted with Dunning R-3327 AT-1 prostatic adenocarcinoma (38). After 6 weeks of therapy, the tumor volume and tumor weight in nude mice with DU-145 prostate cancers treated sc with 40 μg/day of MZ-4-71 were significantly decreased, when compared with controls (38). A similar inhibition of tumor growth was obtained in nude mice bearing PC-3 cancers, in which the sc treatment with MZ-4-71 for 4 weeks diminished the tumor volume and weight (38). Therapy with MZ-4-71 also significantly increased tumor doubling time of PC-3 and DU-145 prostate cancers. Serum levels of GH and IGF-I were significantly decreased in animals treated with GHRH antagonist. In PC-3 tumor tissue, the concentration of IGF-I and IGF-II was reduced to nondetectable values after therapy with MZ-4-71, and liver IGF-I levels were decreased. The growth of androgen-independent Dunning R-3327 AT-1 tumors in rats was also significantly inhibited after 3 weeks of intraperitoneal treatment with 100 μg MZ-4-71/day, as shown by a reduction in tumor volume and weight (38).

Specific high-affinity binding sites for IGF-I were found on the membranes of DU-145, PC-3, and Dunning R-3327 AT-1 tumors. These results indicate that GHRH antagonist MZ-4-71 inhibits growth of PC-3, DU-145, and Dunning AT-I androgen-independent prostate cancers, by decreasing GH release and the secretion of hepatic IGF-I, or through mechanisms involving a suppression of tumor IGF-I and IGF-II production (38). In view of these results, GHRH antagonists might be considered for the development of new approaches to the therapy of patients with prostatic carcinoma who have relapsed following conventional androgen deprivation treatment (38).

D. Inhibition of In Vivo Proliferation of Caki-I Renal Adenocarcinoma by GHRH Antagonist MZ-4-71

Renal cell carcinoma is the most common renal tumor, and in the United States, an estimated 12,000 deaths per year are due to this neoplasm (39). For the advanced stages of renal cell carcinoma, no effective therapy is available (39). There is evidence that GH and IGF-I might have an effect on the function and growth of this carcinoma (39). In view of findings that GH and IGF-I may play a role in the development of renal cell carcinoma, we investigated the effects of GHRH antagonist MZ-4-71 on the proliferation of the human renal adenocarcinoma cell line Caki-I in vitro and in vivo (39). Male nude mice bearing xenografts of human Caki-I renal cell carcinoma were treated for 4 weeks with MZ-4-71 injected sc twice daily at a dose of 20 µg/animal (39). After 4 weeks of therapy, the final volume of Caki-I tumors in nude mice treated with MZ-4-71 was significantly decreased, compared with controls. Treatment with GHRH antagonist also significantly reduced tumor weight, serum levels of GH and IGF-I, liver concentrations of IGF-I, and tumor levels of IGF-I and IGF-II (39).

Our findings demonstrate that GHRH antagonists such as MZ-4-71 can significantly inhibit the growth of Caki-I renal cell carcinoma (39). The GHRH antagonists may suppress growth of renal cell carcinomas by inhibiting GH release from the pituitary and diminishing the secretion of IGF-I from the liver, or by decreasing the production of IGF-I and II by the tumors (39). These findings support the merit of further investigations aimed at evaluating the use of GHRH antagonists for the therapy of advanced renal cell carcinoma.

E. Other Oncological Studies with GHRH Antagonists

The GHRH antagonists, such as MZ-4-71 and MZ-5-156, inhibit the growth of other IGF-I- or IGF-II-dependent tumors, including various human cancer cell lines transplanted into nude mice or cultured in vitro. The tumors that can be inhibited by GHRH antagonists include MXT mouse mammary tumors, human mammary cancers, human pancreatic cancer, and nitrosamine (BOP)-induced pancreatic cancers in golden hamsters, human colon cancer, and human glioblastomas (malignant astrocytomas) (10). The results accumulated so far strongly suggest that GHRH antagonists could provide new approaches to the treatment of IGF-I- and IGF-II-dependent tumors.

V. CLINICAL STUDIES WITH GHRH ANTAGONISTS

Only limited clinical studies were carried out with GHRH antagonists. In the initial study, a partial suppression of nocturnal GH secretion was obtained when subjects were treated with a single iv dose of GHRH antagonist [N-Ac-Tyr^1-D-Arg^2]GHRH(1–29) (400 µg/kg) at 2200 h (40). In a subsequent study, it was shown that nocturnal GH secretion can be eliminated by infusion of this GHRH antagonist (41). Eight men were given a 400 µg/kg iv bolus of the GHRH antagonist [N-Ac-Tyr^1, D-Arg^2]GHRH(1–29) at 2300 h, followed by a 50 µg/kg h^{-1} iv infusion of GHRH antagonist between 2300–0700 h. An iv bolus of GHRH (1 µg/kg) was given at 0500 h. GHRH antagonist suppressed GH secretion by an average of 89% and inhibited the response to GHRH by 79% (41).

Clinical efficacy of GHRH antagonists in conditions such as acromegaly remains to be demonstrated (11). Similarly, no oncological studies have been performed so far in cancer patients.

VI. POTENTIAL THERAPEUTIC USES OF GHRH ANTAGONISTS

A. Acromegaly and Other Conditions

The GHRH antagonists could be used in treatment of conditions caused by excess GH, such as acromegaly (10,11). Other examples would be diabetic retinopathy, which is the main cause of blindness in patients with diabetes, in which vascular damage to the eye is thought to be due to GH, and diabetic nephropathy (glomerulosclerosis). GH and IGF-I may be involved in the pathophysiology of diabetic nephropathy; therefore, this condition may be amenable to therapy with GHRH antagonists, based on inhibition of mesangial cell growth.

B. Therapeutic Uses of GHRH Antagonists in Oncology

The main applications of GHRH antagonists would be in the field of cancer (10). This is based on the following considerations: GHRH antagonists are designed to block the binding and, thereby, the action of GHRH (10,23,24). GHRH stimulates the secretion of GH. In turn, GH increases the production of IGF-I in the liver and other tissues and serum IGF-I levels (42). Autocrine or paracrine production of IGF-I by various tumors could also be under control of GH (42). This is suggested by inhibition of tumor IGF-I levels by GHRH antagonists (29,30,38,39). The GHRH antagonists also appear to inhibit the production of IGF-II in diverse tumors (30,38,39). Both IGF-I and IGF-II have been implicated in malignant transformation of cells, tumor progression, and metastases of various cancers (9,10,12,13,29,30,34–39,42–44). The involvement of IGF-I or IGF-II, or both, in breast cancer, SCLC, non-SCLC, prostate cancer, colon cancer, pancreatic cancer, osteosarcomas, brain tumors, renal cancers, and other malignancies is established or suspected (9,10,12, 13,29,30,34–39,42–52). Inhibition of growth of experimental tumors by GHRH antagonists can be linked to a suppression of IGF-I and IGF-II levels or their secretion (10,29,30,38,39).

The receptors for IGF-I are present in primary human breast cancers (45–48,51), lung cancers (SCLC and non-SCLC; 30,34,35), human brain tumors (49), human osteogenic sarcomas (12,13,29), human colon cancers (45) and pancreatic cancers (52), and human prostate cancers (36–38). The mRNAs for IGF-I and IGF-II have been found in human prostate cancers, various sarcomas, benign leiomyomas, neuroblastomas, hepatomas, breast cancer, human colorectal cancers, and human lung cancers (SCLC and non-SCLC lines) (35,37,38,42–44,53). IGF-II modulates tumor growth through the receptor for IGF-I (type I IGF-receptor), although IGF-II mannose-6-phosphate (M-6-P) receptor also exists (43).

The presence of IGF-I and IGF-II and IGF-I receptors in these tumors appears to be related to proliferation, malignant transformation, and progression of these cancers (10,43,44). IGF-I may act as an endocrine growth factor, and both IGF-I and IGF-II as paracrine or autocrine growth factors for various human cancers (10,42–44). By suppressing GH secretion, GHRH antagonists lower the production of hepatic and tumor IGF-I, which should lead to tumor growth inhibition. In addition, GHRH antagonists appear to lower paracrine or autocrine production of IGF-II by the tumors, which should also produce an inhibition of cancer proliferation (30,38,39). The relative importance of both mechanisms is being investigated in various tumors.

The GHRH antagonists could be given alone or in combination with somatostatin analogs (10,12,13,29,30). A combination could achieve a more complete suppression of IGF-I levels. The advantage of GHRH antagonists over somatostatin analogs would be that a GHRH antagonist could be used for suppression of tumors that do not express somatostatin receptors; for example, human osteogenic sarcomas (12,13). The mRNA for SSTR-2 (sub-type 2 of somatostatin receptors) also appears to be absent in human pancreatic and colorectal cancers (54). Tumor growth inhibition and the reduction in IGF-I and IGF-II levels in tumor tissue after therapy with GHRH antagonists was obtained in nude mice bearing transplanted human lung cancers, prostate cancers, osteosarcomas, renal cancers, and other tumors (29,30,38,39). Further work with these GHRH antagonists could lead to the development of a new class of antitumor agents.

ACKNOWLEDGMENTS

Some experimental work described in this paper was supported by the Medical Research Service of the Veterans Affairs Department.

REFERENCES

1. Rivier J, Spiess J, Thorner M, Vale W. Characterization of a growth hormone-releasing factor from a human pancreatic islet tumour. Nature 1982; 300:276–278.
2. Guillemin R, Brazeau P, Bohlen P, Esch F, Ling N, Wehrenberg WB. Growth hormone-releasing factor from a human pancreatic tumor that caused acromegaly. Science 1982; 218:585–587.
3. Ling N, Esch F, Bohlen R, Brazeau P, Wehrenberg WB, Guillemin R. Isolation, primary structure, and synthesis of human hypothalamic somatocrinin: growth hormone-releasing factor. Proc Natl Acad Sci USA 1984; 81:4302–4306.
4. Coy DH, Murphy WA, Lance VA, Heiman ML. Strategies in the design of synthetic agonists and antagonists of growth hormone releasing factor. Peptides 1986; 7:49–52.
5. Felix AM, Heimer EP, Wang C-T, Lambros TJ, Fournier A, Mowles TF, Maines S, Campbell RM, Wegrzynski BB, Toome V, Fry D, Madison VS. Synthesis, biological activity and conformation analysis of cyclic GRF analogs. Int J Pept Protein Res 1988; 32:441–454.
6. Zarandi M, Serfozo P, Zsigo J, Bokser L, Janaky T, Olsen DB, Bajusz S, Schally AV. Potent agonists of growth hormone-releasing hormone. Int J Pept Protein Res 1992; 39:211–217.
7. Kubiak TM, Kelly CR, Krabill LF. In vitro metabolic degradation of a bovine growth hormone-releasing factor analog Leu27-bGRF(1–29)NH$_2$ in bovine and porcine plasma. Drug Metab Dispos 1989; 17:393–397.
8. Schally AV, Comaru-Schally AM. Agonistic analogs of growth hormone-releasing hormone (GHRH); endocrine and growth studies. Chapter 9, this volume.

9. Pollak MN, Polychronakos C, Guyda H. Somatostatin analogue SMS 201-995 reduces serum IGF-I levels in patients with neoplasm potentially dependent on IGF-I. Anticancer Res 1989; 9:889–891.

10. Schally AV, Comaru-Schally AM. Hypothalamic and other peptide hormones. In: Holland JF, Frei E III, Bast RC Jr, Kufe DE, Morton DL, Weichselbaum R, eds. Cancer Medicine, 4th ed. Baltimore, MD: Williams & Wilkins, 1997:1067–1085.

11. Gelato MC. Growth hormone-releasing hormone: clinical perspectives. Endocrinologist 1994; 4:64–68.

12. Pollak MN, Polychronakos C, Richard M. Insulin-like growth factor I: a potent mitogen for human osteogenic sarcoma. J Natl Cancer Inst 1990; 82:301–305.

13. Pollak M, Sem AW, Richard M, Tetenes E, Bell R. Inhibition of metastatic behavior of murine osteosarcoma by hypophysectomy. J Natl Cancer Inst 1992; 84:966–9711.

14. Schally AV. Oncological applications of somatostatin analogs. Cancer Res 1988; 48:6977–6985.

15. Coy D, Murphy WA, Sueiras-Diaz J, Coy EJ, Lance VA. Structure–activity studies on the N-terminal region of growth hormone releasing factor. J Med Chem. 1985; 28:181–185.

16. Robberecht P, Coy D, Waelbroeck M, Heiman M, deNeef P, Camus J-C, Christophe J. Structural requirements for the activation of rat anterior pituitary adenylate cyclase by growth hormone-releasing factor (GRF): discovery of (N-Ac-Tyr^1, D-Arg^2)-GRF(1–29)-NH$_2$ as a GRF antagonist on membranes. Endocrinology 1985; 117:1759–1764.

17. Lumpkin MD, Mulroney SE, Haramati A. Inhibition of pulsatile growth hormone (GH) secretion and somatic growth in immature rats with a synthetic GH-releasing factor antagonist. Endocrinology 1989; 124:1154–1159.

18. Lumpkin MD, McDonald JK. Blockade of growth hormone-releasing factor (GRF) activity in the pituitary and hypothalamus of the conscious rat with a peptidic GRF antagonist. Endocrinology 1989; 124:1522–1531.

19. Sato K, Hotta M, Kageyama J, Chiang T-C, Hu H-Y, Dong M-H, Ling N. Synthesis and in vitro bioactivity of human growth hormone-releasing factor analogs substituted with a single D-amino acid. Biochem Biophys Res Commun 1987; 149:531–537.

20. Sato K, Hotta M, Kageyama J, Hu H-Y, Dong M-H, Ling N. Synthetic analogs of growth hormone-releasing factor with antagonistic activity in vitro. Biochem Biophys Res Commun 1990; 167:360–366.

21. Ling N, Sato K, Hotta M, Chiang T-C, Hu H-Y, Dong M-H. Growth hormone-releasing factor analogs with potent antagonist activity. In: Marshall GR, ed. Peptides: Chemistry and Biology. Leiden: ESCOM 1988:484–486.

22. Coy DH, Hocart SJ, Murphy WA. Human growth hormone-releasing hormone analogues with much improved in vitro growth hormone-releasing potencies in rat pituitary cells. Eur J Pharmacol 1991; 204:179–185.

23. Zarandi M, Horvath JE, Halmos G, Pinski J, Nagy A, Groot K, Rekasi Z, Schally

AV. Synthesis and biological activities of highly potent antagonists of growth hormone-releasing hormone. Proc Natl Acad Sci USA 1994; 91:12298–12301.

24. Zarandi M, Kovacs M, Horvath J, Toth K, Halmos G, Groot K, Nagy A, Kele Z, Schally AV. Synthesis and in vitro evaluation of new potent antagonists of growth hormone-releasing hormone (GH-RH). Peptides 1997; 18:423–430.

25. Rekasi Z, Schally AV. A method for evaluation of activity of antagonistic analogs of growth hormone-releasing hormone in a superfusion system. Proc Natl Acad Sci USA 1993; 90:2146–2149.

26. Vigh S, Schally AV. Interaction between hypothalamic peptides in a superfused pituitary cell system. Peptides 1984; 5:241–247.

27. Halmos G, Rekasi Z, Szoke B, Schally AV. Use of radioreceptor assay and cell superfusion system for in vitro screening of analogs of growth hormone-releasing hormone. Receptor 1993; 3:87–97.

28. Kovacs M, Schally AV, Zarandi M, Groot K. Inhibition of GH release in rats by new potent antagonists of growth hormone-releasing hormone (GH-RH). Peptides 1997; 18:431–438.

29. Pinski J, Schally AV, Groot K, Halmos G, Szepeshazi K, Zarandi M, Armatis P. Inhibition of growth of human osteosarcomas by antagonists of growth hormone-releasing hormone. J Natl Cancer Inst 1995; 87:1787–1794.

30. Pinski J, Schally AV, Jungwirth A, Groot K, Halmos G, Armatis P, Zarandi M, Vadillo-Buenfil M. Inhibition of growth of human small-cell and non–small-cell lung carcinomas by antagonists of growth hormone-releasing hormone (GH-RH). Int J Oncol 1996; 9:1099–1105.

31. Kovacs M, Zarandi M, Halmos G, Groot K, Schally AV. Effects of acute and chronic administration of a new potent antagonist of growth hormone-releasing hormone in rats: mechanisms of action. Endocrinology 1996; 137:5364–5369.

32. Horvath JE, Zarandi M, Groot K, Schally AV. Effect of long-acting antagonists of growth hormone (GH)-releasing hormone on GH and cyclic adenosine 3′,5′-monophosphate release in superfused rat pituitary cells. Endocrinology 1995; 136:3849–3855.

33. Goth MI, Lyons CE, Canny BJ, Thorner MO. Pituitary adenylate cyclase activating polypeptide, growth hormone (GH)-releasing peptide and GH-releasing hormone stimulate GH release through distinct pituitary receptors. Endocrinology 1992; 130:939–944.

34. Macauley VM, Everard MJ, Teale, JD, Trott PA, Van Wyk MM, Smith IE, Millar JL. Autocrine function for insulin-like growth factor I in human small cell lung cancer cell lines and fresh tumor cells. Cancer Res 1990; 50:2511–2517.

35. Reeve JG, Brinkman A, Hughes S, Mitchell J, Schwander J, Bleehen NM. Expression of insulin-like growth factor (IGF) and IGF-binding protein genes in human lung tumor cell lines. J Natl Cancer Inst 1992; 84:628–634.

36. Macaulay VM. Insulin-like growth factors and cancer. Br J Cancer 1992; 65:311–320.

37. Pietrzkowski Z, Mulholland G, Gomella L, Jameson BA, Wernicke D, Baserga R. Inhibition of growth of prostatic cancer cell lines by peptide analogues of insulin-like growth factor 1. Cancer Res 1993; 53:1102–1106.

38. Jungwirth A, Schally AV, Pinski J, Halmos G, Groot K, Armatis P, Vadillo-Buenfil M. Inhibition of in vivo proliferation of androgen-independent prostate cancers by an antagonist of growth hormone-releasing hormone (GH-RH). Br J Cancer 1997; 75:1585–1592.

39. Jungwirth A, Schally AV, Pinski J, Groot K, Halmos G. Growth hormone-releasing hormone (GH-RH) antagonist MZ-4-71 inhibits in vivo proliferation of Caki-I renal adenocarcinoma. Proc Natl Acad Sci USA 1997; 94:5810–5813.

40. Jaffe CA, Friberg RD, Barkan AL. Suppression of growth hormone secretion by a selective GH-releasing hormone antagonist: direct evidence for involvement of endogenous GHRH in the generation of GH pulses. J Clin Invest 1993; 92:695–701.

41. Ocampo-Lim B, Guo W, DeMott-Friberg R, Barkan AL, Jaffe CA. Nocturnal growth hormone (GH) secretion is eliminated by infusion of GH-releasing hormone antagonist. J Clin Endocrinol Metab 1996; 81:4396–4399.

42. Daughaday WH. [Editorial]. The possible autocrine/paracrine and endocrine roles of insulin-like growth factors of human tumors. Endocrinology 1990; 127:1–4.

43. Toretsky JA, Helman LJ. Involvement of IGF-II in human cancer. J Endocrinol 1996; 149:367–372.

44. Westley BR, May FEB. Insulin-like growth factors: the unrecognised oncogenes. Br J Cancer 1995; 72:1065–1066.

45. Pollak MN, Perdue JF, Margolese RG, Baer K, Richard M. Presence of somatomedin receptors in primary human breast and colon carcinomas. Cancer Lett 1987; 38:223–230.

46. Pekonen F, Partanen S, Makinen T, Rutanen E-M. Receptors for epidermal growth factor and insulin-like growth factor I and their relation to steroid receptors in human breast cancer. Cancer Res 1988; 48:1343–1347.

47. Foekens JA, Portengen H, Janssen M, Klijn JGM. Insulin-like growth factor-1 receptors and insulin-like growth factor-1-like activity in human primary breast cancer. Cancer 1989; 63:2139–2147.

48. Foekens JA, Portengen H, van Putten WLJ, Trapman AMAC, Reubi JC, Alexieva-Figusch J, Klijn JGM. Prognostic value of receptors insulin-like growth factor 1, somatostatin, and epidermal growth factor in human breast cancer. Cancer Res 1989; 49:7002–7009.

49. Gammeltoft A, Ballotti R, Kowalski A, Westermark B, van Obberghen E. Expression of two types of receptors for insulin-like growth factors in human malignant glioma. Cancer Res 1988; 48:1233–1237.

50. Goustin AS, Loef EB, Shipley GS, Moses HL. Growth factors and cancer. Cancer Res 1986; 46:1015–1029.

51. Lippman ME, Dickson RB, Gelmann EP, Rosen N, Knabbe C, Bates S, Bronzert D, Huff K, Kasid A. Growth regulatory peptide production by human breast carcinoma cells. J Steroid Biochem 1988; 29:79–88.

52. Ohmura E, Okada M, Onoda N, Kamiya Y, Murakami H, Tsushima T, Shizume K. Insulin-like growth factor I and transforming growth factor α as autocrine growth factors in human pancreatic cancer cell growth. Cancer Res 1990; 50:103–107.

53. Angelloz-Nicoud P, Binoux M. Autocrine regulation of cell proliferation by the

insulin-like growth factor (IGF) and IGF binding protein-3 protease system in a human prostate carcinoma cell line (PC-3). Endocrinology 1995 136:5485–5492.

54. Buscail L, Saint-Laurent N, Chastre E, Vaillant J-C, Gespach C, Capella G, Kalthoff J, Lluis F, Vaysse N, Susini C. Loss of *sst2* somatostatin receptor gene expression in human pancreatic and colorectal cancer. Cancer Res 1996; 56:1823–1827.

11

Clinical Evidence for Extrahypophyseal Action Site of Growth Hormone Secretagogues

Eva Maria Carro, Manuel Pombo Arias, Carlos Dieguez, Roberto Peinó, and Felipe F. Casanueva
University of Santiago de Compostela, Santiago de Compostela, Spain

Dragan Micić and Vera Popović
Institute of Endocrinology, Diabetes, and Diseases of Metabolism, Belgrade, Yugoslavia

Alfonso Leal-Cerro
Hospital Virgen del Rocio, Seville, Spain

I. INTRODUCTION

The excruciating process of developing the nonclassic growth hormone (GH) secretagogues (GHS) in the early 1980s was guided by the in vitro GH-releasing capabilities of the synthetic compounds (1). with this selective process, peptides that were able to release GH in vivo, but were scarcely effective in vitro were discarded, a process that favored the selection of compounds for which the action was mainly at the pituitary level. Interestingly, cumulative evidence suggests that GHSs also act at a central, probably hypothalamic site, and that this alternative point of activation is as relevant as the pituitary one.

II. HYPOTHALAMIC ACTIONS OF GHSs

Receptors for GHSs were initially detected both in the pituitary and in the hypothalamus (2). However, the concept of a relevant participation of the hypothalamic structures on the GHS-mediated GH secretion came from the observation that the GHSs were more effective in vivo than in vitro (3,4). In vivo GHS and GH-releasing hormone (GHRH) also show striking synergistic activity on GH release (i.e., the amount of GH released by the combined administration of GHRH and GH-releasing peptide (GHRP)-6 is significantly greater than the arithmetic sum of GHRH-mediated GH release plus GHRP-6-mediated GH release (4,5). On the contrary, in vitro GHS and GHRH merely show additive activity (6–8), although some controversy surrounds these observations (9,10). A more direct demonstration of their central action is that GHS administration enhances c-*fos* expression and electrical activity in hypothalamic neurons (11).

The GHSs behave similar to a hypothalamic neurohormone. In living rats with surgical hypothalamic destruction (deprived of hypothalamic hormones), but with intact pituitaries, the GHRP-6-mediated GH release shows two phases. In the first week after hypothalamic ablation, the GH released by GHS is greater than in sham-operated rats, whereas 15 days later it is lower (Fig. 1; 12). This phenomenon of early supersensitivity and late subsensitivity is typical of a re-

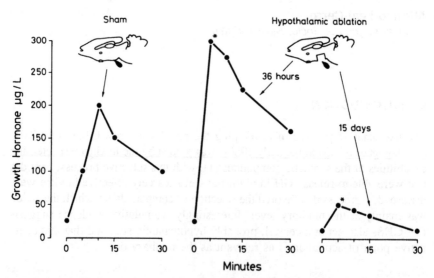

Figure 1   GH secretion in both control rats and rats with surgical ablation (destruction) of the hypothalamus, after stimulation with GHRP-6 (10 µg/kg iv). Rats with hypothalamic ablation had functioning pituitaries, although deprived of any hypothalamic signal and were studied immediately after (36 h) and 15 days after surgery. (From Ref. 12.)

ceptor activated by a neurohormone, and it was also observed for GHRH. Similarly, in hypophysectomized rats that have had a pituitary transplanted under the kidney capsule, the daily administration of GHRP-6 exerts tropic influences over the transplanted somatotrophs, preserving their function (12).

III. CLINICAL EVIDENCE FOR A HYPOTHALAMIC SITE OF ACTION OF GHS

To understand the precise point of action of GHS in releasing GH in humans we studied several adult patients with hypothalamic–pituitary disconnection. These patients harbored hypothalamic lesions, most of them of tumoral origin, which disrupted the communication between the hypothalamus and the pituitary, whereas the pituitary tissue remained intact. In fact, they were selected following the criteria of clinical and magnetic resonance-imaging (MRI) studies, absence of pituitary response to stimuli operating at the hypothalamic level, and preserved, although delayed, pituitary response to exogenously administered hypothalamic factor. In such a model, any factor acting at the hypothalamic level would be inoperative. The GHRH-induced GH secretion in these patients with pituitary–hypothalamic disconnection was roughly similar to the control group, although it was clearly delayed. This responsiveness to exogenous GHRH was not at all surprising, considering that it was part of the basis for the patients' inclusion in this study. Interestingly enough, when the stimulus administered was the GHS GHRP-6, which in normal controls is more efficacious than GHRH, the patients with hypothalamic–pituitary disconnection showed no response at all. This was the first indication that GHS releases GH in humans by acting at a hypothalamic level and that the pituitary action was ancillary (13). An obvious criticism of these conclusions is that for years the lack of stimulation of the pituitary receptor for the endogenous GHS (14) had led to its desensitization. This is unlikely considering that no desensitization was observed for GHRH, and it would be strange that the two hypothalamic compounds have such different properties. Also in these patients with hypothalamic–pituitary disconnection, the GHRH–GHRP-6-mediated GH release was no different from the action of GHRH alone, with an absence of synergism (13). Similar findings have been reported in children with neonatal pituitary stalk transection from perinatal damage, with the ensuing complete GH deficiency and panhypopituitarism of variable degrees (15). In these patients, and in comparison with controls, the GHRH-mediated GH release was minimal, suggesting a reduction in the somatotroph cell dotation. However, again the GHS action was lower than GHRH, pointing to the absence of hypothalamic factors as the cause (Fig. 2).

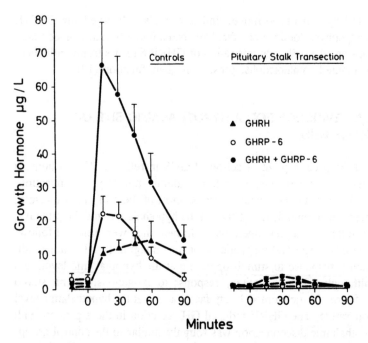

Figure 2 GH secretion elicited by the administration of GHRH, GHRP-6, and GHRH plus GHRP-6 in a group of patients who suffered neonatal pituitary stalk transection and permanent hypopituitarism and matched controls. (From Ref. 15.)

When these children were stimulated with GHRH plus GHRP-6 no synergistic action was evident (15). The most plausible explanation of both studies is that a GHS needs an operative hypothalamus for release of GH, and even the synergistic action of GHRH plus GHS implies the activation of structures located at the hypothalamic level.

To further clarify the GHS site of action, several patients with prolactinomas were studied, either before or after tumor reduction with bromocriptine (16). In patients with microprolactinoma (i.e., with no impairment in the arrival of hypothalamic signals to the remaining normal pituitary tissues), GH secretion was practically normal (Fig. 3). In fact, the GH response to either GHRH, GHRP-6, or GHRH plus GHRP-6 was no different from that observed in normal control subjects. After the disappearance of the microadenoma and normalization of prolactin (PRL) values with bromocriptine, again the GH responses to the three stimuli were preserved, although a nonsignificant increase in GH release was observed after the combined stimulus. These results suggest that independently of the PRL values, the microadenoma did not interfere with the hypothalamic–pituitary connection, allowing GHS to exert its

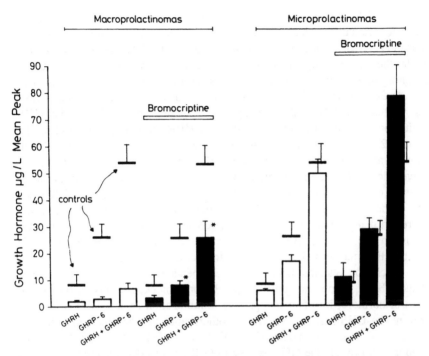

Figure 3 GH secretion in the patients with either micro- or macroprolactinoma before and after tumor reduction and normalization of prolactin levels. These patients were stimulated on different days with GHRH, GHRP-6, or GHRH plus GHRP-6. The repetitive horizontal boxes indicate the GH response after the same stimuli in normal control subjects. *$p < 0.05$ versus pretreatment data with the same stimulus. (From Ref. 16.)

GH-stimulating properties. Conversely, when similar studies were undertaken in patients with macroprolactinoma (i.e., with hypothalamic–pituitary disconnection) a reduction was observed for each of the three stimuli used: GHRH, GHRP-6, and GHRH plus GHRP-6 (see Fig. 3). When both the tumoral mass and the PRL levels were normalized by using bromocriptine, the GHRH-mediated GH response was not enhanced relative to the pretreatment study, suggesting permanent damage to part of the somatroph cells in some patients. Although reduced from the response observed in controls, both the GHRP-6- and the GHRH plus GHRP-6-mediated GH secretion was enhanced relative to the pretreatment situation (16). These findings suggest that when the permeability of the portal vessels and the hypothalamic–pituitary connection are reestablished by the pharmacological reduction in the tumor volume, the GHS again becomes operational.

As GHSs are more effective owing to their hypothalamic action, there are four, nonmutually exclusive explanations for their mechanism of action: (1) exogenous GHSs induce the release of endogenous GHRH, (2) exogenous GHSs inhibit the release of hypothalamic somatostatin, (3) exogenous GHSs, through an autocrine stimulatory mechanism, enhance the release of endogenous GHS, and (4) exogenous GHSs release another unknown hypothalamic factor (Bowers' U-factor) with GH-releasing capability. There is some evidence that GHSs elicit a parallel discharge of hypothalamic GHRH through a nonopioid mechanism (1,2). In sheep, GHSs elevate GHRH levels in hypophyseal portal blood, without altering somatostatin levels (17). In rats, systemic or intracerebroventricular (icv) administration of GHRP-6 or L-692,585 enhances the expression of c-*fos*, as well as the electrical activity on the putative GHRH neurons in the rat arcuate nucleus (11). Although considerable speculation followed the report that pretreatment with anti-GHRH serum blocked GHS-mediated GH release in rats, passive immunization against GHRH blocks GH secretion elicited by all stimuli tested in rats, including somatostatin withdrawal. Thus, the only conclusion available from these experiments is that after acute impairment of the GHRH action, rat somatotrophs become blocked to any further stimulus. Furthermore, in infant rats passive immunization against GHRH does not prevent the GHS action (18). On the other hand, the data indicating that the action of GHS is independent of hypothalamic GHRH release can be summarized as follows: (1) the GH response to GHS is considerably greater than that of GHRH (19); (2) GHSs are able to potentiate GH release in response to a maximal-stimulating dose of exogenous GHRH (4,5); (3) in rat hypothalamic explants, GHs are unable to release GHRH (20); (4) hours of infusion of GHRH block the GHRH-mediated GH release without altering the action of GHRP-6 (21,22). According to the foregoing reports, it appears that no participation of endogenous GHRH seems necessary to explain the GHS's mechanism of action.

Although the GHS-mediated GH release is partially insensitive to the inhibitory action of somatostatin (23), there appears to be little doubt that the secretagogues do not work by inhibiting hypothalamic somatostatin's release (20). In a physiological model, the putative endogenous ligand of the GHS receptor would be envisioned as a secreted factor that would act first at the hypothalamus, before acting on the pituitary. The presence of GHS receptors in hypothalamic structures and the evidence that GHS-elicited GH secretion is not mediated by changes in endogenous GHRH or somatostatin, but needs an operative hypothalamus, suggest that exogenous GH secretagogues could induce the release of another hypothalamic factor that has GH-releasing capabilities (U-factor; 1). Although convoluted, this working hypothesis is supported by inferential evidence and needs further testing.

III. CONCLUSIONS

Given the present evidence, the widely accepted dual control of GH release, which implicates GHRH and somatostatin, must be changed to a trinity that incorporates the GH secretagogues. The CNS actions of GHS, either hypothalamic or extrahypothalamic, and the probable existence of different GHS-receptor subtypes, will provide a further insight into the physiological and pathological regulation of GH secretion in humans within the next few years.

ACKNOWLEDGMENTS

This work was supported by grants from Fundacion Ramon Areces, Fundacion Salud 2000 and Asociacion Española Contra el Cancer. The expert technical assistance of Ms Mary Lage and Ms Veronica Piñeiro is gratefully acknowledged.

REFERENCES

1. Bowers CY. Xenobiotic growth hormone secretagogues: growth hormone releasing peptides. In: Bercu BB, Walker RF, eds. Growth Hormone Secretagogues. New York: Springer-Velag, 1996:9–28.
2. Codd EE, Shu AYL, Walker RF. Binding of a growth hormone releasing hexapeptide to specific hypothalamic and pituitary binding sites. Neuropharmacology 1989; 28:1139–1144.
3. Bowers CY, Sartor AO, Reynolds GA, Badger TM. On the actions of the growth hormone-releasing hexapeptide, GHRP-6. Endocrinology 1991; 128:2027–2035.
4. Bowers CY, Reynolds GA, Durham D, Barrera CM, Pezzoli SS, Thorner MO. Growth hormone (GH)-releasing peptide stimulates GH release in normal men and acts synergistically with GH-releasing hormone. J Clin Endocrinol Metab 1990; 70:975–982.
5. Peñalva A, Carballo A, Pombo M, Casanueva FF, Dieguez C. Effect of growth hormone (GH)-releasing hormone (GHRH), atropine, pyridostigmine or hypoglycemia on GHRP-6-induced GH secretion in man. J Clin Endocrinol Metab 1993; 76:168–171.
6. Sartor O, Bowers CY, Chang D. Parallel studies of His-D Trp-Ala-Trp-D Phe-Lys-NH$_2$ and human pancreatic growth hormone-releasing factor-44-NH$_2$ in rat primary pituitary cell monolayer culture. Endocrinology 1985; 116:952–957.
7. Blake AD, Smith RG. Desensitization studies using perifused rat pituitary cells show that growth hormone-releasing hormone and His-D Trp-Ala-Trp-D Phe-Lys-NH$_2$ stimulate growth hormone release through distinct receptor sites. J Endocrinol 1991; 129:11–19.
8. Wu D, Chen C, Katoh K, Zhang J, Clarke IJ. The effect of GH-releasing pep-

tide-2 (GHRP-2 or KP 102) on GH secretion from primary cultured ovine pituitary cells can be abolished by a specific GH-releasing factor (GHF) receptor
antagonist. J Endocrinol 1994; 140:R9–R13.

9.   Chen C, Wu D, Clarke IJ. Signal transduction systems employed by synthetic GH-
     releasing peptides in somatotrophs. J Endocrinol 1996; 148:381–386.
10. Cheng K, Chan WWS, Butler B, Wei L, Schoen R, Wyvratt MJ, Fisher MH, Smith
     RG. Stimulation of growth hormone release from rat primary pituitary cells by L-
     692,429, a novel non-peptidyl GH secretagogue. Endocrinology 1993; 132:2729–
     2731.
11. Dickson SL, Leng G, Dyball REJ, Smith RG. Central actions of peptide and non-
     peptide growth hormone secretagogues in the rat. Neuroendocrinology 1995; 61:36–
     43.
12. Mallo F, Alvarez CV, Benitez L, Burguera B, Coya R, Casanueva FF, Dieguez
     C. Regulation of His-D Trp-Ala-Trp-D Phe-Lys-NH$_2$ (GHRP-6)-induced GH secre-
     tion in the rat. Neuroendocrinology 1993; 57:247–256.
13. Popovic V, Damjanovic S, Micic D, Djurovic M, Dieguez C, Casanueva FF.
     Blocked growth hormone-releasing peptide (GHRP-6)-induced GH secretion and
     absence of the synergic action of GHRP-6 plus GH-releasing hormone in patients
     with hypothalamopituitary disconnection: evidence that GHRP-6 main action is
     exerted at the hypothalamic level. J Clin Endocrinol Metab 1995; 80:942–947.
14. Howard AD, Feighner SD, Cully DF, Arena JP, Liberator PA, Rosenblum CI,
     Hamelin M, Hreniuk DL, Palyha OC, Anderson J, Paress PS, Diaz C, Chou M,
     Liu KK, Mckee KK, Pong S-S, Chaung L-Y, Elbrecht A, Dashkevicz M, Heav-
     ens R, Rigby M, Sirinathsinghji DJS, Dean DC, Melillo DG, Patchett AA, Nargund
     R, Griffin PR, DeMartino JA, Gupta SK, Schaeffer JM, Smith RS, Van der Ploeg
     LHT. A receptor in pituitary and hypothalamus that functions in growth hormone
     release. Science 1996; 273:974–977.
15. Pombo M, Barreiro J, Peñalva A, Peino R, Dieguez C, Casanueva FF. Absence
     of growth hormone (GH) secretion after the administration of either GH-releasing
     hormone (GHRH), GH-releasing peptide (GHRP-6), or GHRH plus GHRP-6 in
     children with neonatal pituitary stalk transection. J Clin Endocrinol Metab 1995;
     80:3180–3184.
16. Popovic V, Simic M, Ilic L, Micic D. Growth hormone (GH) and prolactin se-
     cretion elicited by GHRH, GHRP-6 and GHRH plus GHRP-6 in patients with
     prolactinoma. Clin Endocrinol. In press.
17. Guillaume V, Magnan E, Cataldi M, Dutour A, Sauze N, Renard M, Razafindraibe
     H, Conte-Devolx B, Deghenghi R, Lenaerts V, Oliver C. GH-releasing hormone
     secretion is stimulated by a new GH-releasing hexapeptide in sheep. Endocrinol-
     ogy 1994; 135:1073–1076.
18. Locatelli V, Torsello A, Grilli R, Ghigo MC, Cella SG, Deghenghi R,
     Weherenberg WB, Muller EE. GHRP-6 stimulates GH secretion and synthesis
     independently from endogenous GHRH and SRIF in the infant rat. J Endocrinol
     Invest 1993; 16:OP27.
19. Ilson BE, Jorkasky DK, Curnow RT, Stote RM. Effect of a new synthetic
     hexapeptide to selectively stimulate growth hormone release in healthy human
     subjects. J Clin Endocrinol Metab 1989; 69:212–214.

20. Korbonits M, Grossman A. Growth hormone-releasing peptide and its analogues: novel stimuli to growth hormone release. Trends Endocrinol Metab 1995; 6:43–49.
21. Robinson BM, Friberg RD, Bowers CY, Barkan AL. Acute growth hormone (GH) response to GH-releasing hexapeptide in humans is independent of endogenous GH-releasing hormone. J Clin Endocrinol Metab 1992; 76:1121–1124.
22. Jaffe CA, Ho PJ, Demott FR, Bowers CY, Barkan AL. Effect of prolongued growth hormone (GH)-releasing peptide infusion on pulsatile GH secretion in normal men. J Clin Endocrinol Metab 1993; 77:1641–1647.
23. Arvat E, Gianotti L, Di Vito L, Imbimbo BP, Lenaerts V, Deghenghi R, Camanni F, Ghigo E. Modulation of growth hormone-releasing activity of hexarelin in man. Neuroendocrinology 1995; 61:51–56.

12

Electrophysiological Studies of the Arcuate Nucleus: Central Actions of Growth Hormone Secretagogues

Gareth Leng and Alex R. T. Bailey
University Medical School, Edinburgh, Scotland

Suzanne L. Dickson
University of Cambridge, Cambridge, England

Roy G. Smith
Merck Research Laboratories, Merck and Co., Inc., Rahway, New Jersey

I. INTRODUCTION: ELECTROPHYSIOLOGICAL STUDIES OF NEUROSECRETORY NEURONS

The secretion of growth hormone (GH) in the rat is sexually dimorphic. In both sexes secretion is pulsatile, but in males the pulses are larger, less frequent, and arise from a lower interpulse baseline than in females (1,2). The sexually dimorphic patterns of growth hormone secretion in the rat appear to derive from sexually dimorphic behavior of the somatostatin neurons, possibly reflecting the sexually dimorphic expression of androgen receptors by these neurons (3). In the male rat, GH-releasing factor (hormone) (GRF) and somatostatin are probably released alternately to produce peaks and troughs of growth hormone release (1,2,4), respectively, whereas in the female somatostatin is released more continuously.

These characteristics would lead us to suspect that there may be sexually dimorphic interactions between hypothalamic somatostatin neurons and GRF neurons, but the evidence is largely circumstantial. Our understanding of the neural control of anterior pituitary hormone secretion from direct observations is as yet rudimentary. By contrast, our detailed knowledge of the oxytocin- and vasopressin-secreting cells of the supraoptic and paraventricular nuclei makes these among the most fully characterized and best-understood neurons in the brain. The electrical activity of individual, identified oxytocin and vasopressin neurons has been extensively studied in a wide range of physiological and experimental circumstances in vivo, and the neuronal behavior is related, on the one hand, to hormonal secretion in the circulation and its biological effects and, on the other hand, to the detailed cell and membrane properties of the neurons (5,6). The reasons for the disparity are not difficult to see; whereas the neuroendocrine cell populations that regulate anterior pituitary secretion are distributed sparsely across relatively large regions of the hypothalamus, and dispersed among several other neuronal cell types, the supraoptic nucleus is a densely packed, homogeneous nucleus comprising essentially only large cells, each of which synthesizes either oxytocin or vasopressin. However, more importantly for the in vivo electrophysiologist, every neuron in the supraoptic nucleus projects to the neural lobe of the pituitary, and this makes it possible to "antidromically identify" supraoptic neurons during an experiment.

The principle of antidromic identification is straightforward: a stimulus pulse applied to any region of the brain will induce action potentials in the axons passing through that region, and those action potentials will propagate orthodromically toward the nerve terminals and also antidromically toward the cell body. Thus, if stimulus pulses are applied to the neural stalk of the pituitary, an electrode that is recording the electrical activity of a supraoptic neuron will detect an antidromically conducted action potential that occurs at a constant latency (depending on conduction velocity) following each stimulus pulse. These antidromic action potentials can be readily distinguished from orthodromically conducted action potentials on the basis of a few well-established electrophysiological criteria. This simple procedure removes most of the uncertainty surrounding cell identification: in the hypothalamus, virtually every cell thus antidromically identified as projecting to the neural stalk is either an oxytocin cell or a vasopressin cell.

II. ELECTROPHYSIOLOGICAL IDENTIFICATION OF GRF NEURONS

In the arcuate nucleus it is similarly possible to identify neurosecretory neurons antidromically by their projections to the median eminence. In our experiments, we expose the ventral surface of the hypothalamus by a transpharyngeal surgical approach, and place a bipolar stimulating electrode directly on the surface of the median eminence, with the two poles aligned sagitally. A recording microelectrode is then introduced into the arcuate nucleus, under direct visual control, at a point opposite the midpoint of the two poles of the stimulating electrode.

Although the arcuate nucleus is immediately adjacent to the median eminence, stimulation of the median eminence induces antidromic action potentials in a surprisingly few arcuate neurons [13.5% reported by Renaud (7); approximately one in six in our experiments]. These antidromically identified neurosecretory cells of the arcuate nucleus will include GRF neurons, but will also include tuberoinfundibular dopamine neurons and possibly other neurosecretory cell types. Vasopressin cells may be distinguished readily from oxytocin cells on the basis of their spontaneous firing patterns, and we had hoped that similar distinguishing characteristics might readily identify distinct subpopulations among these antidromically identified neurosecretory neurons. A subpopulation of arcuate neurons does fire spontaneously in a readily recognized firing pattern comprising brief high-frequency bursts of action potentials (8); unfortunately, however, none of these bursting cells appears to project to the median eminence. The neurosecretory cells, at least in the anesthetized rat, appear to be show very low spontaneous-firing rates (about half are spontaneously active at 0.2 spikes per second or less).

In an attempt to distinguish between these neurosecretory cell types we then studied the effects of stimuli applied to the basolateral amygdala, because stimulation of this site evokes the release of GH, but of no other anterior pituitary hormone (8). About half of the neurosecretory cells in the arcuate nucleus proved to be excited by stimulation of the basolateral amygdala (9); disappointingly, a similar proportion of the nonneurosecretory cells tested were also excited by this stimulation, so clearly this stimulus alone cannot be used to discriminate GRF neurons; however, it is reasonable to conclude that the GRF neurons are among the population of neurosecretory cells that are activated by basolateral amygdala stimulation. Finally, we stimulated the periventricular nucleus of the hypothalamus—the origin of the somatostatinergic innervation of the median eminence.

There is good reason to suspect that an interaction between somatostatin neurons and GRF neurons underlies the reciprocally alternating secretion of GRF and somatostatin in the male rat: 55–60% of the GRF neurons visualized either by immunocytochemistry (10) or in situ hybridization (11) express somatostatin receptors, and somatostatin receptors appear to be particularly abundant on axon terminals in the nucleus (12). In vivo, it is difficult to distinguish the inhibitory effects of systemic administration of somatostatin on GH release at the level of the pituitary from possible central actions of somatostatin. At the pituitary, somatostatin suppresses both spontaneous GH release (13) and release induced either by GRF (14) or by GH secretagogues (GHSs; 15,16). However, the suppression of GH release is likely to reflect, at least partly, a central action, because octreotide (a long-acting somatostatin analog) inhibits GHS-induced GH release when administered icv (17).

In any event, it proved very easy to find arcuate neurons that were inhibited following stimulation of the periventricular nucleus—disappointingly easy, because it appeared that a very high proportion of both neurosecretory and nonneurosecretory cells were inhibited (9). However, some of the neurosecretory cells showed a rebound activation following the end of periventricular nucleus stimulation, and these neurons, in particular, attracted our attention. Prolonged electrical stimulation of the periventricular nucleus (18), and prolonged infusions of somatostatin (19), are both followed by a large rebound secretion of GH that is thought to be partly mediated by a rebound release of GRF. Thus, by combining these criteria, we selected a population of neurosecretory cells within the arcuate nucleus that were excited following stimulation of the basolateral amygdala, that were inhibited by stimulation of the periventricular nucleus of the hypothalamus, and that showed a rebound activation following maintained stimulation of the periventricular nucleus. We felt reasonably confident that these cells, comprising about 1 in 5 of the neurosecretory neurons in the arcuate nucleus and about 1 in 30 of all arcuate neurons, were highly likely to be GRF cells.

III. ACTIVATION OF PUTATIVE GRF NEURONS BY GHRP

After having identified what we thought likely to be GRF cells, we then tested their responsiveness to iv injection of GH-releasing peptide (GHRP-6). Our expectations at this point were mixed; mainly we were anticipating that the GRF cells would probably be subject to negative feedback from GH released by GHRP from the pituitary, but Iain Robinson, with whom we were collaborating, had obtained some results that led him to suspect otherwise: the GH release in response to GHRP-6 was attenuated by prior administration of GRF antiserum, implying that part of the GH release might be a consequence of increased GRF release (20). In any event, the outcome was clear; each putative GRF cell that we tested showed a clear and sustained excitation in response to GHRP-6 (21). Either we had been completely misled in our cell identification, or GHRP-6 was a potent activator of GRF neurons; in either event it was clear that GHRP-6 had a marked action on central neurons following peripheral administration. The cells that were activated showed a consistent profile of response: the firing rate accelerated shortly after iv administration to a maximum level after about 5 min, which was sustained thereafter for often an hour or more. We subsequently found very similar responses to nonpeptidyl secretagogues (Fig. 1), the principal difference being that the latency to maximum response was longer with other secretagogues than with GHRP-6 (22).

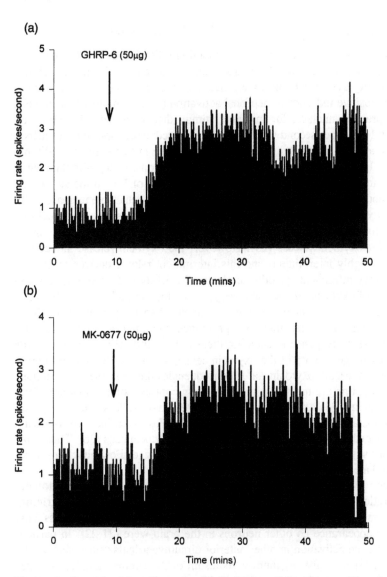

Figure 1 Records of the action potential discharge activity of two antidromically identified arcuate neurons in urethane anesthetized male rats. The arcuate nucleus was exposed by ventral surgery, and a stimulating electrode placed on the median eminence for antidromic identification. Systemic injection of (a) 50 μg GHRP-6 and (b) 50 μg MK-0677 elicits a similar electrophysiological response; each produces a sustained increase in firing rate after a 5- to 10-min latency.

IV. Induction of c-*fos* Following GHRP-6 Administration

We turned to look for evidence of Fos induction following GHRP-6. Fos is the protein product of the immediate-early gene c-*fos*, and in the magnocellular neurosecretory system we and others had shown that it was a robust, consistent, and sensitive indicator of neuronal activation (23,24). In the oxytocin cells of the supraoptic nucleus, for instance, stimuli that increased cell activity by as little as 1 spike per second would induce increased expression of c-*fos* mRNA in as little at 10 min, resulting in detectable increase in Fos expression 90 min later (24). In supraoptic cells, with only two exceptions, every stimulus that is known to induce oxytocin release induces expression of Fos in the supraoptic nucleus, and the proportion of neurons in which Fos is detectable increases with the intensity of the stimulus. The two exceptions are interesting: first, suckling is not a strong stimulus for Fos induction, despite being a major physiological stimulus for oxytocin release (25). However, the oxytocin release induced by suckling is highly intermittent, and this intermittent release derives from very brief and very infrequent episodes of electrical activation. In response to suckling, pulses of oxytocin are released every 5–10 min, and each pulse is the result of just 1–2 seconds of intense activity (at about 50 spikes per second). Moreover, following each burst, there is a prolonged inhibition of basal firing rate, so that overall in response to suckling there is very little if any change in the average discharge activity of the oxytocin cells. The bursts, though, are highly efficient for releasing oxytocin—at high frequencies the processes of stimulus-secretion-coupling are massively facilitated. By contrast, stimuli that are effective in inducing Fos expression, although they may be less intense, are all relatively sustained. The second exception is that Fos is not induced following even prolonged, intense antidromic stimulation of oxytocin cells (26). Thus spike activity per se in oxytocin cells is not sufficient to trigger Fos induction—a receptor-mediated mechanism may be essential.

When we mapped the pattern of Fos induction in the brain following administration of GHRP-6 and other secretagogues, the outcome was striking: a large number of cells in the arcuate nucleus were activated to express Fos, but at first appearance no other neurons in the brain were (21,22). In particular there was no activation in other anterior circumventricular sites, such as the subfornical organ or the organum vasculosum of the laminae terminalis—sites lacking in a blood–brain barrier that readily show Fos expression in response to many humoral signals and that might be anticipated to be sites vulnerable to nonspecific actions of a blood-borne peptide, and there was no activation in the periventricular nucleus (although subsequently we have found activation in some brain stem sites [see Chap. 17]). Central administration of GHRP-6, at doses ineffective when given peripherally, reproduced the same selective pattern of Fos induction in the hypothalamus, indicating that the selective distribution of

Fos did not reflect selectivity of access, but specificity of action. By contrast, administration of GRF, either centrally or peripherally, does not lead to any induction of Fos in the hypothalamus; hence, the neuronal activation induced by GHRP-6 is not a secondary consequence of GH release from the pituitary, nor is GHRP-6 acting as a mimic of GRF within the brain.

The nonpeptidyl secretagogues L-692,429 and L-692-585 (22) are less effective than GHRP-6 in inducing Fos expression in the arcuate nucleus, although the degree of electrical activation of responsive cells appears similar. It may be that the nonpeptidyl secretagogues are more selective than GHRP-6 in their central action, activating fewer arcuate cells, or that they have more restricted access to central sites. Whereas, at the pituitary, stimulation of GH release by GHRP-6 appears to require a functional GRF receptor, this appears not to be true of central actions of GHRP-6 (27), which is effective at inducing Fos expression in the arcuate nucleus of the lit/lit mouse that lacks functional GRF receptors. At the pituitary, the nonpeptidyl GHSs appear to be at least as potent as GHRP-6 for stimulating GH release (28), and only one GHS receptor has as yet been identified, which is present in both the hypothalamus and the pituitary (29).

V. GHRP-6 SELECTIVELY ACTIVATES NEUROSECRETORY NEURONS IN THE ARCUATE NUCLEUS

The specificity of action of GHRP-6 within the arcuate nucleus became apparent with more extensive electrophysiological studies. Of the spontaneously active cells that were antidromically identified as neurosecretory cells, about half were activated following injection of GHRP-6—a proportion probably too high to be exclusively GRF cells, but when cells in the arcuate nucleus that were not antidromically identified were tested, very few were excited, although a significant proportion were inhibited (22). This population is likely to include some neurosecretory cells, for only those neurosecretory cells that projected to the region of the median eminence between the stimulating electrodes would be identifiably neurosecretory.

Recent studies employing in situ hybridization histochemistry, and localized Fos protein and peptide mRNA have confirmed that, in the arcuate nucleus, a high proportion of GRF neurons express Fos after systemic administration of GHRP-6 (30) or the peptidyl GHS, KP-2 (31), but showed, in addition, that a significant population of neurons that synthesize neuropeptide Y (NPY) also express Fos following GHRP-6 administration. Arcuate neurons that synthesize β-endorphin, somatostatin, or high levels of tyrosine hydroxylase do not seem to be significantly activated (see Chap. 18).

VI. DIRECT ACTIONS OF GHRP-6 IN THE ARCUATE NUCLEUS

That GHRP-6 selectively activates neurosecretory neurons in the arcuate nucleus does not imply that these cells are the direct targets of action of GHRP-6. In particular, it is possible that the direct central actions of GHRP-6 are inhibitory, perhaps in the arcuate nucleus, but also perhaps elsewhere, and the arcuate neurosecretory cells are activated indirectly as a result of disinhibition. This point has yet to be resolved, but recordings from rat arcuate neurons in vitro have shown depolarizing effects of bath-applied GHRP-6 that probably do reflect a direction action at the recorded cells and that certainly reflect an action within or very close to the arcuate nucleus (Fig. 2). If the transduction mechanisms coupled to the central GHRP-6 receptor are the same as those coupled to the pituitary GHRP-6 receptor, then we should certainly expect the primary

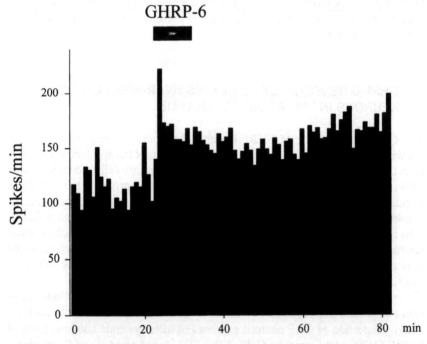

Figure 2 Record of the action potential discharge activity of a rat arcuate nucleus neuron recorded in a hypothalamic slice preparation in vitro. GHRP-6 (15 µM) was added to the extracellular perfusion medium for 10 min (bar). As did many cells tested, this neuron showed a prompt increase in electrical activity that outlasted the perfusion. (From unpublished experiments in collaboration with REJ Dyball.)

central action to be depolarizing, and we should also expect that Fos expression should be a reliable indicator of that activation because cAMP is a strong inducer of c-*fos* expression.

VII. CENTRAL INTERACTIONS BETWEEN SOMATOSTATIN NEURONS AND GRF NEURONS AND THE CENTRAL ACTIONS OF GHRP-6

Classically, the secretion of GH is controlled by the GRF neurons of the arcuate nucleus and the periventricular somatostatin neurons, which stimulate and inhibit GH secretion, respectively. It appears probable that, in the male rat, GRF and somatostatin are released alternately to produce alternate peaks and troughs of GH release. Interestingly, in the male rat, prolonged infusion of somatostatin leads to a sustained inhibition of GH release, followed by a dramatic rebound secretion of GH after the end of somatostatin infusion. Although this rebound is partly generated at the level of the pituitary, it also appears to reflect a large rebound secretion of GRF (19). Similar rebound secretion of GRF follows electrical stimulation of the periventricular nucleus (18). The periventricular nucleus appears to provide a direct inhibitory projection to neurons in the arcuate nucleus (9).

Hence, it is possible that the reciprocity in the output of GRF and somatostatin during pulsatile GH secretion reflects neuronal interactions between the GRF and somatostatin cells. Our electrophysiological studies in vivo support this hypothesis, because arcuate cells excited by GHRP-6 or nonpeptidyl secretagogues were also inhibited during electrical stimulation of the periventricular nucleus (21), and such secretagogue-responsive cells are also inhibited following iv injection of somatostatin or octreotide (Fig. 3). By contrast, cells that are not responsive to secretagogues are mainly unaffected by somatostatin injections. In addition, the central actions of GHSs to induce expression of Fos in the arcuate nucleus can be attenuated by systemic or central administration of octreotide. Thus, it would appear that a subpopulation of the arcuate cells activated by GH secretagogues are also inhibited by central somatostatin action. In recording from antidromically identified arcuate neurons, we have also seen examples of cells that are activated electrophysiologically following the end of an infusion of somatostatin, although these appear to be rare in the anesthetized rat (Fig. 4).

Suppression by octreotide of the GHS-induced increase in the expression of Fos in the arcuate nucleus is likely to be mediated by a direct central action of this peptide, for injection of a very low dose of octreotide (2 µg) was as effective as iv injection of 100 µg. Indeed, this also suggests that octreotide is able to gain access to central sites when administered by the iv route. Octreotide

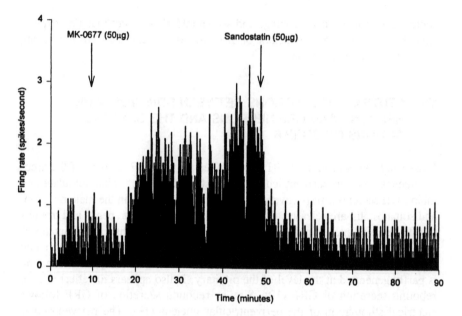

Figure 3 Record of the action potential discharge activity of an antidromically identified arcuate neuron recorded in an urethane-anesthetized male rat. Systemic injection
of MK-0677 (50 μg; iv) elicited a sustained increase in firing rate that was promptly attenuated by injection of the specific somatostatin analog octreotide (Sandostatin; 50 μg,
iv).

injection alone did not induce Fos protein expression in the arcuate nucleus, or
in any other hypothalamic or forebrain structure studied. However, although it
seems likely that octreotide acts through somatostatin receptors on GRF neurons, it is also possible that it acts through an afferent pathway to these cells.
Indeed, it is not possible to determine whether the cells that are the direct target for the action of the secretagogues are also the direct target for the action
of octreotide.

Thus, interactions between somatostatin neurons and GRF neurons may
underly the pulsatile patterning of GH secretion; it is also possible that negative-feedback actions of GH on GH release are mediated by activation of the
somatostatin neurons or by inhibition of GRF neurons, or by both (32), and
that this feedback plays a key role in determining the interval between pulses
of GH secretion. There are somatostatin-containing neurons within both the
arcuate and periventricular nuclei. Although GH receptors are present in the
arcuate nucleus, very few of the somatostatin neurons at this site appear to
express GH receptor mRNA (33). However, GH receptors are present in the
periventricular nucleus (33), and somatostatin neurons at this site do appear to

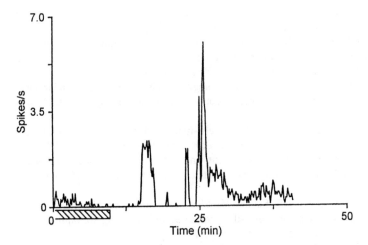

Figure 4 Record of the action potential discharge activity of an antidromically identified arcuate neuron recorded in a pentobarbitone-anesthetized male rat. In these experiments, recordings were made from single cells during prolonged iv infusion with somatostatin (10 μg/h). After a control period in the presence of somatostatin, the infusion
was then terminated, and subsequent changes in activity were observed. Activation such
as this was observed in a few such antidromically identified neurons (three out of nine
cells). (From unpublished experiments in collaboration with A. Ison.)

be sensitive to GH feedback because administration of GH receptor antisense
decreases periventricular somatostatin expression (34), and in hypophysectomized rats, GH activates periventricular somatostatin neurons to express c-*fos*
(31). The induction of Fos in the periventricular nucleus following MK-0677
would be expected to follow from the effects of endogenous GH release. Such
induction has not been observed previously in intact rats. However, in intact
rats, rather than in hypophysectomized rats, GH itself may be less effective in
inducing Fos expression, suggesting that central GH receptors may be down-
regulated by the feedback actions of GH.

REFERENCES

1. Tannenbaum G. Genesis of episodic growth hormone secretion. J Paediatr
    Endocrinol 1993; 6:273–282.
2. Robinson ICAF. The growth hormone secretory pattern: a response to neuroen-
    docrine signals. Acta Paediatr Scand Suppl 1991; 372:70–78.
3. Herbison AE. Sexually dimorphic expression of androgen receptor immunoreac-

tivity by somatostatin neurones in rat hypothalamic periventricular nucleus and bed nucleus of the stria terminalis. J Neuroendocrinol 1995; 7:543–554.

4. Kasting NW, Martin JB, Arnold MA. Pulsatile somatostatin release from the median eminence of the unanaesthetized rat and its relationship to plasma growth hormone levels. Endocrinology 1981; 190:1739–1745.

5. Bourque CW, Renaud LP. Electrophysiology of mammalian magnocellular vaso-pressin and oxytocin neurosecretory neurons. Front Neuroendocrinol 1990; 11:183–212.

6. Hatton GI. Emerging concepts of structure–function dynamics in adult brain: the hypothalamo-neurohypophysial system. Prog Neurobiol 1990; 34:437–504.

7. Renaud LP. Influence of medial preoptic–anterior hypothalamic area stimulation on the excitability of mediobasal hypothalamic neurones in the rat. J Physiol 1977; 264:541–564.

8. Koibuchi N, Kakagawa T, Suzuki M. Electrical stimulation of the basolateral amygdala (ABL) elicits only growth hormone secretion among six pituitary hormones. J Neuroendocrinol 1991; 3:685–687.

9. Dickson SL, Leng G, Robinson ICAF. Electrical activation of the rat periventricular nucleus influences the activity of hypothalamic arcuate neurones. J Neuroendocrinol 1994; 6:359–367.

10. Epelbaum J, Moyse E, Tannenbaum GS, Kordon C, Beaudet A. Combined auto-radiographic and immunohistochemical evidence for an association of somatostatin binding sites with growth hormone-releasing factor-containing nerve cell bodies in the rat arcuate nucleus. J Neuroendocrinol 1989; 1:109–115.

11. Bertherat J, Dournaud P, Berod A, Normand E, Bloch B, Rostene W, Kordon C, Epelbaum J. Growth hormone-releasing hormone-synthesizing neurons are a sub-population of somatostatin receptor-labeled cells in the rat arcuate nucleus—a combined in situ hybridization and receptor light-microscopic autoradiographic study. Neuroendocrinology 1992; 56:25–31.

12. Dournand P, Gu YZ, Schonbrunn A, Mazella J, Tannenbaum G, Beaaudet A. Localization of the somatostatin receptor SST2A in rat brain using a specific antipeptide antibody. J Neurosci 1996; 16:4468–4478.

13. Brazeau P, Vale W, Burgus R, Ling N, Butcher M, Rivier J, Guillemin R. Hypothalamic polypeptide that inhibits the secretion of immunoreactive pituitary growth hormone. Science 1973; 179:77–79.

14. Fukata J, Diamond DJ, Martin JB. Effects of rat growth hormone (rGH)-releasing factor and somatostatin on the release and synthesis of rGH in dispersed pituitary cells. Endocrinology 1985; 117:457–467.

15. Bowers CY, Momany FA, Reynolds GA, Hong A. On the in vitro and in vivo activity of a new synthetic hexapeptide that acts on the pituitary to specifically release growth hormone. Endocrinology 1984; 114:1537–1545.

16. Bowers CY, Sartor AO, Reynolds TM, Badger TM. On the actions of the growth hormone- releasing hexapeptide, GHRP. Endocrinology 1991; 128:2027–2035.

17. Fairhall KM, Mynett A, Robinson ICAF. Central effects of growth hormone-releasing hexapeptide (GHRP-6) on growth hormone release are inhibited by central somatostatin action. J Endocrinol 1995; 144:555–560.

18. Okada K, Wakabayashi I, Sugihara H, Minami S, Kitamura T, Yamada J. Electrical stimulation of hypothalamic periventricular nucleus is followed by a large rebound secretion of growth hormone in unanaesthetized rats. Neuroendocrinology 1991; 53:306–312.

19. Clark RG, Carlsson LMS, Rafferty YB, Robinson ICAF. The rebound release of growth hormone (GH) following somatostatin infusion in rats involves hypothalamic growth hormone-releasing factor release. Endocrinology 1988; 119:397–40479.

20. Clark RG, Carlsson LMS, Trojnar J, Robinson ICAF. The effects of a growth hormone-releasing peptide and growth hormone-releasing factor in conscious and anaesthetized rats. J Neuroendocrinol 1989; 1:249–255.

21. Dickson SL, Leng G, Robinson ICAF. Systemic administration of growth hormone-releasing peptide (GHRP-6) activates hypothalamic arcuate neurones. Neuroscience 1993; 53:303–306.

22. Dickson SL, Leng G, Dyball REJ, Smith RG. Central actions of peptide and nonpeptide growth hormone secretagogues in the rat. Neuroendocrinology 1995; 61:36–43.

23. Hoffman GE, Smith MS, Verbalis JG. c-*fos* and related immediate early gene products as markers of activity in neuroendocrine systems. Front Neuroendocrinol 1993; 14:173–213.

24. Hamamura M, Leng G, Emson PC, Kiyama H. Electrical activation and c-*fos* mRNA expression in rat neurosecretory neurones after systemic administration of cholecystokinin. J Physiol 1991; 444:51–63.

25. Fenelon VS, Poulain DA, Theodosis DT. Oxytocin neuron activation and Fos expression: a quantitative immunocytochemical analysis of the effect of lactation, parturition, osmotic and cardiovascular stimulation. Neuroscience 1993; 53:77–89.

26. Luckman SM, Dyball REJ, Leng G. Induction of c-*fos* expression in hypothalamic magnocellular neurones requires synaptic activation and not simply spike activity. J Neurosci 1994; 14:4825–4830.

27. Dickson SL, Doutrelant-Viltart O, Leng G. GH-deficient dw/dw rats and lit/lit mice show increased Fos expression in the hypothalamic arcuate nucleus following systemic injection of GH-releasing peptide 6. J Endocrinol 1995; 146:519–526.

28. Patchett AA, Nargund RP, Tata JR, Chen MH, Barakat KJ, Johnston DBR, Cheng K, Chan WWS, Butler B, Hickey G, Jacks T, Schleim K, Pong S-S, Chaung LYP, Chen HY, Frazier E, Leung KH, Chiu SHL, Smith RG. Design and biological-activities of L-163,191 (MK-0677)—a potent, orally-active growth-hormone secretagogue. Proc Natl Acad Sci USA 1995; 92:7001–7005.

29. Howard AD, Feighner SD, Cully DF, Arena JP, Liberator PA, Rosenblum CI, Hamelin M, Hrenluk DL, Palyha OC, Anderson J, Paress PS, Diaz C, Chou M, Liu KK, McKee KK, Pong S-S, Chaung L-Y, Elbrecht A, Dashkevicz M, Heavens R, Rigby M, Sirinathsinghji DJS, Dean DC, Melillo DG, Patchett AA, Nargund R, Griffin PR, DeMartino JA, Gupta SK, Schaeffer JM, Smith RG, Van der Ploeg LHT. A pituitary gland and hypothalamic receptor that functions in growth hormone release. Science 1996; 273:974–977.

30. Dickson SL, Luckman SM. Induction of c-*fos* messenger ribonucleic acid in neuropeptide Y and growth hormone (GH)-releasing factor neurones in the rat arcu-

ate nucleus following systemic injection of the GH secretagogue, GH-releasing peptide-6. Endocrinology 1977; 138:771–777.

31. Kamegai J, Hasegawa O, Minami S, Sugihara H, Wakabayashi I. The growth hormone-releasing peptide KP-102 induces c-*fos* expression in the arcuate nucleus. Mol Brain Res 1996; 39:153–159.

32. Sato M, Chihara K, Kita T, Kashio Y, Okimura Y, Kitajima N, Fujita T. Physiological role of somatostatin-mediated autofeedback regulation for growth hormone: importance of growth hormone in triggering somatostatin release during a trough period of pulsatile growth hormone release in conscious male rats. Neuroendocrinology 1989; 50:139–151.

33. Burton KA, Kabigting EB, Clifton DK, Steiner RA. Growth hormone receptor message ribonucleic acid distribution in the adult male rat brain and its colocalisation in hypothalamic somatostatin neurones. Endocrinology 1992; 131:958–963.

34. Pellegrini E, Bluet-Pajot MT, Mounier F, Bennett P, Kordon C, Epelbaum J. Central administration of a growth hormone (GH) receptor mRNA antisense increases GH pulsatility and decreases hypothalamic somatostatin expression in rats. J Neurosci 1996; 16:8140–8148.

13

Effects of a Growth Hormone-Releasing Peptide-like Nonpeptidyl Growth Hormone Secretagogue on Physiology and Function in Aged Rats

Richard F. Walker
University of South Florida, Tampa, Florida

Barry B. Bercu
University of South Florida College of Medicine, Tampa, and All Children's Hospital, St. Petersburg, Florida

I. INTRODUCTION AND BACKGROUND

Progressive physical deterioration characterized by specific changes in body composition with decreased functional capacity is one of the most obvious characteristics of normal aging. Compared with the young, old individuals experience reduced muscle mass, increased fat mass, and decreased strength (1–3). For example, the average muscle strength of men in their 70s is about half of that of men in their 20s. As a result of these physical decrements, the risk for falls, fractures, and frailty increases with aging (4,5).

Pituitary secretion of growth hormone (GH), which is both anabolic and lipolytic (6–8), decreases during aging (9). Circulating concentrations of insulin-like growth factor I (IGF-I), which mediates the action of GH on peripheral tissues, also decline (10), and GH deficiency is associated with changes in body composition that resemble those that accompany the aging process. Furthermore, GH administration to elderly men reversed some of their age-related changes in body composition (i.e., lean body mass increased 9% and adipose tissue mass decreased by 15%; 11). These data were used to support the hypothesis that diminished GH secretions at least partly, contributes to the somatic changes of aging. However, in a recent study, functional ability on selected tests of physical and mental performance did not improve in elderly men administered GH for 6 months (12).

Except for two reports (13,14), it is generally agreed that GH secretion in response to GH-releasing hormone (GHRH) administration in vivo declines with advancing age (15–23). Although age-related hyposensitivity of the pituitary to GHRH has been associated with increased somatostatin influence (21), it is also accompanied by reduced hypothalamic GHRH secretion, immunoreactivity, and mRNA concentrations (24–26). In light of potential endogenous GHRH deficiency, age-related pituitary hyposensitivity may be compounded by diminished stimulation of the gland. This view is supported by the fact that hormones often induce their own receptors, and that GH release in response to GHRH administration in the elderly improves with repeated dosing (17), although youthful responses are not totally restored. Pituitary hyposensitivity may also result from reduced stimulation by GH secretagogues (GHSs), other than GHRH, that complement GHRH activity. This hypothesis derives from our finding that GH hypersecretion occurred in old rats coadministered GHRH and GH-releasing hexapeptide (GHRP; His-D-Trp-Ala-Trp-D-Phe-Lys-NH$_2$), a xenobiotic GH secretagogue (27). The GHRH and GHRP are functional complements, working through independent receptors and second messengers (28–35). Old rats were hyposensitive to individually administered GHRP (36) as well as to individually administered GHRH, but were hyperresponsive to the coadministered peptides (27). Because other investigators reported a "potential physiological role for hitherto putative hypophysiotropic factors" (34), our data suggested that reduced efficacy of GHRH in old animals was at least partly due to the absence or reduced concentrations of GHRH as well as other, yet undefined, endogenous cosecretagogues.

Although the results of our acute study (27) demonstrated that the mechanism for stimulated GH release does not irreversibly decay, the data were insufficient to determine whether:

1. Pituitary reserve in old rats is sufficient to sustain daily episodes of high-amplitude GH secretion when the gland is repeatedly stimulated
2. Endogenous GH released in response to secretagogue administration is bioactive
3. Pituitary GH concentrations change in response to chronic stimulation
4. Physical performance improves in old rats as a function of increased secretion of endogenous GH in those administered the GH secretagogues

Thus, the present study was designed to answer these questions using GHRH and a nonpeptidyl GHRP analog, Merck L-692,585.

II. METHODS

A. Animal Husbandry

Fischer 344 female rats (20 months of age) were purchased from the NIA colony at Harlan Industries (Indianapolis, IN) and acclimated to local conditions for 14 days before being used in the present study. Female rats were chosen for study because they respond more consistently and reproducibly to GHRP than do conscious males (37). The rats were housed under alternating 12-h periods of light and darkness (0600–1800 h), $60 \pm 5\%$ relative humidity, $22 \pm 2°C$ ($72 \pm 2°F$) ambient temperature, and provided with a constant source of food (Purina Rat Chow; Purina Ralston Company, Kalamazoo, MI) and tap water. The animal facility at All Children's Hospital in which this study was performed is a fully accredited, institutional member of the American Association of Laboratory Animal Sciences and complies with federal statutes and regulations for animal use in research.

B. Study Design

Sixty rats were randomly assigned to four groups (15 rats per group) designated to receive daily, subcutaneous (sc) injections of L-692,585 (GHRP; 1 mg/kg), GHRH (10 μg/kg), GHRP, and GHRH (1 mg/kg and 10 μg/kg, respectively) or vehicle (saline). The decision to coadminister GHRP and GHRH was based on our prior findings with GHRH and GHRP in which the two different classes of GH secretagogues potentiated each other's action and, thereby, simulated youthful profiles of GH secretion in old rats (27). Solutions (1 mL/kg) were administered at approximately 0900 h for 120 consecutive days. Rats were given physical examinations during the period of drug administration. Each animal was weighed, visually examined for lesions, palpated for masses, and observed for any unusual aspects of posture and locomotion. Physical performance tests were performed before the last day of the study. On the final study day, the rats were anesthetized with pentobarbital (Nembutal; 35 mg/kg, ip) and administered a provocative dose of GHRP (0.5 mg/kg, ip). All rats received GHRP regardless of prior treatment to determine whether previous exposure to the same, to different, or to no compound, affected the GH secretory response to the nonpeptidyl secretagogue. A lower dose of GHRP was used during the provocative test because pentobarbital anesthesia suppresses somatostatin secretion and, thereby, enhances the response to pituitary stimulation (38). The rats were decapitated 15 min after administration of GHRP (time to peak stimulated GH release) and trunk blood was collected in heparinized centrifuge tubes for subsequent analysis of plasma hormone concentrations. Rats awaiting decapitation were isolated in an adjacent room to avoid potential stress resulting from sounds and smells associated with the procedure. After euthanasia, the calvarium of each rat was opened, the brain removed, and the pituitary examined visually in situ before removal from the sella turcica. Thereafter, the pituitaries were removed, weighed, and immediately processed for determination of GH concentrations. Hypothalamic tissues were similarly processed for GHRH and somatostatin concentrations.

C. Physical Performance Tests

Tests of physical performance were designed to determine whether long-term administration of GHRP was associated with increased activity, neuromuscular reflexes, strength, endurance, or coordination. Preferred physical tests were those that required no conditioning or training. They included running wheel (spontaneous activity), negative geotaxis (48° inclined plane; neuromuscular reflex; 58° inclined plane; upper body strength, neuromuscular reflex), vertical screen (coordination and strength), balance bar (coordination and strength), and swimming (coordination and endurance).

Although a prior study failed to demonstrate increased functional ability in elderly men administered GH, it was not clear whether the results accurately evaluated the complex events associated with performance of tasks (12). For example, none of the tests administered to study participants controlled for motivation. It is possible that the return to more youthful body composition in the study subjects was not accompanied by increased enthusiasm or willingness to perform each task. In the present study, one test of physical performance was chosen because it specifically excluded motivation as a factor for performing each test. Whereas running wheel, negative geotaxis, vertical screen, and balance bar allowed the experimental animals a choice of whether or not to give full effort, or even to continue the test, forced swimming required continuous and total effort to remain above water.

To test spontaneous activity, rats were housed for 24 h in cages fitted with running wheels that were constantly available for use. Rotation of the wheel triggered electronic data collection that could provide information on the amount of activity (total distance), temporal patterns of running (circadian rhythmicity), maximum speed, and average speed (Fig. 1).

To test complex neuromuscular reflex and strength, rats were placed head down on inclined planes of 48° or 58° to the horizontal (Fig. 2). The inclines were covered with sandpaper to provide traction against gravity. Time to reversal from a head-down to a head-up position was recorded with a stopwatch. The lesser incline favored testing the reflex because less strength was required to resist slipping. The greater incline tested upper limb strength, which was needed to prevent slipping.

To test coordination and strength, rats were placed head-down on a vertical screen and released (Fig. 3). The animals would reflexively attempt to

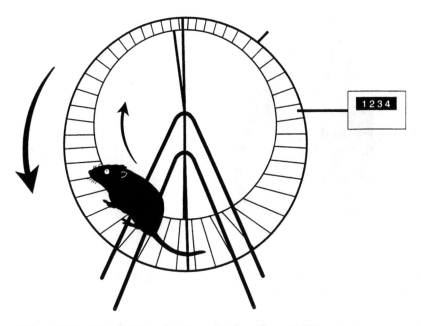

Figure 1   Running wheel used to determine the effects of GH secretagogues on spontaneous activity in old female rats.

reverse their position while supporting total body weight on the screen. After reversal, they remained on the screen, resisting gravity until lowering themselves to the table below. A balance bar was also used to test coordination and strength. Each rat was placed on a 1½-in (3.8-cm)– diameter wooden rod projected from

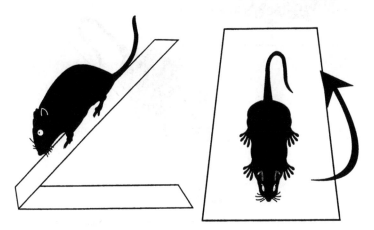

Figure 2   Inclined plane used to determine the effects of GH secretagogues on complex neuromuscular reflexes and strength in old female rats.

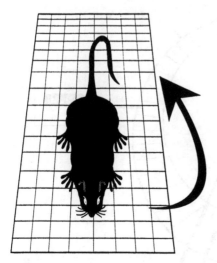

Figure 3 Hanging or vertical screen used to determine the effects of GH secretagogues on coordination and strength in old female rats.

a table top, above a platform (Fig. 4). Because the animals were facing away from the table (i.e., into space), they attempted to turn and return to the table. They usually slipped in the process of turning, but were able to grasp the bar and either return to a balancing position or hold the bar until they tired and fell

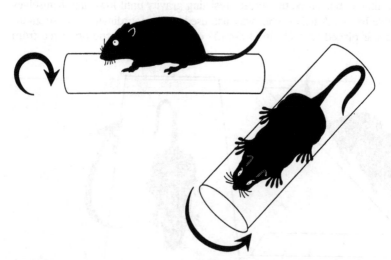

Figure 4 Balance bar used to determine the effects of GH secretagogues on coordination, balance, and strength in old female rats.

to the platform below. Ability to successfully balance or hang from the bar were measured and compared as indicators of coordination and strength.

To test endurance and coordination, rats were placed in a 3-ft– (91-cm)–deep, glass, paper chromatography tank that was three-quarters filled with room-temperature water (Fig. 5). The rats began swimming instantly and sustained their efforts to remain above water as long as they were in the tank. Depending on their physical condition, they gradually assumed a vertical position and finally submerged, nose up. To prevent their inhaling water, the rats were removed from the tank at the instant of first reverse submersion, and total time spent in the water to that point was recorded with a stopwatch.

D. Assays, Chemicals, and Reagents

The GHRP was provided by Merck Research Laboratories (Rahway, NJ) and rat GHRH was purchased from Peninsula Laboratories (Belmont, CA). Controls received saline administered in equal volume and by the same route as the peptides. Blood samples were centrifuged and plasma was collected for radio-immunoassay (RIA) analysis of GH and IGF-1 concentrations. Pituitaries were homogenized in 0.1 N NaOH–saline, diluted, and aliquots were analyzed by RIA. The GH concentrations were estimated in duplicate by RIA and expressed in terms of the reference preparation recombinant GH receptor protein-2 (rGH-

Figure 5 Swimming tank used to determine the effects of GH secretagogues on endurance and coordination in old female rats.

RP-2) from the NIDDK. Radiolabeled rat GH was purchased from Chemicon International, Inc. (Temecula, CA) and precipitating antibody (goat antimonkey $\gamma\gamma$ [P-4] for GH was purchased from Antibodies Inc.; Davis, CA) and used at a concentration of 1:10. Intra-assay and interassay coefficients of variation were less than 5 and 8%, respectively. IGF-1 concentrations were estimated after acid ethanol extraction from plasma using methods provided in a commercially available kit (Nichols Institute; Diagnostics; San Juan Capistrano, CA). Hypothalamic somatostatin and GHRH were measured by RIA using commercially available kits purchased from Incstar Corp. (Stillwater, MN) and Peninsula Laboratories, Inc. (Belmont, CA), respectively. Pituitary and hypothalamic protein concentrations were measured using a Bio-Rad Protein Assay macromethod against a standard of bovine albumin (Bio-Rad Chemical Division, Richmond, CA).

E. Data Analysis

Data were compared among the groups of old rats using analysis of variance with Hochberg's GT2 method for multiple comparisons. Hochberg's test was used because it is the most robust statistical method for comparing multiple groups having unequal sample sizes. Results were expressed as means plus–minus standard errors of the means (m $\pm$ SEM). A p value less than 0.05 was used to define statistically significant differences.

III. RESULTS

Eleven rats died during the course of this study. Four died in the saline group and four in the GHRP group. One died in the GHRH group and two died in the GHRP plus GHRH group. Deaths were associated with diencephalic compression caused by pituitary enlargement (three rats), brain stem hemorrhage (one rat), splenomegaly and wasting (two rats), diffuse hepatic or renal tumors (three rats), undefined causes (four rats). No pattern of disease and death was apparent among the different groups. It is concluded that the deaths were related to advanced age and independent of treatment.

Body weights are presented in Fig. 6. The data show that there were no statistically significant differences among the groups, regardless of treatment.

Spontaneous activity, as measured by running wheel rotations, was so low in the old rats that it was essentially nonexistent. As seen in Figs. 7–10, administration of GHRP alone or in combination with GHRH, or GHRH alone had no effect on any test of physical performance, except swimming. The data presented in Figure 10 show that the mean time until first submersion of the head was significantly increased in rats administered GHRP. Although mean

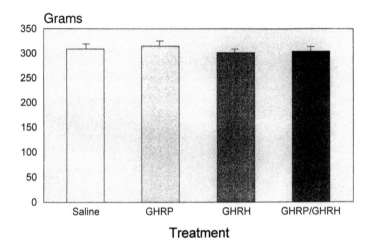

Figure 6 Lack of effect of GH secretagogue administration on body weight in old female rats.

swimming time was also increased in rats administered GHRH or GHRP plus GHRH, the differences were not statistically significant ($p < 0.08$) compared with saline-treated controls.

Increased swimming performance in GHRP-treated rats correlated with high plasma GH concentrations 15 min after a provocative dose of the GH

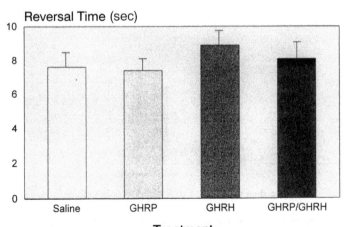

Figure 7 Lack of effect of GH secretagogue administration on inclined plane (negative geotaxis 48°) righting reflex in old female rats.

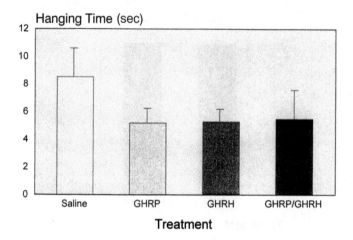

Figure 8 Lack of effect of GH secretagogue administration on ability of old female rats to hang on vertical screen.

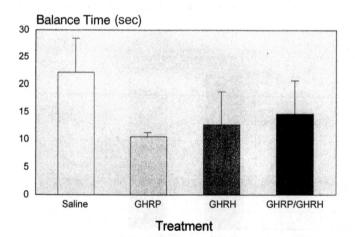

Figure 9 Lack of effect of GH secretagogue administration on ability of old female rats to balance on a rotating rod.

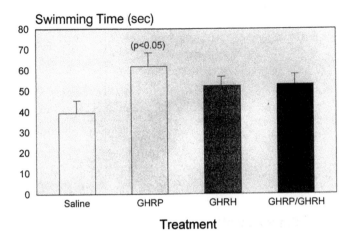

Figure 10   Increased duration of forced exercise in old female rats administered GH secretagogues.

secretagogue administered on the last day of the study. Before decapitation, all rats received the same provocative test to determine whether the different prior treatments would influence the response to a dose of GHRP. As seen in Figure 11, the highest concentrations of plasma GH in response to the GHRP challenge occurred in the control rats previously administered saline (2545 ± 466 ng/mL). Basal values of plasma GH under anesthesia in old, saline-treated rats are less than 100 ng/mL (historical data; aging studies 1991–1994). Mean plasma concentrations of stimulated GH in rats previously administered GHRP, were significantly higher than those previously administered GHRH (1535 ± 285 vs. 379 ± 68 ng/mL; $p < 0.01$). Mean plasma concentrations of stimulated GH in rats previously treated with GHRP were also higher than those previously treated with GHRP and GHRH, but the differences were not statistically significant. The data presented in Figure 11 also show that repeated administration of GHRP reduced the response to a provocative dose of GHRP, but that even greater inhibition occurred after repeated administration of GHRH, an unexpected finding in light of presumed potentiation of GHRP activity by GHRH.

The data presented in Figure 12 compare pituitary concentrations of GH in 24-month-old rats after administration of saline, GHRP, GHRH, or GHRP

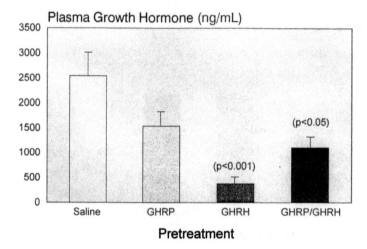

Figure 11 Effect of GH secretagogue administration on episodic GH release resulting from a provocative dose of GHRP.

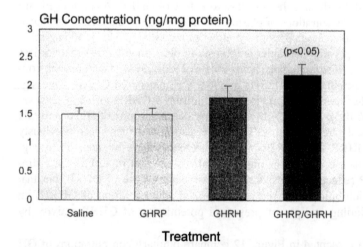

Figure 12 Effect of extended administration of GH secretagogues on pituitary GH concentrations.

plus GHRH for 120 consecutive days. In contrast with the effects of plasma GH, administration of GHRP was associated with the lowest pituitary GH concentrations. Pituitary GH concentrations in rats administered GHRP were significantly different from those administered GHRP plus GHRH. Although GHRP apparently produced episodes of robust GH secretion each day, as indicated by the amplitude of GH secretion on the last day of study, pituitary accumulation of the hormone was not changed when compared with saline-treated rats. On the other hand, GHRH increased pituitary accumulation of GH, but stimulated less release of the hormone than did GHRP. These data suggest that the two GH secretagogues have differential effects on GH secretion and production. Quantitation of pituitary GH mRNA should provide insight into the possibility of GHRP and GHRH having different mechanisms on somatotroph physiology and molecular biology.

Plasma IGF-1 concentrations were significantly higher in all groups of rats administered GH secretagogues than in the group administered saline (Fig. 13). As seen in Figure 13, plasma IGF-I concentrations were significantly higher in GHRP-rats than in saline-treated controls (533 $\pm$ 16 vs. 654 $\pm$ 32 ng/mL; p < 0.01). Although mean IGF-I concentrations were not as high in rats administered GHRP and GHRH (636 $\pm$ 21 ng/mL) or GHRH (623 $\pm$ 21 ng/mL), the values were still significantly higher than those of controls. However, swimming times were not significantly increased in these groups, suggesting that

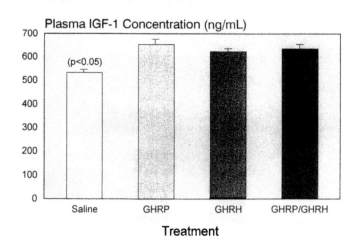

Figure 13 Effect of extended administration of GH secretagogues on plasma IGF-I concentrations in old female rats.

plasma IGF-I concentrations may not have been the only factor contributing to increased physical performance in the old rats.

As seen in Figures 14 and 15, hypothalamic somatostatin concentrations and hypothalamic GHRH concentrations, respectively, were no different among treatment groups. Several GHRH samples were lost owing to technical error during the RIA. However, those remaining had highly variable concentrations, supporting the view that differences in tissue concentrations of the hormones were not changed by administration of the GH secretagogues.

IV. DISCUSSION

The results of the experiments described in the foregoing provide answers to the questions proposed in the introduction to this study and support the following conclusions:

1. High peak plasma GH concentrations in old female rats administered GHRP on the final day of the study demonstrate sufficient pituitary reserve to sustain high-amplitude GH secretion on stimulation of the gland with GH secretagogues each day.
2. Statistically significant increases in plasma GH of old rats administered GH secretagogues demonstrate that endogenous GH is bioactive.
3. The pituitary glands of old female rats are capable of increased GH synthesis in response to stimulation with certain GH secretagogues or their combinations.

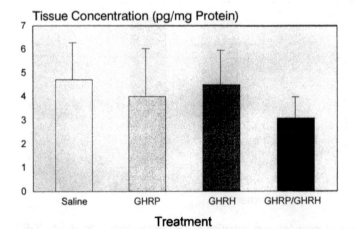

Figure 14 Effect of extended administration of GH secretagogues on hypothalamic GHRH concentrations in old female rats.

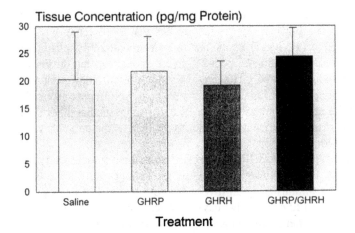

Figure 15 Effect of extended administration of GH secretagogues on hypothalamic somatostatin concentrations in old female rats.

4. Increased capability to perform physical tasks can be demonstrated in old female rats in certain tests.

To begin discussion of these findings, GHRP presumably improved swimming endurance in old animals by an anabolic effect resulting from increased endogenous GH secretion and elevated plasma IGF-I concentrations. These endocrine effects are similar to those previously observed in rats administered GHRH and GHRP (27,36). Although the increase in GHRP-stimulated GH secretion in the present study was comparable with that following GHRH plus GHRP-stimulated GH secretion in the previous study, its effect on IFG-I was not as great. This difference is attributed to the fact that rats used in the present study were six months older than those in the previous study (18 months vs. 24 months). The older rats were chosen for use in the present study because it was assumed that a lower basal concentration of IGF-I in the older rats would amplify the potential effects of GHRP. To the contrary, basal IGF-I concentrations in the 24-month-old saline-treated rats in the present study were comparable with those in 18-month-old saline-treated rats in the prior study (497 ± 27 vs 533 ± 16 ng/mL), but the mean increase after administration of the GH secretagogue was less (344 vs 121 ng/mL). The data are insufficient to determine whether the difference is due to age, treatment, or both factors. However, the finding that IGF-I concentrations are as low at 18 months as at 24 months, supports the choice of using the younger age for future studies. This suggestion is based on the greater incidence of intrinsic disease in the older rats, increasing their risk of dying and perhaps attenuating responses that occur more readily at the younger age, before irreversible physical deterioration of senescence proceeds too far. Further support for the fact that the older group may provide less reliable data derives from the fact that attrition at 24 months of age was 18% (from 60 to 49), compared with 2% in the younger group during the prior study. Although stimulated GH secretion in the present study was very robust, demonstrating normal functional capacity of the pituitary despite the advanced age of the rats, the modest increase in IGF-I suggests that other components of the GH neuroendocrine axis are not as resilient as the pituitary when subjected to repeated stimulation.

The provocative effect of GHRP on GH secretion in the present study was differentially affected by prior treatment with GH secretagogues. An unexpected finding resulted from administering a provocative dose of GHRP to rats previously given GHRH for 120 consecutive days. The response to the GHRP challenge was significantly attenuated, suggesting that prior exposure to GHRP in some way negatively altered the mechanism for GHRP-mediated GH release. This observation was somewhat surprising because age-related hyposensitivity of the pituitary to GHRH has been associated with reduced hypothalamic GHRH secretion, immunoreactivity, and mRNA concentrations (24–26). This potential endogenous GHRH deficiency suggested that age-related pituitary hyposensitivity may partly result from diminished stimulation of the gland and, conversely, GHRH administration could restore pituitary sensitivity. This hypothesis is supported by the facts that hormones often induce their own receptors, and that GH release in response to GHRH administration in the elderly improves with repeated dosing (17), although youthful responses are not totally restored. However, although GHRH replacement can restore responses to GHRH stimulation, it did not seem to have the same effect on GHRP stimulation. To the contrary, in the present study, daily administration of GHRH diminished GHRP efficacy. It is not likely that up-regulation of GHRH sensitivity by prolonged administration of the peptide caused a reciprocal down-regulation of GHRP sensitivity that was expressed in the GHRH-treated rats in the present study. More likely, when GHRH was given over an extended period and independent of GHRP, GHRH concentrations at the pituitary level were probably reduced or absent when GHRP was administered. This hypothesis is based on the fact that daily administration of GHRH suppresses release of endogenous GHRH from hypothalamic neurosecretory neurons by a negative-feedback process. However, the data from this study on hypothalamic concentrations of somatostatin and GHRH were insufficient to evaluate possible feedback effects of GHRP and GHRH. Nonetheless, we previously showed that passive immunization against GHRH grossly attenuated responses to GHRP, indicating that local GHRH concentrations were essential for expression of GHRP efficacy (39). This view is further supported by the fact that coadministration of GHRH and GHRP to old rats caused GH hypersecretion, presumably because GHRH and GHRP are functional complements working through independent receptors and second messengers (28–35). Apparently to express their synergism, both secretagogues must be present at the same time as during GHRH and GHRP coadministration. Thus, it seems that GHRH administration reduced local concentrations of GHRH that, in turn, caused GHRP to be less efficacious than in control rats. Interestingly, Bercu and Walker (see Chap. 20) recently found that in middle age, men become hyposensitive to GHRH but not GHRP. This age-related decrement could partly result from a progressive decline in availability of a putative endogenous analog of GHRP. This interesting possibility, which deserves further study to help achieve a better understanding of the physiological mechanisms subserving the interactive secretory potential of GHRP and GHRH, is discussed in greater detail by Bercu and Walker (see Chap. 20).

V. SUMMARY AND CONCLUSIONS

The purpose of this study was to determine whether long-term administration of L-692,585 (GHRP) to old rats would restore endocrine parameters of a more youthful GH neuroendocrine axis and also improve performance of certain physical tasks. Four groups of rats were administered GHRP, rat GHRH, GHRP plus GHRH, or saline for 120 consecutive days.

The GH secretion in response to a provocative dose of GHRP was tested in the four treatment groups on the last day of study. Plasma GH concentrations 15 min after administration of GHRP were highest in naive rats (saline-treated) and significantly different from those previously administered GHRH or GHRP plus GHRH. Differences between saline-treated and GHRP-treated rats, as well as between GHRP-treated and GHRP plus GHRH-treated rats were not statistically significant. The data suggest that prolonged administration of GHRH opposed the brief stimulatory effect of GHRP on GH release through a negative-feedback effect that reduced pituitary concentrations of endogenous GHRH.

Pituitary GH concentrations in GHRH plus GHRP-treated rats were significantly higher than those in GHRP-treated rats. The data suggest that GHRP stimulated GH secretion but not GH synthesis.

Plasma IGF-1 concentrations were significantly higher in rats administered GH secretagogues over extended periods than in those administered saline. Hypothalamic concentrations of GHRH and somatostatin were not affected by a treatment.

Physical performance as measured by time of swimming until exhaustion was significantly improved in rats administered GHRP. In contrast to performance in a forced exercise paradigm, long-term administration of GH secretagogues had no effect on performance in tests that gave animals the choice of whether or not to participate (i.e., vertical screen, balance bar, running wheel, and inclined plane).

The results of this study demonstrate that although long-term administration of GH secretagogues stimulate youthful changes in endocrine physiology, the functional correlates of those changes are more difficult to demonstrate. Physical performance in old rats was improved by chronic administration of GHRP. However, significant improvement in physical performance was observed only when the rats were forced to participate. This observation suggests that, whereas physical capacity was improved, attitude or motivation to perform were not increased by treatment. Increased swimming time correlated best with high plasma GH concentrations as well as high plasma IGF-I concentrations, suggesting that quantitative changes in both hormones, not IGF-I alone, contributed to improved physical performance.

In conclusion, the following summary is offered:

1. Daily administration of GHRP increased physical endurance in old rats as tested by swimming capability.
2. GHRP stimulated high-amplitude GH release in old, saline-treated rats, but the response to GHRP was significantly reduced by extended pretreatment with GHRH. This effect is attributed to a negative-feedback effect that reduced local GHRH concentrations and, thus, its complementary effect on GHRP expression.
3. The GH released in response to GHRP administration was bioactive, as indicated by a statistically significant increase in plasma IGF-I.
4. GHRP did not increase pituitary GH concentrations, whereas they were increased by GHRP plus GHRH. These data suggest different effects of the two secretagogues on pituitary GH synthesis, storage, or secretion, or some combination thereof.

REFERENCES

1. Rudman D. Growth hormone, body composition and aging. J Am Geriatr Soc 1985; 33:800–807.
2. Novak LP. Aging, total body potassium, fat free mass and cell mass in males and females between ages 15 and 85 years. J Gerontol 1972; 27:438–443.
3. Vandervoot AA, Hayes KC, Belanger AY. Strength and endurance of skeletal muscle in the elderly. Physiother Can 1986; 38:162–173.
4. Ory MD, Schectman KB, Miller JP, Hadley EC, Flatarone MA, Province MA. Frailty and injuries in later life. The FICSIT trials. J Am Geriatr Soc 1993; 41:283–296.
5. Tinnetti ME, Williams TF, Mayewski R Fall risk index for elderly patients based on the number of chronic disabilities. Am J Med 1986; 80:429–434.

6. Flaim KE, Li JB, Jefferson LS. Protein turnover in rat skeletal muscle; effects of hypophysectomy and growth hormone. Am J Physiol 1978; 234:E38–E43.
7. Beck JC. Primate growth hormone studies in man. Metabolism 1960; 9:699–737.
8. Bergenstal DM, Lipsett MB. Metabolic effects of human growth hormone and growth hormone of other species in man. J Clin Endocrinol Metab 1960; 20:1427–1436.
9. Rudman D, Kutner MH, Rogers CM, Lubin MP, Fleming GA, Bain RP. Impaired growth hormone secretion in the adult population; relation to age and adiposity. J Clin Invest 1981; 67:1361–1369
10. Corpas E, Harman SM, Blackman MR. Human growth hormone and human aging. Endocr Rev 1993; 14:20–30.
11. Rudman D, Feller AG, Hoskote SN, Gergans GA, Lalitha PY, Goldberg AF, Schlenker RA, Chon L, Rudman IW, Mattson DE. Effects of human growth hormone in men over 60 years old. N Engl J Med 1990; 323:1–6.
12. Papadakis MA, Grady D, Black D, Tierney MJ, Gooding GAW, Schambelan M, Grunfeld C. Growth hormone replacement in healthy older men improves body composition but not functional ability. Ann Intern Med 1996; 124:708–716.
13. Wehrenberg WB, Ling N. The absence of an age-related change in the pituitary response to growth hormone-releasing factor in rats. Neuroendocrinology 1983; 37:463–466.
14. Pavlov EP, Harman SM, Merriam GR, Gelato MC, Blackman MR. Responses of growth hormone (GH) and somatomedin-C to GH-releasing hormone in healthy aging men. J Clin Endocrinol Metab 1986; 62:595–600.
15. Ceda GP, Valenti G, Butturini U, Hoffman AR. Diminished pituitary responsiveness to growth hormone-releasing factor in aging male rats. Endocrinology 1986; 118:2109–2114.
16. Ghigo E, Goffi S, Arvat E, Nicolosi M, Procopio M, Bellone J, Imperiale E, Mazza E, Baracchi G, Camanni F. Pyridostigmine partially restores the GH responsiveness to GHRH in normal aging. Acta Endocrinol 1990; 123:169–174.
17. Iovino M, Monteleone P, Steardo L. Repetitive growth hormone-releasing hormone administration restores the attenuated growth hormone (GH) response to GH-releasing hormone testing in normal aging. J Clin Endocrinol Metab 1989; 69:910–913.
18. Lang I, Schernthaner G, Pietschmann P, Kurz R, Stephenson JM, Templ H. Effects of sex and age on growth hormone response to growth hormone-releasing hormone in healthy individuals. J Clin Endocrinol Metab 1987; 65:535–540.
19. Lang I, Kruz R, Geyer G, Tragl KH. The influence of age on human pancreatic growth hormone releasing hormone stimulated growth hormone secretion. Horm Metab Res 1988; 20:574–578.
20. Shibasaki T, Shizume K, Nakahara M, Masuda A, Jibiki K, Demura H, Wakabayashi I, Ling N. Age-related changes in plasma growth hormone response to growth hormone-releasing factor in man. J Clin Endocrinol Metab 1984; 58:212–214.
21. Sonntag WE, Gough MA. Growth hormone releasing hormone induced release of growth hormone in aging male rats: dependence on pharmacological manipulation of endogenous somatostatin release. Neuroendocrinology 1988; 47:482–488.

22. Sonntag WE, Steger RW, Forman LJ, Meites J. Decreased pulsatile release of growth hormone in old male rats. Endocrinology 1980; 107;1875–1879.

23. Sonntag WE, Hylka VW, Meites J. Impaired ability of old male rats to secrete growth hormone in vivo but not in vitro in response to hpGRF (1–44). Endocrinology 1983; 113:2305–2307.

24. De Gennaro Colonna V, Zoli M, Cocchi D, Maggi A, Marrama P, Agnati LF, Muller EE. Reduced growth hormone releasing factor (GHRF)-like immunoreactivity and GHRF gene expression in the hypothalamus of aged rats. Peptides 1989; 10:705–708.

25. Morimoto N, Kawakami F, Makino S, Chihara K, Hasegawa M, Ibata Y. Age-related changes in growth hormone releasing factor and somatostatin in the rat hypothalamus. Neuroendocrinology 1988; 47:459–464.

26. Ono M, Miki N, Shizume D. Release of immunoreactive growth hormone releasing factor (GRF) and somatostatin from incubated hypothalamus in young and old male rats. Neuroendocrinology 1986; 43:111.

27. Walker RF, Yang S-W, Bercu BB. Robust growth hormone (GH) secretion in aged female rats co-administered GH-releasing hexapeptide (GHRP) and GH releasing hormone (GHRH). Life Sci 1991; 49:1499–1504.

28. Abribat T, Boulanger L, Gaudreau P. Characterization of ["251-Tyr"0] human growth hormone-releasing factor (1–44) amide binding to rat pituitary: evidence for high and low affinity classes of sites. Brain Res 1990; 528:291–299.

29. Blake AD, Smith RG. Desensitization studies using perfused pituitary cells show that growth hormone-releasing hormone and His-D-Trp-Ala-Trp-D-Phe-Lys-NH$_2$ stimulate growth hormone release through distinct receptor sites. J Endocrinol 1991; 129:11–19.

30. Bowers CY, Sartor AO, Reynolds GA, Badger TM. On the actions of growth hormone-releasing hexapeptide, GHRP. Endocrinology 1991; 128:2027–2035.

31. Cheng K, Chan WW-S, Barreta A, Convey DM, Smith RG. The synergistic effects of His-D-Trp-Ala-Trp-D-Phe-Lys-NH$_2$ on growth hormone (GH)-releasing factor stimulated GH release and intracellular adenosine 3'5'-monophosphate accumulation in rat primary pituitary cell culture. Endocrinology 1989; 124:2791–2798.

32. Cheng K, Chan WW-S, Butler B, Barreto A Jr, Smith RG. Evidence for a role of protein kinase-C in His-D-Trp-Ala-Trp-D-Phe-Lys-NH$_2$-induced growth hormone release from rat primary pituitary cells. Endocrinology 1991; 129:3337–3342.

33. Codd EE, Shu AYL, Walker RF. Binding of a growth hormone releasing hexapeptide to specific hypothalamic and pituitary binding sites. Neuropharmacology 1989; 28:1139–1144.

34. Goth MI, Lyons CE, Canny BJ, Thorner MO. Pituitary adenylate cyclase activating polypeptide, growth hormone (GH)-releasing peptide and GH-releasing hormone stimulate GH release through distinct pituitary receptors. Endocrinology 1992; 130:939–944.

35. Sethumadhavan K, Veeraragavan K, Bowers CY. Demonstration and characterization of the specific binding of growth hormone-releasing peptide to rat anterior pituitary and hypothalamic membranes. Biochem Biophys Res Commun 1991; 178:31–37.

36. Walker RF, Yang S-W, Masuda R, Hu C-S, Bercu BB. Effects of growth hormone releasing peptides on stimulated growth hormone secretion in old rats In: Bercu BB, Walker RF, eds. Growth Hormone II: Basic and Clinical Aspects. New York: Springer-Verlag, 1994.

37. Bercu BB, Weidman CA, Walker RF. Sex differences in growth hormone secretion by rats administered growth hormone-releasing hexapeptide (GHRP-6). Endocrinology 1991; 129:2592–2598.

38. Chihara K, Arimura A, Schally V. Immunoreactive somatostatin in rat hypophyseal portal blood: effects of anesthetics. Endocrinology 1979; 104:1434–1441.

39. Bercu BB, Yang S-W, Masuda R, Walker RF. Role of selected endogenous peptides in growth hormone releasing hexapeptide (GHRP) activity: analysis of GHRH, TRH and GnRH. Endocrinology 1992; 130:2579–2586.

14

Anterior Pituitary Function in Prolonged Critical Illness: Effects of Continuous Infusion of Growth Hormone Secretagogues

Greet Van den Berghe and Francis de Zegher
University of Leuven, Leuven, Belgium

Roger Bouillon
Katholieke Universiteit Leuven, Leuven, Belgium

Johannes D. Veldhuis
University of Virginia, Charlottesville, Virginia

Cyril Y. Bowers
Tulane University Medical Center, New Orleans, Louisiana

I. INTRODUCTION

In prolonged critical illness (vital organ function supported with intensive care for weeks), feeding is unable to reverse ongoing wasting of protein, whereas fat is preserved or even further stored (1,2). This feeding-resistant loss of protein is mainly caused by its activated degradation together with suppressed synthesis; thereby, dramatically reducing the protein content of vital tissues. The residual protein content in skeletal muscle of critically ill patients apparently correlates inversely with the duration, instead of the severity, of illness, and is mainly determined by the (impaired) capacity to synthesize protein (2). The mechanisms underlying these inappropriate metabolic responses in the fed, critically ill state remain incompletely understood.

The metabolic response is associated with endocrine changes, including elevated serum cortisol levels, blunted growth hormone (GH) secretion, low insulin-like growth factor I (IGF-I) concentrations, and low circulating thyroid hormone levels in the presence of low-normal thyrotropin (TSH; 3). The contribution of these endocrine changes to the catabolic state is conceivable. Administration of exogenous GH is being intensely investigated as a strategy to reverse the catabolism of critical illness.

In the prolonged critically ill state, endogenous GH is readily released in response to a single bolus of GH-releasing peptide-2 (GHRP-2), even more so to the combined administration of GHRP-2 and GH-releasing hormone (GHRH), and paradoxically, also to thyrotropin-releasing hormone (TRH). TSH is released in reduced amounts in response to administration of a TRH bolus, and it is paradoxically secreted in response to a GHRH bolus (4,5). These findings suggest an altered hypothalamic control of pituitary function in the catabolic phase of prolonged critical illness.

In an attempt to gain further insight into this altered hypothalamic–pituitary function, we dynamically studied the somatotropic, thyrotropic, and lactotropic axes and cortisol secretion in this condition, together with the effects exerted by continuous infusion of GHRH or GHRP-2, or their combination (6,7).

II. PATIENTS AND METHODS

Twenty-six critically ill adults (mean ± SEM age: 63 ± 2 years), who were dependent on intensive care support, including mechanical ventilation for 18 ± 1 (range 7–35) days at the time of inclusion, were studied during 2 consecutive nights (9 PM–6 AM) (Table 1). Patients were selected for the duration of critical illness (1 week at least), rather than the type or severity of the underlying disease (2); consequently, they represent a rather heterogeneous population, with the critical condition occurring mainly as a complication of major surgery or trauma. Cerebral trauma or disease, preexisting endocrine disease, as well as concomitant treatment with drugs that have known effects on hypothalamic–pituitary function (e.g., dopamine, Ca^{2+}-reentry blockers, amiodarone, clonidine, and others) were exclusion criteria. A proportion of the patients received major analgesia with opioids, sedation with benzodiazepines, or cardiovascular support with nondopaminergic catecholamines, but nevertheless, did not respond differently from those who were not treated with these substances, as we have previously shown (3). The studied patients ultimately remained in the ICU for 41 ± 4 (range 16–107) days, and 73% (19/26) survived. During the studied episode of 45 h, the concomitant ICU therapy remained virtually unaltered, including the continuously administered, parenteral or enteral artificial nutritional support.

According to a weighted randomization schedule, patients received one of four combinations of infusions within a strictly randomized, crossover design for each combination: placebo (1 night) and GHRP-2 (other night) (n =

Table 1 Clinical Patient Data

Random	Gender	Age	B.M.I.	Diagnosis	Apache II adm.	Total ICU stay	Outcome	Incl. day	Type feeding	Opioids	Benzod.	Exog. insul.	Exog. catech.
Placebo/GHRP-2	M	64	20	AFG + ischemic GI bleeding	19	33	Died	20	TPN	y	n	n	y
Placebo/GHRP-2	M	67	30	Polytrauma	20	26	Home	19	TPN	y	y	n	n
Placebo/GHRP-2	M	59	29	AAA + resp. insufficiency	23	48	Home	7	PN + enteral	y	y	y	y
Placebo/GHRP-2	M	69	29	Bricker + sepsis	21	33	Home	19	TPN	n	n	n	y
Placebo/GHRP-2	M	52	29	Necrotising pancreatitis	17	44	Home	31	TPN	y	y	y	n
Placebo/GHRP-2	M	65	23	Infected AFG	22	30	Died	21	TPN	y	y	n	y
GHRP-2/Placebo	M	56	28	Guillain/Barre	15	103	Home	35	Enteral	y	y	y	n
GHRP-2/Placebo	F	67	35	CABG + sepsis	13	30	Home	16	TPN	y	y	y	n
GHRP-2/Placebo	F	57	33	CABG + resp. insufficiency	19	35	Home	24	PN + enteral	n	n	y	y
GHRP-2/Placebo	M	66	28	Hemicolectomy + resp. insuff.	19	50	Home	8	TPN	y	y	y	n
Placebo/GHRH	F	69	35	CABG + shock	11	45	Home	18	PN + enteral	n	y	y	y
Placebo/GHRH	M	57	27	Biliary peritonitis	13	33	Home	11	PN + enteral	y	y	y	y
GHRH/Placebo	M	74	26	CABG + resp. insufficiency	11	54	Home	9	Enteral	y	n	n	n
GHRH/Placebo	M	76	29	Abdominal trauma	17	49	Home	18	TPN	n	n	n	n
GHRP-2/GHRH	M	31	31	Gastrectomy + resp. insufficiency	14	16	Home	13	Enteral	y	y	y	n
GHRP-2/GHRH	M	62	20	Polytrauma	19	20	Home	13	TPN	y	y	y	n
GHRP-2/GHRH	M	79	17	Lobectomy + resp. insufficiency	20	107	Died	35	BN	n	n	n	n
GHRH/GHRP-2	M	71	23	Necrotizing pancreatitis	33	19	Died	14	TPN	y	y	y	y
GHRH/GHRP-2	M	73	26	CABG + shock	7	20	Home	9	TPN	y	y	y	n
GHRH/GHRP-2	M	74	24	Peritonitis	17	41	Home	11	TPN	n	n	n	y
GHRP-2/G+G	F	76	25	CABG + mediastinitis	13	71	Died	21	PN + enteral	y	n	y	n
GHRP-2/G+G	M	39	24	Facial hemangioma, resp. insuffic.	17	22	Home	17	TPN	y	y	n	n
GHRP-2/G+G	M	56	19	Peritonitis	20	45	Died	15	TPN	y	y	n	y
G+G/GHRP-2	M	69	23	CABG + pneumonia	7	32	Home	23	TPN	y	y	n	y
G+G/GHRP-2	M	47	31	Peritonitis + mediastinitis	24	22	Home	17	TPN	y	y	y	y
G+G/GHRP-2	M	71	22	Esophageal resection + multiple organ failure	21	41	Died	24	TPN	y	y	y	y

Randomization group; gender; age (years); BMI (kg/m^2); diagnosis (AFG, complicated aortofemoral graft; CABG, complicated coronary artery bypass grafts; AAA, complicated abdominal aortic aneurysm repair; GI, gastrointestinal); Apache II score after the first 24 h of intensive care; total ICU stay (days); ultimate outcome; day of stay in the ICU at the time of inclusion; type of continuous feeding (TPN, total parenteral nutrition; PN, parenteral nutrition; PN + EN, parenteral and enteral nutrition; EN, full enteral nutrition); the concomitant administration of exogenous opioids, benzodiazepines, insulin, nondopaminergic catecholamines. *Source:* Ref. 13.

10); placebo and GHRH ($n = 4$); GHRH and GHRP-2 ($n = 6$); GHRP-2 and GHRH plus GHRP-2 ($n = 6$). The peptide infusions (over 21 h) were started after a bolus of 1 µg/kg at 9 AM and infused (1 µg/kg per hour) until 6 AM. Serum concentrations of GH, TSH, and prolactin (PRL) were measured every 20 min, serum cortisol concentrations were determined every hour, and serum IGF-I, thyroxine (T_4), triiodothyronine (T_3), and reverse T_3 levels were determined in the first and the last sample of each night's profile.

The dynamic characteristics of GH, TSH, and PRL secretion were calculated with deconvolution [this technique allows determination of the occurrence and magnitude of secretory hormone bursts, and quantifies the pulsatile as well as the basal secretion of the hormones (8)]. Also, it was specified as to what extent the pulsatile release of the studied anterior pituitary hormones was synchronized, by calculation of the amount of cross-correlation of the GH and TSH with PRL concentration profiles within the same patient (9). Furthermore, hormonal secretion was characterized by determination of the approximate entropy (A_pE_n), a model-independent statistic for assessing regularity of time-series, with larger values corresponding to greater irregularity or apparent process randomness (10,11).

III. RESULTS

The GH, TSH, and PRL nightly mean serum concentrations during placebo infusion ($n = 14$) were low-normal: mean GH concentration was 1.5 ± 0.24 µg/L; mean TSH level was 1.25 ± 0.42 mIU/L; and mean PRL concentration was 9.4 ± 0.9 µg/L. In contrast, the fraction of hormone released in a pulsatile fashion (determined as the product of number and mass of the secretory bursts [pulsatile production], relative to the total [pulsatile plus basal] production) was consistently reduced: GH $51 \pm 6\%$ versus normal mean 99%, TSH $32 \pm 6\%$ versus normal 65%, and PRL $16 \pm 3\%$ versus normal 48% (all $p < 0.0001$; Figs. 1–3). The normal nocturnal surges of GH, TSH, and PRL were absent, independent of concomitant sleep. The reduced pulsatile GH and TSH production over 9 h nightly correlated positively with, respectively, low circulating IGF-I (106 ± 11 µg/L; $R^2 = 0.35$; $p = 0.02$) and low T_3 (0.64 ± 0.06 nmol/L; $R^2 = 0.34$; $p = 0.03$; see Fig. 1). In contrast with the normal partial synchronization between TSH and PRL release (12), there were no significant cross-correlations among the concentration profiles of either GH or TSH and PRL in the critically ill patients infused with placebo (Fig. 4).

The GHRH infusion elicited a two- to three-fold increase of basal ($p = 0.03$) and pulsatile ($p = 0.007$) GH secretion and, paradoxically, of pulsatile TSH release ($p = 0.03$; see Figs. 2 and 3). GHRH infusion did not significantly alter serum IGF-I but did increase serum T_3 by 21% ($p = 0.03$). GHRH

% PULSATILE SECRETION **(M ± 95% CI)**

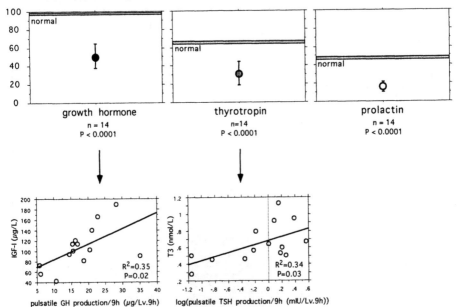

Figure 1 (*Upper panel*) In the presence of low-normal nightly mean serum concentrations (GH, 1.5 ± 0.24 μg/L; TSH, 1.25 ± 0.42 mIU/L; PRL, 9.4 ± 0.9 μg/L), the fraction of hormone released in a pulsatile fashion was consistently reduced for all three hormones: GH $51 \pm 6\%$ versus normal mean 99%; TSH $32 \pm 6\%$ versus normal 65%; and PRL $16 \pm 3\%$ versus normal 48% (all $p < 0.0001$). (*Lower panel*) The reduced pulsatile GH and TSH production over 9 h nightly correlated positively with, respectively, low circulating IGF-I (106 ± 11 μg/L; $R^2 = 0.35$; $p = 0.02$) and low T_3 (0.64 ± 0.06 nmol/L; $R^2 = 0.34$; $p = 0.03$). CI, confidence intervals. (Adapted from Refs. 13 and 14.)

increased the GH A_pE_n score by 22% (1.027 ± 0.03 vs. 0.842 ± 0.047; $p = 0.009$; meaning a more irregular secretory pattern; Fig. 5), but did not alter the absence of synchronization among the secretory patterns of GH, TSH, and PRL (see Fig. 4).

The GHRP-2 infusion provoked a four- to six-fold increase of basal ($p = 0.0007$) and pulsatile ($p = 0.002$) GH secretion (see Fig. 2), associated with a $61 \pm 13\%$ increase of serum IGF-I within 24 h ($p = 0.02$). In contrast, GHRP-2 suppressed pulsatile TSH secretion ($p = 0.02$; see Fig. 3); whereas it exclusively stimulated the basal component of PRL release ($p = 0.02$). In addition, GHRP-2 infusion partially coupled the secretory patterns of GH and of TSH with PRL ($p < 0.01$ for significant cross-correlation; see Fig. 4), without significantly altering the A_pE_n scores.

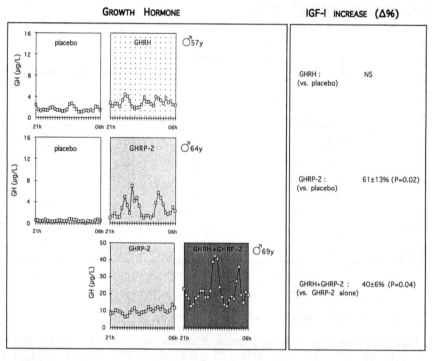

Figure 2 (*Left panel*) Three representative serum GH concentration profiles (sampling every 20 min between 9 PM and 6 AM) during placebo are shown, and the effects of continuous GHRH infusion (1 µg/kg per hour), GHRP-2 infusion (1 µg/kg per hour), and GHRH plus GHRP-2 (1 + 1 µg/kg per hour) are illustrated. (*Right panel*) IGF-I delta increase for the different groups of patients. The group data revealed that (1) GHRH infusion elicited a two- to threefold increase of basal ($p = 0.03$) and pulsatile ($p = 0.007$) GH secretion, without an effect on serum IGF-I; (2) GHRP-2 infusion provoked a four- to sixfold increase of basal ($p = 0.0007$) and pulsatile ($p = 0.002$) GH secretion, associated with a 61 ± 13% increase of serum IGF-I within 24 h ($p = 0.02$); (3) GHRH plus GHRP-2 infusion, as compared with GHRP-2 alone, elicited a further twofold increase of GH secretion ($p = 0.02$), and an additional 40 ± 6% rise in serum IGF-I ($p = 0.04$). (Adapted from Ref. 13.)

The GHRH plus GHRP-2 infusion, as compared with GHRP-2 alone, elicited a further two-fold increase of mainly basal GH secretion ($p = 0.02$; see Fig. 2), and an additional 40 ± 6% rise in serum IGF-I ($p = 0.04$). The addition of GHRP-2 abolished the amplification of pulsatile TSH release by GHRH alone (see Fig. 3), whereas it suppressed, but *did not abolish* the GHRH-induced increase in GH A_pE_n (see Fig. 5). During GHRH plus GHRP-2 infusion GH, but not TSH, secretion was coupled to PRL ($p < 0.01$; see Fig. 4).

THYROTROPIN

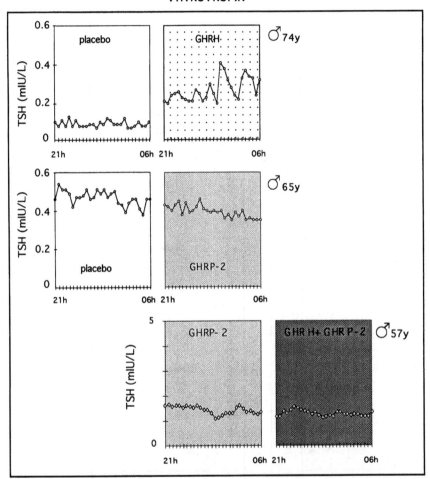

Figure 3 Three representative serum TSH concentration profiles (sampling every 20 min between 9 PM and 6 AM) during placebo are shown, and the effects of continuous GHRH infusion (1 μg/kg per hour), GHRP-2 infusion (1 μg/kg per hour) and GHRH plus GHRP-2 (1 + 1 μg/kg per hour) are illustrated. The group data revealed that (1) GHRH infusion paradoxically doubled pulsatile TSH release ($p = 0.03$), eliciting a 21% rise in serum T_3 ($p = 0.03$); (2) GHRP-2 infusion reduced the pulsatile fraction of TSH release ($p = 0.02$); (3) GHRP-2 addition to GHRH infusion abolished the amplification of pulsatile TSH release by GHRH alone. (Adapted from Ref. 14.)

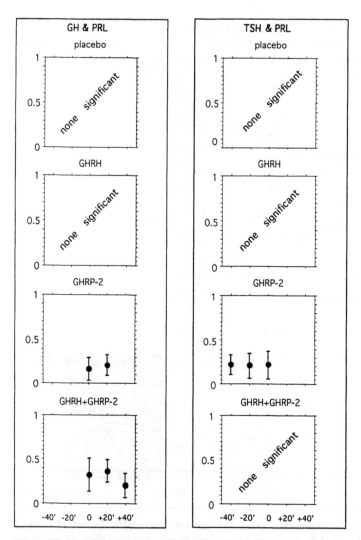

Figure 4 Cross-correlation of GH and TSH to PRL concentrations: In contrast with the normal partial synchronization of TSH and PRL release (12), there were no significant cross-correlations between the serum profiles of TSH or GL and PRL during placebo in the critically ill patients. GHRH did not alter the abnormal dyssynchronization (absence of cross-correlation) among the nocturnal profiles of TSH, GH, and PRL. GHRP-2 infusion partially coupled the secretory patterns of TSH and of GH with PRL ($p < 0.01$ for significant cross-correlation). During GHRH plus GHRP-2 infusion GH but not TSH secretion was partly coupled to PRL ($p < 0.01$). Significant R is $p < 0.01$, mean $\pm$ 95% CI.

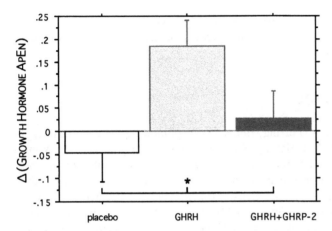

Figure 5 ΔGH A_pE_n in the various clinical groups: The GH A_pE_n, an indicator of irregularity of GH secretion, is rather high during critical illness. GHRP-2 infusion did not alter this A_pE_n. GHRH alone induces a rise in A_pE_n. This rise is attenuated, but not abolished, by the addition of GHRP-2. Data are presented as mean $\pm$ SEM, and are analyzed by ANOVA. *$p = 0.02$. (Adapted from Ref. 13.)

Cortisol levels were continuously elevated in all patients, and were not significantly altered by any of the peptide infusions.

IV. CONCLUSIONS

In the presence of low-normal mean nightly serum concentrations, the pulsatile fractions of GH, TSH, and PRL secretion are uniformly reduced during prolonged critical illness. In addition, the normally occurring nocturnal surges, as well as the partial synchronization between TSH and PRL release are absent. Circulating IGF-I and T_3 levels are low and correlate positively with respectively decreased pulsatile GH and TSH release, suggesting that they at least partly have a neuroendocrine origin. The suppressed pulsatile GH secretion can be substantially amplified by the continuous infusion of GHRP-2, with or without GHRH, generating a striking IGF-I response. The paradoxical TSH release induced by GHRH infusion as well as the increase of more irregularity in GH secretion by GHRH, both suppressed by the addition of GHRP-2, suggests the presence of particular linkages between the somatotropic and the thyrotropic axes in this condition. Together with the observation that exogenous GHRP-2 has a synchronizing effect on the pulsatile release of GH and TSH with PRL, these findings may point to a coordinating role for the putative endogenous GHRP-like ligand in pituitary hormone release. The substantial IGF-I responses within

24 h generated by GHRP-2 with or without GHRH infusion, open therapeutic perspectives for GH-secretagogues as potential antagonists of catabolic state in critical care medicine.

ACKNOWLEDGMENTS

We thank Dr. Mehuys (Ferring, Belgium) for organizing the generous supply of GHRH, and Mr. Jean Hellings (Baxter, Belgium) for providing the Vamp-systems for undiluted blood sampling. We also thank Professor Norbert Blanckaert and Dr. Jaak Billen for the TSH and thryroid hormone determinations. We thank Mrs. Viviane Celis, Mrs. Myriam Smets, Mrs. Christiane Eyletten, Mrs. Marianne Aerts, and Mr. Mathieu Gerits for expert technical assistance. The authors wish to thank the medical and nursing staff of the intensive care unit for their cooperation. Supported by research grants from the National Science Foundation Center for Biological Timing (JDV), the Belgian Fund for Scientific Research (G.0162.96), the University of Leuven (OT 95/24), and the Belgian Study Group for Pediatric Endocrinology (GVdB and FdZ). Dr. Van den Berghe is a Clinical Research Investigator of the Belgian Fund for Scientific Research (G.3c05.95N).

REFERENCES

1. Streat SJ, Beddoe AH, Hill GL. Aggressive nutritional support does not prevent protein loss despite fat gain in septic intensive care patients. J Trauma 1987; 27:262–266.
2. Gamrin L, Essén P, Forsberg AM, Hultman E, Wernerman J. A descriptive study of skeletal muscle metabolism in critically ill patients: free amino acids, energy-rich phosphates, protein, nucleic acids, fat, water, and electrolytes. Crit Care Med 1996; 24:575–583.
3. Van den Berghe G, de Zegher F. Anterior pituitary function during critical illness and dopamine treatment. Crit Care Med 1996; 24:1580–1590.
4. Van den Berghe G, de Zegher F, Vlasselaers D, Schetz M, Verwaest C, Ferdinande P, Lauwers P. Thyrotropin releasing hormone in critical illness: from a dopamine-dependent test to a strategy for increasing low serum triiodothyronine, prolactin, and growth hormone. Crit Care Med 1996; 24:590–595.
5. Van den Berghe G, de Zegher F, Bowers CY, Wouters P, Muller P, Soetens F, Vlasselaers D, Schetz M, Verwaest C, Lauwers P, Bouillon R. Pituitary responsiveness to growth hormone (GH) releasing hormone, GH-releasing peptide-2 and thyrotropin releasing hormone in critical illness. Clin Endocrinol 1996; 45:341–351.

6. Bowers CY, Sartor AO, Reynolds GA, Badger TM. On the actions of the growth hormone-releasing hexapeptide, GHRP. Endocrinology 1991; 128:2027–2035.

7. Frohman LA, Jansson JO. Growth hormone-releasing hormone. Endocr Rev 1986; 7:223–253.

8. Veldhuis JD, Johnson ML. Deconvolution analysis of pulsatile hormone data. Methods Enzymol 1992; 210:539–575.

9. Veldhuis JD, Johnson ML, Seneta E. Analysis of copulsatility of anterior pituitary hormones. J Clin Endocrinol Metab 1991; 73:569–576.

10. Pincus SM. Approximate entropy as a measure of system complexity. Proc Natl Acad Sci USA 1991; 88:2297–2301.

11. Pincus SM, Gladstone IM, Ehrenkranz RA. A regularity statistic for medical data analysis. J Clin Monit 1991; 7:335–345.

12. Samuels MH, Veldhuis JD, Ridgway EC. Copulsatile release of thyrotropin and prolactin in normal and hypothyroid subjects. Thyroid 1995; 5:369–372.

13. Van den Berghe G, de Zegher F, Veldhuis JD, Wouters P, Awouters M, Verbruggen W, Schetz M, Verwaest C, Lauwers P, Bouillon R, Bowers CY. The somatotropic axis in critical illness: effect of continuous GHRH and GHRP-2 infusion. J Clin Endocrinol Metab 1997; 82:590–599.

14. Van den Berghe G, de Zegher F, Veldhuis JD, Wouters P, Gouwy S, Stockman W, Weekers F, Schetz M, Lauwers P, Bouillon R, Bowers CY. Thyrotropin and prolactin release in prolonged critical illness: dynamics of spontaneous secretion and effects of growth hormone secretagogues. Clin Endocrinol 1997; 47:599–612.

15

Hepatic Extraction of Hexarelin, a New Peptidic Growth Hormone Secretagogue, in the Isolated Perfused Rat Liver

Marie Roumi, Sylvie Marleau, Patrick du Souich, and Huy Ong
University of Montreal, Montreal, Quebec, Canada

Muny Franklin Boghen
Pharmacia & Upjohn, Milan, Italy

Magnus H. L. Nilsson
Pharmacia & Upjohn, Stockholm, Sweden

Romano Deghenghi
Europeptides, Argenteuil, France

I. INTRODUCTION

Among the GH-releasing class of compounds, a novel and potent growth hormone-releasing peptide (GHRP) analog, hexarelin* [His-D-Trp(2Me)Ala-Trp-D-Phe-Lys-NH$_2$] has recently been developed. The compound displays high stability owing to the substitution of the D-tryptophan(2ME) position, and is more potent than GHRP-6 in vitro (1). Preclinical studies of hexarelin in rats and dogs following intravenous (iv) and subcutaneous (sc) administration have demonstrated that the peptide elicits a long-lasting GH release and is more effective than GHRP-6 (2,3). In humans, intravenous (iv) administration of hexarelin results in substantial dose-dependent increases in plasma GH concentrations (4). Following intravenous, subcutaneous, intranasal, and oral administration of hexarelin to humans, Ghigo et al. have demonstrated that all routes of administration elicit significant GH response (5).

---

*Hexarelin is a trademark for examorelin (I.N.N.).

Despite the low bioavailability of hexarelin, satisfactory GH-releasing activity is observed following a 20-mg oral dose of this drug in humans (5). The results are comparable with a GH response similar to that observed after iv administration of 1 μg/kg hexarelin. The low bioavailability of hexarelin stems from either of two possible sources: (1) an elevated hepatic extraction, or (2) a limitation in drug absorption from the gastrointestinal (GI) tract. Hence, investigation of whether the low bioavailability of hexarelin is due to an extensive effect of hepatic first-pass, or low intestinal absorption is imperative.

Because little is known about the disappearance rate of hexarelin in the liver, the objective of this study was to assess the extraction of hexarelin at 5, 50, and 500 ng/mL by an isolated perfused rat liver model. The pharmacokinetic parameters, dose dependency, and extraction ratio of the peptide were documented. In addition, the binding of hexarelin to plasma proteins was documented in vitro using plasma of rats, dogs, porcine, and humans.

II. MATERIALS AND METHODS

A. Isolated Perfused Liver

Male Sprague–Dawley rats (Charles River, St. Constant, Quebec), weighing 210–230 g, were used as liver donors. Rats were anesthetized by intraperitoneal injection of sodium pentobarbital (1 mL/kg). The portal vein, the common bile duct, and the vena cava above the hepatic veins were cannulated as described by Ross (6). The rat was then transferred to the perfusion system where the liver, maintained at a temperature of 37°C, was perfused in situ in a recirculating circuit using an MX/AMBEC two/ten perfuser (MX International Aurora, CO).

The perfusion medium consisted of a Krebs–Henseleit buffer containing 20% (v/v) washed bovine erythrocytes, 1% bovine serum albumin (BSA), and 1 g/L dextrose. The perfusate was adjusted to pH 7.4, saturated with a mixture of 95% oxygen and 5% carbon dioxide, and thermostatically controlled at a temperature of 37°C. The recirculating perfusate volume was 200 mL. The perfusion rate was fixed at 16 mL/min and verified volumetrically following diversion of venous outflow at the end of each experiment.

Following equilibrium, the circuit was switched to the reservoir containing a fixed final concentration of hexarelin. Samples of the medium, 0.5 mL from the reservoir (C_{in}) and 0.5 mL from a catheter of the caval vein (C_{out}), were taken at 0, 1, 3, 5, 7, 10, 15, 20, 30, 60, 90, and 120 min and kept on ice during the study. Bile samples were collected at 10-min intervals until the end of the perfusion. Perfusate samples were centrifuged and the plasma obtained was immediately stored with the bile samples at –20°C until analysis. Hexarelin in the perfusate samples (C_{out}) was measured using the established hexarelin radioimmunoassay method, as described by Roumi et al. (7). The perfusate medium concentration data were analyzed by a nonlinear least-squares program (PCNONLIN, Scientific Consulting Inc., Apex, NC).

B. Experimental Design

Two sets of experiments were conducted: (1) liver perfusion ($n = 4$) at three hexarelin dose levels (5, 50, and 500 ng/mL); (2) control liver perfusion ($n = 5$) without hexarelin.

C. In Vitro Plasma Protein-Binding Studies

The protein binding of hexarelin in plasma was determined by the method of ultrafiltration. Stock solutions consisting of a fixed amount of [^{3}H]hexarelin and various concentrations of unlabeled compound, diluted in a 0.054 M phosphate buffer (pH 7.4), were mixed with plasma to yield final concentrations of 0.003, 0.01, 0.03, 0.1, 0.3, 1, and 3 μM. The samples (in triplicate) were capped and incubated at 37°C for 30 min. One-milliliter aliquots were transferred to the sample reservoir of micropartition centrifuge tubes (Centrifree Micropartition Centrifuge Tube, Amicon, Beverly, MA) that had been pretreated with 2% polyethyleneimine for 24 h. The tubes were centrifuged at 37°C for 20 min at $1000 \times g$. All samples were analyzed by liquid scintillation counting. Because the binding studies were conducted using frozen plasma, the effects of freezing on the extent of plasma binding was examined.

D. Statistical Methods

Statistical analysis of clearance, half-life, and extraction ratio parameters were conducted using one-way ANOVA to verify difference between dose groups ($p < 0.05$).

III. RESULTS

A. Liver Perfusion

For all the studies, livers appeared macroscopically intact. They were not swollen, nor did they contain patchy areas that might suggest malperfusion or congestion. Liver function parameters monitored throughout the experiment for the control group and the three hexarelin concentrations indicated that the metabolic integrity of the liver was maintained.

Figure 1 illustrates the disappearance profiles of hexarelin from the recirculating perfusate over time following the addition of 5, 50, and 500 ng/mL

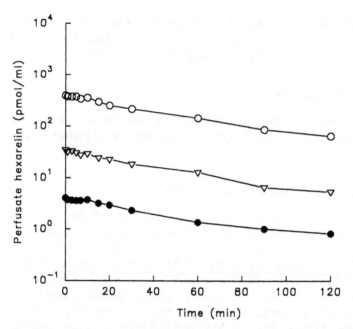

Figure 1 Disappearance of hexarelin from the recirculating perfusate (mean ± SEM) following the addition of 5 ng/mL (●), 50 ng/mL (∇), and 500 ng/mL (○) to the medium (n = 4).

hexarelin to the perfusion medium. As presented in Table 1, the pharmacokinetics of hexarelin are concentration-independent. This conclusion is attributed to the fact that a linear increase in area under the curve (AUC) was observed when increasing doses of the peptide were added to the perfusion medium. Consequently, the clearance values (0.345–0.401 nL/min per gram liver) are not significantly different over the dose range studied, as are the half-life values (55.20, 45.06, and 49.93 min). The determination of the extraction ratio indicates that hexarelin displays an extraction coefficient of 20%.

B.   In Vitro Binding Studies

The extent of binding of hexarelin to plasma proteins is summarized in Table 2. There were no appreciable differences found among rat, porcine, and human plasma, all of which exhibited a significant degree of binding ($\beta \geq 64\%$); binding of hexarelin to dog plasma protein was slightly higher ($\beta \approx 79\%$). For the concentration range studied, all species demonstrated a concentration-independent degree of binding. The suitability of the ultrafiltration method using frozen plasma samples indicated that there was approximately a 6% increase

Table 1 Pharmacokinetic Parameters of Hexarelin Following Administration of Single Doses of 5, 50, or 500 ng/mL in the Perfusion Medium[a] [$n = 4$]

Parameter	5 ng/mL	50 ng/mL	500 ng/mL
K (min^{-1})	0.013 ± 0.0004	0.016 ± 0.002	0.014 ± 0.0002
$t1/2$ (min)	55.20 ± 1.904	45.06 ± 5.71	49.93 ± 0.81
$AUC_{0-\infty}$ (min pmol/mL)	269.85 ± 12.55	1992.62 ± 116.64	24334.35 ± 1492.75
Cl (mL/min/g liver)	0.345 ± 0.012	0.401 ± 0.042	0.369 ± 0.023
E (%)	20.14 ± 0.88	20.95 ± 1.93	18.86 ± 0.67
AUC/D (min/mL)	0.33 ± 0.017	0.31 ± 0.029	0.32 ± 0.022

[a]Abbreviations of mean (±SEM) pharmacokinetic parameters: K, disappearance rate constant; $t1/2$, terminal half-life; $AUC_{0-\infty}$, area under the curve from time zero to infinity; Cl, total clearance per gram of liver; E, hepatic extraction ratio.

Table 2 Percentage Fraction of [^{3}H]Hexarelin Bound to Plasma Proteins (β) ($n = 3$)

Plasma concentration (μg/M)	Rat[a]	Dog	Porcine	Human
0.003	70.73 ± 0.72	80.73 ± 0.62	64.70 ± 0.98	65.67 ± 0.40
0.01	68.22 ± 0.97	77.99 ± 0.30	67.70 ± 0.91	63.20 ± 1.18
0.03	70.65 ± 0.55	80.78 ± 0.61	65.29 ± 0.46	65.20 ± 1.06
0.1	68.93 ± 1.25	78.63 ± 1.02	66.68 ± 2.07	63.10 ± 1.97
0.3	67.92 ± 0.74	79.22 ± 1.17	68.28 ± 1.51	65.47 ± 0.38
1	64.33 ± 0.45	76.08 ± 1.13	69.34 ± 0.79	67.33 ± 0.48
3	69.86 ± 0.98	77.63 ± 0.05	68.86 ± 0.62	66.35 ± 1.24

[a]There is approximately a 6% increase in β for *fresh* rat plasma versus *frozen* plasma.

in β when fresh rat plasma was used versus the frozen plasma. It is assumed that similar results would be expected using porcine, dog, or human plasma.

The plasma/blood ratio (C_p/C_b), estimated from studies conducted to investigate the uptake of hexarelin by red blood cells, was greater than 1.5 for the concentration range studied. This indicates that the bulk of hexarelin was present in plasma, and there was no uptake of the peptide by red blood cells.

IV. DISCUSSION

In this study, two main conclusive findings on the elimination of hexarelin by the liver have been made. First, at physiological pH and temperature, hepatic clearance of hexarelin obeys first-order kinetics over the concentration range studied. Evidence for dose linearity stems from the following two parameters: (1) the AUC increases in a first-order fashion, and (2) the clearance values remain constant independent of dose. Second, according to the present in situ model, hexarelin displays a low hepatic extraction ratio (20%).

Although there are no data on the hepatic extraction of GHRP analogs in the literature, the clearance of hexarelin may be compared with other neuropeptides, such as cholecystokinin (CCK-8) and somatostatin (SS), for which hepatic extractions have been documented (8,9). Whereas the two forms of somatostatin, SS-14 and SS-28, display first-order kinetics, CCK-8 exhibits concentration-dependent kinetics. The hepatic extraction of CCK-8, in the concentration range between 5 and 20 ng/mL, is significantly high (40–60%) following a single-pass perfusion. These values contrast with the low extraction ratio of hexarelin documented after a 2-h recirculating perfusion study. The hepatic extraction ratio of the active form of somatostatin was reported to be 35%.

Among the several implications associated with a poorly extracted drug, the relation between protein binding and drug clearance becomes notably important, for protein binding may be a rate-limiting factor. As a next step in this study, plasma-binding studies were conducted to assess the degree of binding of hexarelin to plasma proteins.

The concentration range of hexarelin used in the plasma protein-binding studies corresponds to that found in the perfusion studies. The binding of hexarelin to plasma proteins is concentration-independent, for a slightly greater binding in dog (β ≈ 79%) is observed. Only a minor portion of hexarelin in blood is taken up by blood cells. In view of the foregoing findings, one can conclude that the hepatic clearance of hexarelin could be limited by its binding to plasma proteins, and may suggest that the extraction of hexarelin might be dependent on plasma protein binding.

In conclusion, it has been demonstrated that the isolated perfused rat liver clears hexarelin in its unbound state from the perfusate following first-order kinetic process up to a 500-ng/mL concentration. The hepatic extraction ratio of hexarelin, being low, may be limited by its binding to plasma proteins. As a clinical implication, although the contribution of intestinal extraction on the effect of first-pass merits investigation, the low hepatic extraction ratio of hexarelin is an advantage for the oral administration of the peptide.

ACKNOWLEDGMENTS

The authors wish to thank Drs. M. G. Barré, M. Huet, and F. Varin for their expert advice in the field of liver perfusions and allowing the use of their perfusion equipment and laboratory facilities. This work was supported by Pharmacia Upjohn.

REFERENCES

1. Deghenghi R, Boutignon F, Wuthrich P. GH releasing properties of hexarelin (EP 23905), International Symposium on Growth Hormone II: Basic and Clinical Aspects. 1992: Abstr 31.
2. Deghenghi R, Canazi MM, Torsello A, Battisti C, Muller EE, Locatelli V. GH-releasing activity of hexarelin, a new growth hormone releasing peptide, in infant and adult rats. Life Sci 1994; 54:1321–1328.
3. Cella SG, Locatelli V, Poratelli M, Colonna VDG, Imbimbo BP, Deghenghi R, Muller EE. Hexarelin, a potent GHRP analogue: interactions with GHRH and clonidine in young and aged dogs. Peptides 1995; 16:81–86.
4. Imbimbo BP, Mant T, Edwards M, Dalton DAN, Boutignon F, Lenaerts V, Wuthrich P, Deghenghi R. Growth hormone-releasing activity of hexarelin in humans. Eur J Clin Pharmacol 1994; 46:421–425.
5. Ghigo E, Arvat E, Gianotti L, Imbimbo BP, Lenaerts V, Deghenghi R, Camanni F. Growth hormone-releasing activity of hexarelin, a new synthetic hexapeptide, after intravenous, subcutaneous, intranasal, and oral administration in man. J Clin Endocrinol Metab 1994; 78:693–698.
6. Ross RD. Perfusion Techniques in Biochemistry: A Laboratory Manual in the Use of Isolated Perfused Organs in Biochemical Experimentation. Oxford: Clarendon Press, 1972.
7. Roumi M, Lenaerts V, Boutignon F, Wuthrich P, Deghenghi R, Bellemare M, Adam A, Ong H. Radioimmunoassay for hexarelin, a peptidic growth hormone secretagogue, and its pharmacokinetic studies. Peptides 1995; 16:1301–1306.
8. Gores GJ, LaRusso NF, Miller LJ. Hepatic processing of cholecystokinin peptides.

I. Structural specificity and mechanism of hepatic extraction. Am J Physiol 1986; 250:G344–G349.

9. Seno M, Seno Y, Takemura Y, Nishi S, Ishida H, Kitano N, Imura H, Taminato T, Matsukura S. Comparison of somatostatin-28 and somatostatin-14 clearance by the perfused rat liver. Can J Physiol Pharmacol 1985; 63:62–67.

16

The Effect of Growth Hormone Secretagogues on the Release of Growth Hormone-Releasing Hormone, Somatostatin, Vasopressin, and Corticotrophin-Releasing Hormone from the Rat Hypothalamus In Vitro

Márta Korbonits, John A. Little, Peter J. Trainer, Michael Besser, and Ashley Grossman
St. Bartholomew's Hospital, London, England

Mary L. Forsling
United Medical and Dental Schools, St. Thomas's Campus, London, England

Alfredo Costa
University of Pavia, Pavia, Italy

Giuseppe Tringali and Pierluigi Navarra
Catholic University Medical School, Rome, Italy

I. BACKGROUND

Growth hormone secretagogues (GHSs), growth hormone-releasing peptides (GHRPs) and their pharmacological nonpeptidyl analogs, are small molecules with in vitro and in vivo growth hormone-releasing activity both in animal and human studies. Recently, a specific GHS receptor belonging to the G–protein-coupled receptor family has been cloned (1). The receptor is located in the pituitary and in the hypothalamus as well as in other tissues, and probably activates the protein kinase C (PKC) system; it also has a direct effect on intracellular calcium concentrations (2–4). In spite of the many experimental and clinical studies with different GHSs, the exact mode of their growth hormone-releasing action has not been fully established.

A. The Effects of GHSs

The GHSs have a direct effect on in vitro pituitary growth hormone (GH) release in both animal and human somatotrophs, but a variety of studies have indicated that they are much more potent in vivo and can also act on the hypothalamus, which action seems to be especially important in humans (5–10). All studies agree that GH-releasing hormone (GHRH) potentiates the effect of GHSs on GH release in vivo. In in vitro studies on pituitary cells, some data point to a potentiating effect, whereas others have found only an additive effect on GH release (11–15). The presence of endogenous GHRH is crucial for the effect of GHSs in humans, as established in several clinical studies in patients with pituitary stalk section (8–10). However, some animal studies were still able to demonstrate GH release after the transection of the pituitary stalk (16,17). These discrepancies could be explained by the possibly more complex destruction of the hypothalamic–pituitary area in the patients studied. Furthermore, apart from possible species differences, the patient groups studied were obviously more heterogeneous in terms of age, cause of the pituitary–hypothalamic disconnection, and underlying diseases, than the group of experimental animals.

Surprisingly, the effect of intracerebroventricular (icv) GHS administration on GH release has been studied in only two papers. Fairhall et al. administered GH-releasing peptide-6 (GHRP-6) to guinea pigs in the dose range of 8 ng–1µg and found a clear stimulation of circulating GH levels (18). However, Yagi et al. found *inhibition* of GH levels in rats after the icv administration of a higher dose of GHRP-6 (1, 3, and 10 µg/rat) (19).

The GHSs may also act as physiological somatostatin (SS) antagonists at the pituitary and the hypothalamic level (18); however, the exact mechanism of action of this effect has as yet not been clarified. A direct inhibitory effect of GHSs on hypothalamic SS release is highly unlikely, because no inhibition of SS release has been found in vivo (20,21), and SS neurons do not show electrical activation or c-*fos* expression after GHS administration (22).

Several studies have suggested a possible direct activation of GHRH-secreting neurons in the arcuate nucleus. In an in vivo study in sheep, hexarelin* caused significant elevation of GHRH in petrosal sinus blood, whereas no change was detected in SS levels (20). Intracerebroventricular administration of GHRP-6 and L-692,429 in rats causes electrical activation and c-*fos* gene expression in cells in the arcuate nucleus, the main GHRH-secreting site (6,23). However,

---

*Hexarelin is a trademark for examorelin (I.N.N.).

50% of these cells are neuropeptide Y (NPY) cells, whereas only 23% are GHRH cells (22). Is it possible that GHSs simply stimulate endogenous GHRH release? If so, how could GHSs elevate GH levels to a higher level than the maximal dose of GHRH, and how could we explain the potentiating effect (i.e., the rise in GH levels after GHRH plus a GHS is higher than the sum of the levels after each drug) of GHSs on GH release?

Pituitary somatotrophs are heterogeneous. Electron microscopy enables us to classify somatotrophs according to the size and shape of their secretory granules. Separated by density gradient, sparely granulated (type I) and heavily granulated (type II) cells both respond in a similar fashion to GHRP-6 administration, whereas type II cells were more sensitive and released more GH to GHRH administration. The presence of a GHRH-dependent and -independent cell population had been proposed earlier (24,25). Somatotrophs seem to be heterogeneous in their responsiveness to GHRP-6: when using the single- and double-reverse hemolytic plaque assay, 23.3% of cells responded to GHRH stimulation, but not to GHRP-6; 11.9% to GHRP, but not to GHRH; whereas 7.8% of cells responded to both drugs (15). Different GH responses to GHRH and GHRP-6 might be attributed to a change in the proportion of somatotroph subpopulation under pathological conditions (15).

The GHSs not only stimulate GH release, but also stimulate prolactin (PRL) and activate the hypothalamic–pituitary adrenal (HPA) axis in both animal and human studies (26–31). The effect on the HPA axis occurs probably by the hypothalamus because GHSs do not elevate corticotropin (adrenocorticotropic hormone; ACTH) levels directly in pituitary cell culture (11,32). The two major ACTH stimulators in the hypothalamus are corticotropin-releasing hormone (CRH) and arginine vasporession (AVP). An in vivo study on rats found that the coadministration of GHRP-6 and CRH cause similar ACTH elevation compared with CRH on its own, whereas the coadministration of GHRP-6 and AVP had a larger effect on ACTH release, suggesting that GHRPs interact with AVP in the release of ACTH, possibly by the stimulation of CRH (30).

The GHSs have two further effects that are shared by GHRH: on feeding and on sleep. The effect on feeding is not so surprising in the light of the newly established NPY connection. NPY has one of the strongest appetite-stimulating effects of all the hormones, and inhibition of NPY seems to partly mediate the anorectic effect of the fat cell hormone, leptin (33). GHSs stimulate food intake in rats (34–36); however, no such effect was reported in human studies. In our study on healthy male volunteers, intravenous (iv) administration of high-dose hexarelin caused mild flushing and nausea, and these side effects could be confounding factors in the evaluation (using visual analog scales) of the effect on hunger (29). With subcutaneous administration of different GHRP analogs in rats, a divergence between the GH-releasing effect and the feeding effect was observed with certain analogs, suggesting an independent mechanism of these two effects (36).

The other effect shared with GHRH is on sleep. However, the mechanism of the two peptides to cause sleep is probably different: electroencephalographic (EEG) recordings of healthy volunteers suggest that GHRH stimulates phase 4 sleep, whereas GHRP-6 primarily stimulates phase 2 sleep (37). The mechanism of this effect is unknown.

B. Neuropeptide Y

Recently, a new player has entered the GHS game. Dickson and Luckman reported that 50% of the stimulated neurons in the arcuate nucleus, by far the highest percentage of all the GHS-stimulating neurons according to their experimental technique, are NPY-containing neurons (22). Could NPY have a role in the GHS-stimulated release of GH, ACTH, and prolactin?

Since it was first characterized in 1982, NPY has been described as playing a variety of roles as an integrator of metabolic function (38,39). This neuropeptide, which is widely distributed in the brain, has been demonstrated to be involved in the regulation of food intake, sexual behavior, energy balance, and in the hypothalamic–pituitary system. NPY exerts its diverse actions on at least six different receptor subtypes, denoted as Y_{1-6}. Several fragments and antagonists have been synthesized, with different activity on the different receptor subtypes. For example, it is believed that sites in the COOH-terminal region of NPY are important for the interaction with the Y_1 receptor (40).

Neuropeptide Y has well-known connections to all three axes stimulated by GHSs: NPY colocalizes with GHRH in various hypothalamic cells, although the direct effect of NPY on GHRH release is unknown (41); however, NPY stimulates hypothalamic SS release (42). A possible direct effect on pituitary GH release is controversial. Some papers report a direct stimulatory effect on rat pituitary cells (43), whereas others found inhibition of GH release in human somatotroph tumors (44). Prolonged icv administration of NPY results in decreased GH and insulin-like growth factor-I (IGF-I) levels in the rat (39,45). Further information on the involvement of NPY in the GH axis has been gained from two recent reports on the localization of GH receptors in the arcuate nucleus (46,47). Both papers report that although no GH receptor has been found on GHRH cells, this receptor is expressed by almost all NPY cells; the GH receptor has also been found on SS cells. These data suggest that the negative-feedback effects of GH in the hypothalamus may be by stimulation of SS release and possibly by NPY cell activation. It has been suggested that the effect of NPY on the GH axis is probably through the Y_1 receptor (39). Considering all these data, which suggest mainly an inhibitory effect of NPY on the GH axis, we have wondered: Is that how NPY might be involved in the GH-releasing effect of GHSs?

Neuropeptide Y stimulates prolactin release, probably by the inhibition of the release of dopamine (48), whereas icv administered NPY antibody blocked prolactin secretion (49). This is in line with the stimulating effect of GHSs on prolactin release.

The connection of NPY with CRH and AVP is well characterized. NPY neurons innervate CRH-containing neurons in the paraventricular nucleus (PVN) and stimulate in vitro CRH release from hypothalamic explants (50,51). Direct connections have been shown between NPY neurons from the arcuate nucleus and AVP neurons in the PVN, and several studies have indirectly shown a stimulatory effect of NPY on AVP release from the hypothalamus (52–54). Although no effect of NPY on basal AVP release was reported from the neurointermediate lobe, NPY potentiated potassium-stimulated AVP release in vitro (54). No direct study on hypothalamic AVP release has been published. Recent data, based on studies using different NPY fragments and analogs, suggest, that the effect of icv NPY on the stimulation of plasma ACTH release is through the Y_5 receptor (Meeran K, personal communication, 1997).

Given these reports, we have speculated that some of the effects of GHSs on hypothalamic hormone release might be mediated by NPY. In the present studies, therefore, we have investigated possible direct effects of different GHS analogs on in vitro GHRH, SS, AVP, and CRH release, and the effect of NPY on GHRH and SS release, in an acute rat hypothalamic incubation system.

II. MATERIALS AND METHODS

A. Hypothalamic Incubation

Freshly dissected hypothalamic halves from adult male Wistar rats (200–220 g) were incubated in Earle's balanced salt solution (EBSS, Gibco) supplemented with 0.5% human serum albumin, 100 kIU/mL aprotinin, and 60 µg/mL ascorbic acid, pH 7.4 under 95% O_2 with 5% CO_2. Two half hypothalami from the same animal were incubated in 500 µL incubation solution for the GHRH and SS experiments, in 400 µL for the vasopressin experiments, and 4 half-hypothalami were incubated in 500 µL incubation solution for the CRH experiments. After an 80-min preincubation, a 20-min incubation period with medium alone was given (control period), and the tissue was then exposed to either medium alone (control group) or test solutions containing the different GHRP analogs in various concentrations for one to four 20-min periods. We also tested the effect of the drugs on potassium-stimulated (28 and 40 mM KCl) somatostatin and GHRH. The viability of the system was tested by the effect of 56 mM KCl at the end of each experiment. Several earlier studies by us and by others proved the suitability of this incubation system to study changes in GHRH, SS, AVP, and CRH output by different drugs and hormones (55–64). Media from the incubations were stored at $-20°C$ until assayed for GHRH and SS by enzymoimmunoassay (EIA), and by radioimmunoassay (RIA) for vasopressin and CRH.

B. Immunoassays

The methods for the somatostatin EIA used and the CRH and AVP RIA have been previously published (65–67). A GHRH immunoassay was developed using a two-stage, heterogeneous EIA for rat GHRH. The specificity and cross-reactivity of the rabbit anti-GHRH antibody has been previously assessed in detail (68).

C. GHRP Analogs

The GHRP-6 was purchased from Peninsula (Paisley, Scotland). GHRP-1 and -2 was a gift from Dr. C. Y. Bowers (Tulane University, New Orleans, LA). Hexarelin was a gift from Dr. R. Deghenghi (Europeptides, Argentuile, France), and L-692,429 and L-692,585 were a gift from Dr. R. G. Smith (Merck Sharp & Dohme Research Laboratories, Rahway, NJ).

D. Statistics

Results are expressed as the ratio of treated hormone release to basal release for each animal, and statistical tests were calculated between control animals and treated animals. Data were checked for distribution with the Shapiro–Wilks test. Analysis of variance and Student's t-test were used to calculate statistical significance ($p < 0.05$). Data on the figures are shown as mean $\pm$ standard error; *, **, *** represent $p < 0.05$, $p < 0.01$, and $p < 0.001$ compared with control incubations, respectively.

III. RESULTS

A. The Effect of GHSs

None of the GHRP analogs in the dose range of 10^{-6} to 10^{-4} M had any effect on basal GHRH release even after an 80-min incubation (Figs. 1 and 2). However, at 10^{-3} M concentration a strong inhibitory effect was shown using GHRP-6, hexarelin, and L-692,585 at both 20- and 80-min incubation, whereas the

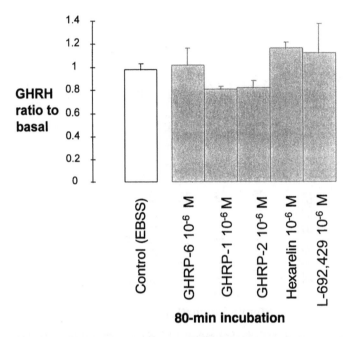

Figure 1   The effect of GHSs on GHRH release at 10^{-6} M concentration after 80 min of incubation ($n = 8$). Data are shown as mean $\pm$ SEM.

less potent L-692,429 showed no significant effect (Figs. 3 and 4). Potassium-stimulated GHRH release was also inhibited by hexarelin 10^{-3} M (Fig. 5).

Neither basal nor potassium-stimulated somatostatin was affected by any of the drugs at 10^{-4} and 10^{-6} M (Fig. 6), whereas at 10^{-3} M, hexarelin and L-692,585 showed a slight, but significant, stimulation of SS release; GHRP-6 10^{-3} M and L-692,429 10^{-3} M showed no effect (Fig. 7). The release of vaso-pressin was dose-dependently stimulated by 10^{-5} to 10^{-3} M concentrations of all the analogs tested (Fig. 8). Changes in CRH release were more variable, but in general, there was no consistent stimulation of CRH by GHSs. Figure 9 shows a representative experiment with no statistically significant changes at 10^{-6} and 10^{-4} M concentrations.

B.   The Effect of NPY on GHRH and SS Release

Neuropeptide Y at 10^{-5}M showed a statistically significant stimulation of SS release and inhibition of GHRH release over 80 min (Figs. 10 and 11).

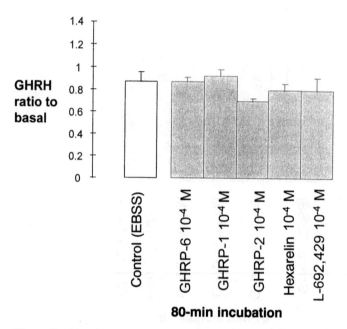

Figure 2 The effect of GHSs on GHRH release at 10^{-4} M concentration ($n = 21-24$).

IV. DISCUSSION

The GHSs neither *stimulated* the release of GHRH nor *inhibited* the release of SS in our in vitro rat hypothalamic incubation system. At millimolar doses they paradoxically inhibited GHRH and stimulated SS release. Neuropeptide Y stimu-

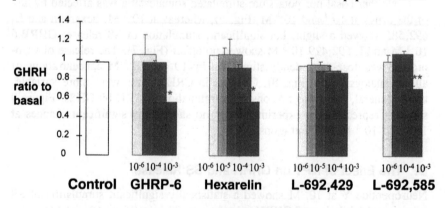

Figure 3 Dose–response curve of GHSs on GHRH release at 20 min incubation ($n = 43-44$).

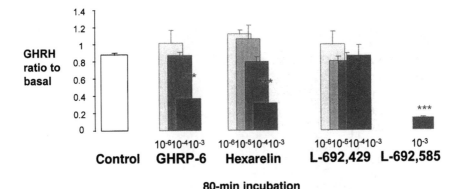

Figure 4 Dose–response curve of GHSs on GHRH release at 80 min incubation ($n = 21$–24).

lated SS release and inhibited GHRH release. In the HPA axis, the analogs consistently stimulated hypothalamic AVP secretion, although showing no consistent stimulation of CRH release.

Studies on electrical activation and c-*fos* expression in hypothalamic cells located in the arcuate nucleus after iv or icv administration of GHSs suggested, along with the in vivo direct GHRH measurement after hexarelin injection in sheep, that GHSs activate GHRH neurons (6,20,23). However, hypophysial portal plasma concentrations of GHRH in sheep after the iv bolus administration of GHRP-6 did not show a coincident release of GHRH, and infusion of GHRP-6 caused no change in GHRH pulse amplitude, but a 50% rise in pulse frequency, suggesting an effect on the frequency of the pulsatile discharge of

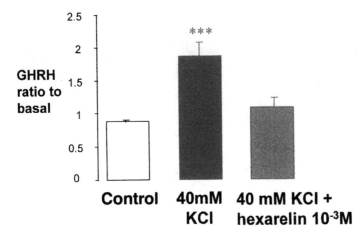

Figure 5 The effect of GHSs on potassium-stimulated GHRH release ($n = 20$–24).

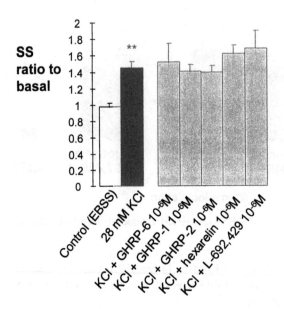

Figure 6   The effect of GHSs on potassium-stimulated SS release ($n = 12$–14).

GHRH (21). Other studies indicated that GHSs inhibit the effect of SS (7,18). However, Bowers and colleagues have long suggested the effects of GHSs on GH could not be explained purely in terms of modulation of GHRH and SS, but required the presence of an unknown endogenous factor to fully explain their

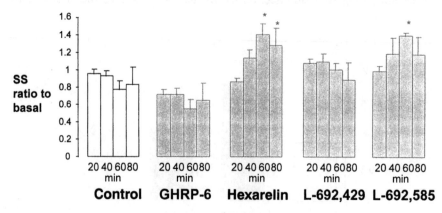

Figure 7   The effect of high-dose GHSs on SS release ($n = 19$–24).

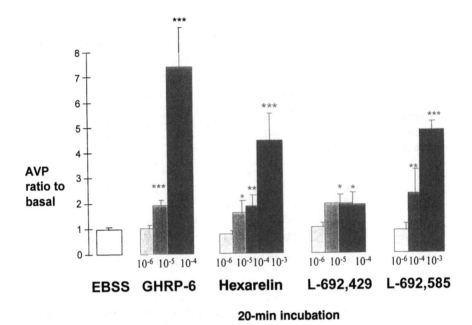

Figure 8   The effect of GHSs on AVP release ($n = 15$–24).

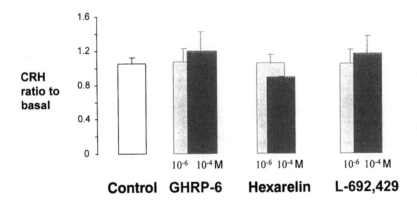

Figure 9   The effect of GHRP-6, hexarelin, and L-692,429 on CRH release ($n = 8$–14).

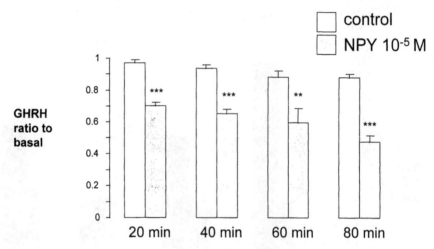

Figure 10 The effect of NYP 10^{-5} M on GHRH release ($n = 4$).

complex effects. Our results are compatible with this speculation. Furthermore, the results of experiments of icv administration of GHRPs on GH levels do not show consistent results. Although in guinea pigs the expected rise was shown (18), in rats a paradoxical decrease of circulating GH levels was observed (19). The dose administered by the icv catheter in the latter study corresponds to a

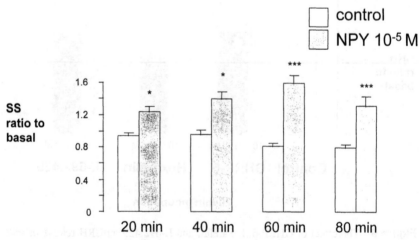

Figure 11 The effect of NYP 10^{-5} M on SS release ($n = 4$).

very high concentration of GHRP-6 in solution (2.3×10^{-3} M). Our finding, that the release of GHRH is inhibited by the presence of similarly high doses of GHS analogues, is compatible with these results in the rat. It is unlikely that our findings are the result of a toxic effect of the high concentration of GHSs: the less potent nonpeptidyl analog L-692,429 showed no change in SS and GHRH release, whereas the others showed inhibition on GHRH and stimulation on SS and AVP release. Some of the analogs were peptides, but the other nonpeptidyl compounds and the analogs were purchased from three different sources, again suggesting that it was unlikely that a common chemical contaminant is responsible for the similar neurosecretory effects. We have found no effect of high concentrations of GHSs in the EBSS solution on immunoassay characteristics. Species differences might explain the difference in guinea pig and rat icv studies. The regulation of the GH axis in the rat is different from that of other species; for example, stress or hypoglycemia causes GH inhibition in rats (58), whereas in guinea pigs or humans, these tests cause stimulation of the GH axis. Thus, certain effects observed in rat experiments cannot be readily extrapolated to other animal species or to humans.

Recent data have suggested that 50% of the neurons expressing c-*fos* mRNA following GHRP-6 administration in the arcuate nucleus are NPY-secreting neurons, and only 23% are GHRH neurons (22). The effect of NPY on GH secretion is inhibitory in various experimental systems (39,42). Although these data do not necessarily imply the involvement of NPY in the GH-releasing effect of GHSs, they are consistent with an effect of high doses of GHSs releasing NPY which, in turn, inhibits GHRH and stimulates SS release.

We have clearly demonstrated that GHSs stimulate AVP release. However, the effect on CRH was variable, suggesting overall a lack of stimulation. In vivo data from the GHRP-6 administration to rats show a variable effect on circulating ACTH release in relation to the preinjection levels of ACTH concentrations (30). Similar factors could explain the variable CRH response to GHSs in our experiments. Alternatively, these effects, stimulation of AVP and no effect or even inhibition of CRH release, may be related to substance P, which is also known to cause stimulation of AVP and inhibition of CRH (69). Substance P antagonists inhibit the effects of GHRP-6 (70).

In conclusion, GHSs neither stimulated the release of GHRH nor inhibited the release of SS in our in vitro rat hypothalamic incubation system. At millimolar doses they paradoxically inhibited GHRH and stimulated SS release. GHSs stimulated hypothalamic AVP secretion while showing no stimulation of CRH release. It is speculated that some of these neuroendocrine effects might be mediated by NPY, but stimulation of GH release may involve other as yet unidentified GH secretagogues.

ACKNOWLEDGMENTS

We thank Dr. C. Y. Bowers (Tulane University, New Orleans, LA) for the gift of GHRP-1 and GHRP-2, Dr. Deghenghi (Europeptides, Argenteuil, France) for the gift of hexarelin and Dr. R. G. Smith (Merck Research Laboratories, Rahway, NJ) for the gift of L-692,429 and L-692,585. We thank Dr. Jennifer Stewart (Virginia Commonwealth University, Richmond, VA) for the gift of the SS antibody and Dr. Millard for the gift of the rat GHRH antibody (University of Florida, Gainesville, FL).

REFERENCES

1. Howard AD, Feighner SD, Cully DF, Liberator PA, Arena JP, Rosenblum CI, Hamelin MJ, Hreniuk DL, Palyha OC, Anderson J, Paress PS, Diaz C, Chou M, Liu K, Kulju McKee K, Pong S, Chaung LY, Elbrecht A, Dashkevicz M, Heavens R, Rigby M, Sirinathsinghji DJS, Dean DC, Melillo DG, Patchett AA, Nargund R, Griffin PR, DeMartino JA, Gupta SK, Schaeffer JM, Smith RG, Van der Ploeg LHT. A receptor in pituitary and hypothalamus that functions in growth hormone release. Science 1996; 273:974–977.
2. Herrington J, Hille B. Growth hormone-releasing hexapeptide elevates intracellular calcium in rat somatotropes by 2 mechanisms. Endocrinology 1994; 135:1100–1108.
3. Cheng K, Chan WW, Butler B, Barreto A Jr, Smith RG. Evidence for a role of protein kinase-C in His-D-Trp-Ala-Trp-D-Phe-Lys-NH$_2$-induced growth hormone release from rat primary pituitary cells. Endocrinology 1991; 129:3337–3342.
4. Lei T, Bucherfelder M, Fahlbush R, Adams EF. Growth hormone-releasing peptide (GHRP-6) stimulates phosphatidylinositol (PI) turnover in human pituitary somatotroph cells. J Mol Endocrinol 1995; 14:135–138.
5. Bowers CY, Sartor AO, Reynolds GA, Badger TM. On the actions of the growth hormone-releasing hexapeptide, GHRP. Endocrinology 1991; 128:2027–2035.
6. Dickson SL, Leng G, Robinson ICAF. Systemic administration of growth hormone-releasing peptide activates hypothalamic arcuate neurons. Neuroscience 1993; 53:303–306.
7. Clark RG, Carlsson L, Trojnar J, Robinson ICAF. The effect of growth hormone-releasing peptide and growth hormone-releasing factor on conscious and anaesthetized rats. J Neuroendocrinol 1989; 1:249–255.
8. Hayashi S, Kaji H, Ohashi S, Abe H, Chihara K. Effect of intravenous administration of growth hormone-releasing peptide on plasma growth hormone in patients with short stature. J Clin Paediatr Endocrinol 1993; 2(suppl 2):69–74.
9. Pombo M, Barreiro J, Penalva A, Peino R, Dieguez C, Casanueva FF. Absence of growth-hormone (GH) secretion after the administration of either GH-releasing hormone (GHRH), GH-releasing peptide (GHRP-6), or GHRH plus GHRP-6 in children with neonatal pituitary-stalk transection. J Clin Endocrinol Metab 1995; 80:3180–3184.

10. Popovic V, Damjanovic S, Micic D, Djurovic M, Dieguez C, Casanueva FF. Blocked growth hormone-releasing peptide (GHRP-6)-induced growth hormone secretion and absence of the synergic action of GHRP-6 plus growth hormone-releasing hormone in patients with hypothalamopituitary disconnection—evidence that GHRP-6 main action is exerted at the hypothalamic level. J Clin Endocrinol Metab 1995; 80:942–947.

11. Cheng K, Chan WW, Barreto A Jr, Convey EM, Smith RG. The synergistic effects of His-D-Trp-Ala-Trp-D-Phe-Lys-NH$_2$ on growth hormone(GH)-releasing factor-stimulated GH release and intracellular adenosine 3'5'-monophosphate accumulation in rat pituitary cell culture. Endocrinology 1989; 124:2791–2798.

12. Blake AD, Smith RG. Desensitization studies using perifused rat pituitary cells show that growth hormone-releasing hormone and His-D-Trp-Ala-Trp-D-Phe-Lys-NH$_2$ stimulate growth hormone release through distinct receptor sites. J Endocrinol 1991; 129:11–19.

13. Bowers CY, Reynolds GA, Durham D, Barrera CM, Pezzoli SS, Thorner MO. Growth hormone (GH)-releasing peptide stimulates GH release in normal men and acts synergistically with GH-releasing hormone. J Clin Endocrinol Metab 1990; 70:975–982.

14. Penalva A, Carballo A, Pombo M, Casanueva FF, Dieguez C. Effect of growth hormone (GH)-releasing hormone (GHRH), atropine, pyridostigmine, or hypoglycemia on GHRP-6-induced GH secretion in man. J Clin Endocrinol Metab 1993; 76:168–171.

15. Mitani M, Kaji H, Abe H, Chiara K. Growth hormone (GH)-releasing peptide and GH-releasing hormone stimulate GH release from subpopulations of somatotrophs in rats. J Neuroendocrinol 1996; 8:825–830.

16. Fletcher TP, Thomas GB, Willoughby JO, Clarke IJ. Constitutive growth hormone secretion in sheep after hypothalamopituitary disconnection and the direct in vivo pituitary effect of growth hormone releasing peptide 6. Neuroendocrinology 1994; 60:76–86.

17. Hickey GJ, Drisko J, Faidley T, Chang C, Anderson LL, Nicholich S, McGuire L, Rickes E, Krupa D, Feeney W, Friscino B, Cunningham P, Fraizer E, Chen H, Laroque P, Smith RG. Mediation by the central nervous system is critical to the in vivo activity of the GH secretagogue L-692,585. J Endocrinol 1994; 148:371–380.

18. Fairhall KM, Mynett A, Robinson ICAF. Central effects of growth hormone-releasing hexapeptide (GHRP-6) on growth-hormone release are inhibited by central somatostatin action. J Endocrinol 1995; 144:555–560.

19. Yagi H, Kaji H, Sato M, Okimura Y, Chihara K. Effects of intravenous or intracerebroventricular injections of His-D-Trp-Ala-Trp-D-Phe-Lys-NH$_2$ on GH release in conscious, freely moving male rats. Neuroendocrinology 1996; 63:198–206.

20. Guillaume V, Magnan E, Cataldi M, Dutuor A, Sauze N, Conte-Delvox B, Oliver C, Lanaerts V, Deghenghi R. GH-releasing hormone secretion is stimulated by a new GH-releasing hexapeptide in sheep. Endocrinology 1994; 135:1073–1076.

21. Fletcher TP, Thomas GB, Clarke IJ. Growth hormone-releasing hormone and somatostatin concentrations in the hypophysial portal blood of conscious sheep

during the infusion of growth hormone-releasing peptide-6. Domest Anim Endocrinol 1996; 13:251–258.

22. Dickson SL, Luckman SM. Induction of c-*fos* messenger ribonucleic acid in neuropeptide Y and growth hormone (GH)-releasing factor neurones in the rat arcuate nucleus following systemic injection of the GH secretagogue, GH-releasing peptide-6. Endocrinology 1997; 138:771–777.

23. Dickson SL, Leng G, Dyball REJ, Smith RG. Central actions of peptide and nonpeptide growth hormone secretagogues in the rat. Neuroendocrinology 1995; 61:36–43.

24. Frawley LS, Neil JD. A reverse hemolytic plaque assay for microscopic visualization of growth hormone release from individual cells: evidence for somatotrope heterogeneity. Neuroendocrinology 1984; 39:484–487.

25. Lin SC, Lin CR, Gukovsky I. Molecular basis of the little mouse phenotype and implications to cell type-specific growth. Nature 1993; 364:208–213.

26. Bowers CY, Momany FA, Reynolds GA, Hong A. On the in vitro and in vivo activity of a new synthetic heptapeptide that acts on the pituitary to specifically release growth hormone. Endocrinology 1984; 114:1537–1545.

27. Ghigo E, Arvat E, Gianotti L, Imbimbo BP, Lenaerts V, Deghenghi R, Camanni F. Growth hormone-releasing activity of hexarelin, a new synthetic hexapeptide, after intravenous, subcutaneous, and oral administration in man. J Clin Endocrinol Metab 1994; 78:693–698.

28. Gertz BJ, Barrett JS, Eisenhandler R, Krupa DA, Wittreich JM, Seibold JR, Schneider SH. Growth hormone response in man to L-692,429, a novel nonpeptide mimic of growth hormone-releasing peptide-6. J Clin Endocrinol Metab 1993; 77:1393–1397.

29. Korbonits M, Trainer PJ, Besser GM. The effect of an opiate antagonist on the hormonal changes induced by hexarelin. Clin Endocrinol (Oxf) 1995; 43:365–371.

30. Thomas GB, Fairhall KM, Robinson ICAF. Activation of the hypothalamo-pituitary-adrenal axis by the growth hormone (GH) secretagogue, GH-releasing peptide-6, in rats. Endocrinology 1997; 138:1585–1591.

31. Korbonits M, Grossman AB. Growth hormone-releasing peptide and its analogues; novel stimuli to growth hormone release. Trends Endocrinol Metab 1995; 6:43–49.

32. Cheng K, Chan WW, Butler B, Wei L, Smith RG. A novel non-peptidyl growth hormone secretagogue. Horm Res 1993; 40:109–115.

33. Mercer JG, Hoggard N, Williams LM, Lawrence CB, Hannah LT, Morgan PJ, Trayhurn P. Coexpression of leptin receptor and preproneuropeptide Y mRNA in arcuate nucleus of mouse hypothalamus. J Neuroendocrinol 1996; 8:733–735.

34. Locke W, Kirgis HD, Bowers CY, Abdoh AA. Intracerebroventricular growth-hormone-releasing peptide-6 stimulates eating without affecting plasma growth-hormone responses in rats. Life Sci 1995; 56:1347–1352.

35. Okada K, Ishii S, Minami S, Sugihara H, Shibasaki T, Wakabayashi I. Intracerebroventricular administration of the growth hormone-releasing peptide KP-102 increases food intake in free-feeding rats. Endocrinology 1996; 137:5155–5158.

36. Locatelli V, Torsello A, Grilli R, Guidi M, Luoni M, Deghenghi R, Müller EE. Growth hormone-releasing peptides stimulate feeding independently of their GH-releasing activity (abstr). Endocrinol Metab 1990; O-063.

37. Frieboes RM, Murck H, Maier P, Schier T, Holsboer F, Steiger A. Growth hormone-releasing peptide-6 stimulates sleep, growth hormone, ACTH and cortisol release in normal man. Neuroendocrinology 1995; 61:584–589.
38. Tatemoto K, Carlquist M, Mutt V. Neuropeptide Y—a novel brain peptide with structural similarities to peptide YY and pancreatic polypeptide. Nature 1982; 296:659–662.
39. Pierroz DD, Catzeflis C, Aebi AC, Rivier JE, Aubert ML. Chronic administration of neuropeptide Y into the lateral ventricle inhibits both the pituitary–testicular axis and growth hormone and insulin-like growth factor I secretion in intact adult male rats. Endocrinology 1996; 137:3–12.
40. Lundberg JM, Modin A, Malström RE. Recent developments with neuropeptide Y receptor antagonists. Trends Pharmacol Sci 1996; 17:301–304.
41. Ciofi P, Croix D, Tramu G. Coexistence of hGHRF and NPY immunoreactivities in neurons of the arcuate nucleus of the rat. Neuroendocrinology 1987; 45:425–428.
42. Rettori V, Milenkovic L, Aguila MC, McCann SM. Physiologically significant effect of neuropeptide Y to suppress growth hormone release by stimulating somatostatin discharge. Endocrinology 1990; 126:2296–2301.
43. Chabot J, Enjalbert A, Pelletier G, Dubois PM, Morel G. Evidence for a direct action of neuropeptide Y in the rat pituitary gland. Neuroendocrinology 1988; 47:511–517.
44. Adams EF, Venetikou MS, Woods CA, Lacoumenta S, Burrin JM. Neuropeptide Y directly inhibits growth hormone secretion by human pituitary somatotropic tumours. Acta Endocrinol (Copenh) 1987; 115:149–154.
45. Catzeflis C, Pierroz DD, Rohner-Jeanrenaud F, Rivier JE, Sizonenko PC, Aubert ML. Neuropeptide Y administered chronically into the lateral ventricle profoundly inhibits both the gonadotropic and the somatotropic axis in intact adult female rats. Endocrinology 1993; 132:224–234.
46. Chan YY, Steiner RA, Clifton DK. Regulation of hypothalamic neuropeptide-Y neurons by growth hormone in the rat. Endocrinology 1996; 137:1319–1325.
47. Kamegai J, Minami S, Sugihara H, Hasegawa O, Higuchi H, Wakabayashi I. Growth hormone receptor gene is expressed in neuropeptide Y neurons in hypothalamic arcuate nucleus of rats. Endocrinology 1996; 137:2109–2112.
48. Harfstrand A, Eneroth P, Agnati L, Fuxe K. Further studies on the effects of central administration of neuropeptide Y on neuroendocrine function in the male rat: relationship to the hypothalamic catecholamines. Regul Pept 1987; 17:169–179.
49. Rettori V, Milenkovic L, Riedel M, McCann SM. Physiological role of neuropeptide Y (NPY) in control of anterior pituitary hormone release in the rat. Endocrinol Exp 1990; 24:37–45.
50. Liposits Z, Sievers L, Paull WK. Neuropeptide-Y and ACTH-immunoreactive innervation of corticotropin releasing factor (CRF)-synthesizing neurons in the hypothalamus of the rat. Histochemistry 1988; 88:227–234.
51. Tsagarakis S, Rees LH, Besser GM, Grossman AB. Neuropeptide-Y stimulates corticotrophin-releasing factor-41release from the rat hypothalamus in vitro. Brain Res 1989; 502:167–170.
52. Iwai C, Ochiai H, Nakai Y. Electron-microscopic immunochemistry of neuropeptide

Y immunoreactive innervation of vasopressin neurons in the paraventricular nucleus of the rat hypothalamus. Acta Anat 1989; 136:279–284.

53. Willoughby JO, Blessing WW. Neuropeptide Y injected into the supraoptic nucleus causes secretion of vasopressin in the unanesthetized rat. Neurosci Lett 1987; 75:17–22.

54. Larsen PJ, Jukes KE, Chowdrey HS, Lightman SL, Jessor DS. Neuropeptide-Y potentiates the secretion of vasopressin from the neurointermediate lobe of the rat pituitary gland. Endocrinology 1994; 134:1635–1639.

55. Berelowitz M, Szabo M, Frohman LA, Firestone S, Chu L, Hintz HL. Somatomedin-C mediates growth hormone negative feedback by effects on both the hypothalamus and the pituitary. Science 1981; 212:1279–1281.

56. Berelowitz M, Firestone S, Frohman LA. Effects of growth hormone excess and deficiency on hypothalamic somatostatin content and release and on somatostatin distribution. Endocrinology 1981; 109:714–719.

57. Berelowitz M, Dudlak D, Frohman LA. Release of somatostatin-like immunoreactivity from incubated rat hypothalamus and cerebral cortex. Effects of glucose and glucoregulatory hormones. J Clin Invest 1982; 69:1293–1301.

58. Lengyel AJ, Grossman AB, Niewenhuyzen-Kruseman AC, Ackland J, Rees LH, Besser GM. Glucose modulation of somatostatin and LHRH release from rat hypothalamic fragments in vitro. Neuroendocrinology 1984; 39:31–38.

59. Aguila MC, McCann SM. Stimulation of somatostatin release in vitro by synthetic human growth hormone-releasing factor by a nondopaminergic mechanism. Endocrinology 1985; 117:762–765.

60. Lengyel AJ, Grossman AB, Bouloux G, Rees LH, Besser GM. Effects of dopamine and morphine on immunoreactive somatostatin and LH-releasing hormone secretion from hypothalamic fragments in vitro. J Endocrinol 1985; 106:317–322.

61. Shibasaki T, Yamauchi N, Hotta M, Masuda A, Imaki T, Demura H, Ling N, Shizume K. In vitro release of growth hormone-releasing factor from rat hypothalamus: effect of insulin-like growth factor-1. Regul Pept 1986; 15:47–53.

62. Tsagarakis S, Ge F, Rees LH, Besser GM, Grossman AB. Stimulation of alpha-adrenoreceptors facilitates the release of growth hormone-releasing hormone from rat hypothalamus in vitro. J Neuroendocrinol 1989; 1:129–133.

63. Honegger J, Spagnoli A, D'Urso R, Navarra P, Tsagarakis S, Besser GM, Grossman AB. Interleukin-1β modulates the acute release of growth hormone-releasing hormone and somatostatin from rat hypothalamus in vitro, whereas tumor necrosis factor and interleukin-6 have no effect. Endocrinology 1991; 129:1275–1282.

64. Aguila MC, Boggaram V, McCann SM. Insulin-like growth factor I modulates hypothalamic somatostatin through a growth hormone releasing factor increased somatostatin release and messenger ribonucleic acid levels. Brain Res 1993; 625:213–218.

65. Korbonits M, Little JA, Camacho-Hübner C, Trainer PJ, Besser GM, Grossman AB. Insulin-like growth factor-I and -II in combination inhibit the release of growth hormone-releasing hormone from the rat hypothalamus in vitro. Growth Regul 1996; 6:110–120.

66. Cunnah D, Jessop DS, Besser GM, Rees LH. Measurement of circulating corticotrophin-releasing factor in man. J Endocrinol 1987; 113:123–131.
67. Forsling ML. Measurement of vasopressin in body fluids. In: Baylis PH, Padfield PL, eds. Posterior Pituitary. New York: Marcel Dekker, 1985:161–192.
68. Cugini CD, Millard WJ, Leidy JW Jr. Signal transduction systems in growth hormone-releasing hormone and somatostatin release from perfused rat hypothalamic fragments. Endocrinology 1991; 129:1355–1362.
69. Jessop DS, Chowdrey HS, Larsen PJ, Lightman SL. Substance P; multifunctioning peptide in the hypothalamo-pituitary system? J Endocrinol 1992; 132:331–337.
70. Bitar KG, Bowers CY, Coy DH. Effect of substance P/bombesin antagonists on the release of growth hormone by GHRP and GHRH. Biochem Biophys Res Commun 1991; 180:156–161.

17
Immunocytochemical Studies Investigating the Central Actions of the Nonpeptidyl Growth Hormone Secretagogue, MK-0677

Alex R. T. Bailey and Gareth Leng
University Medical School, Edinburgh, Scotland

Roy G. Smith
Merck Research Laboratories, Merck and Co., Inc., Rahway, New Jersey

I. INTRODUCTION

Over 4 years ago it was reported that, in addition to eliciting growth hormone (GH) release from the anterior pituitary gland, GHRP-6, specifically induced Fos-like immunoreactivity (Fos-LI) within the rat hypothalamic arcuate nucleus, in the region of the nucleus thought to contain growth hormone-releasing hormone (GHRH) neurons (1). As the expression of Fos-LI, and its progenitor c-*fos*, have been used extensively within neuroendocrine systems to map cellular activation, this observation was taken to indicate that GHRP-6 acts centrally to increase the neuronal activity of arcuate nucleus GHRH neurons. Subsequently, colocalization studies have demonstrated that GHRP-6 induces c-*fos* mRNA in an appreciable proportion of arcuate nucleus GHRH neurons (2). Similarly, systemic injection of GHRP-6 increases the electrical activity of identified neuroendocrine arcuate nucleus neurons and stimulates GHRH release into the hypophysial portal blood (3,4). The possibility that GHRP-6 could influence GH release at both central and peripheral sites gave this secretagogue an enormous therapeutic potential. This potential was tempered by the fact that, as a peptide, GHRP-6 possesses a limited oral bioavailability, and this led to the generation of a range of orally available GH secretagogues (GHSs). Initial attempts at GH secretagogue generation utilized a benzolactam core as a structural template (e.g., L-692,429) and, although these compounds possessed GH secretory activity, their oral efficacy was not greatly enhanced. However, a new structural class of compounds, based on a spiroindoline core, has both GH secretory activity and oral bioavailability. Structural refinement of these compounds has led to the synthesis of the potent, orally active GH secretagogue, MK-0677 (5).

MK-0677 stimulates the release of GH in a rat pituitary cell assay, with an intrinsic activity greater than that of either L-692,429 or GHRP-6 (6). Furthermore, as maximal MK-0677-induced GH release cannot be potentiated by GHRP-6, both GH secretagogues apparently act at the same pituitary receptor (6). However, pituitary cells desensitized to GHRH will release GH after the introduction of GHRP-6 or MK-0677 to the perfusion medium (6), demonstrating that GHRP-6 and MK-0677 act at a different receptor than GHRH. In vivo, MK-0677 has a dose-dependent effect on GH release, with an oral bioavailability of more than 60%, a property ascribed to its lipophilicity and peptidase stability (5) and, although GHRH and MK-0677 are equally efficient at inducing GH release from pituitary cell assays, MK-0677 is more potent in vivo, possibly reflecting an additional nonpituitary site of action. Further evidence for a central site of action has come from recent studies showing specific binding sites for MK-0677 within the ventrolateral regions of the arcuate nucleus (7), although the identity of the endogenous ligand for this receptor remains to be determined.

II. HYPOTHALAMIC ACTIONS OF MK-0677

By using the immunocytochemical localization of Fos-LI as an indicator of cellular activation, we initially set out to determine whether MK-0677 exerted any central actions. In conscious male rats, intravenous (iv) injection of MK-0677 induced a significant increase in Fos-LI in the hypothalamic arcuate nucleus, indicating that this secretagogue can activate a population of arcuate nucleus neurons (Figs. 1,2). A similar amount, and distribution, of Fos-LI was observed in the arcuate nucleus after intracerebroventricular (icv) injection of MK-0677, showing that this secretagogue can act centrally to activate arcuate nucleus neurons (see Fig. 1). Systemic injection of GHRH did not induce Fos-LI in the arcuate nucleus, demonstrating that the observed activation was not secondary to secretagogue-induced GH release. After both central and systemic injection of MK-0677 the vast majority of Fos-positive cells were located in the ventromedial regions of the nucleus in close proximity to the median eminence (see Fig. 2). This distribution is not entirely coincident with the generally accepted location of GHRH neurons, which are thought to be situated in the ventrolateral regions of this nucleus. However, it is becoming increasingly appar-

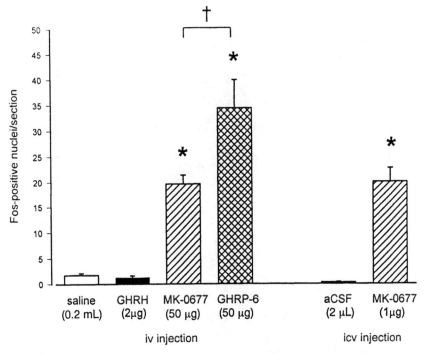

Figure 1   The number of Fos-positive cells in the arcuate nucleus of conscious male rats injected systemically with saline (0.2 mL, $n = 10$), GHRH (2 μg, $n = 6$), GHRP-6 (50 μg, $n = 6$) or MK-0677 (50 μg, $n = 11$) and intracerebroventricularly with artificial cerebrospinal fluid (aCSF, 5 μL, $n = 8$) or MK-0677 (1 μg in 5 μL, $n = 7$). *$p <$ 0.05 compared with vehicle control; †$p < 0.05$ compared with GHRP-6 injection.

ent that GH secretagogues are not necessarily specific in their actions; for example, in situ hybridization studies have shown that, in addition to GHRH neurons, GH secretagogues also activate neuropeptide Y (NPY) arcuate nucleus nucleus neurons (2,8), and such studies may determine the extent to which MK-0677 activates non-GHRH arcuate neurons. Interestingly, the quantity of arcuate nucleus activation induced by MK-0677 was significantly less than that seen after a similar dose of GHRP-6, yet, at the level of the anterior pituitary, MK-0677 is the more potent secretagogue (5). This disparity is unlikely to reflect limited access of the nonpeptide to the CNS because the level of MK-0677-induced arcuate nucleus activation is the same whether the secretagogue is given centrally or systemically. Thus, it is more plausible that MK-0677 has a greater specificity within the hypothalamus. In addition to inducing Fos-LI in GHRH neurons, GHRP-6 also activates several arcuate nucleus NPY neurons (2), which may underlie the increase in feeding observed in rats given this secretagogue

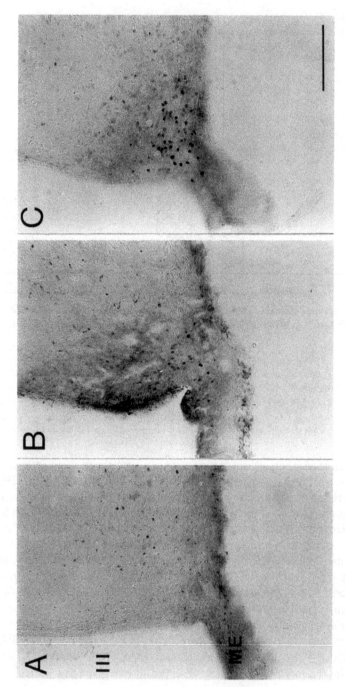

Figure 2 Photomicrographs showing Fos-like immunoreactivity in the rat hypothalamic arcuate nucleus following systemic injection of (A) saline (0.2 mL); (B) GHRH (2 μg); (C) MK-0677 (50 μg). III, third ventricle; ME, median eminence. Scale bar: 0.2 mm.

(9), as central injections of NPY stimulate food intake (10). Unlike most other GH secretagogues, MK-0677 does not alter feeding behavior (9), which may indicate that it does not activate NPY neurons to the same extent as GHRP-6.

A. Other Hypothalamic Actions of MK-0677

While investigating the central actions of MK-0677, we noticed that in some conscious animals this secretagogue activated a small number of cells in the supraoptic nucleus (SON). Although the extent of activation was small in comparison with other stimuli, this observation was unexpected, for neurons in this nucleus project almost exclusively to the posterior pituitary gland where they release oxytocin (OT) or vasopressin (AVP) into the circulation. Additionally, the central actions of other GH secretagogues have been restricted solely to the arcuate nucleus (1). Further studies showed that, in anesthetized rats, systemic injection of MK-0677 had no effect on plasma levels of OT, as measured by radioimmunoassay, or on the firing rate of antidromically-identified SON OT neurons. However, the neuronal activity of some phasically firing AVP neurones did appear to be decreased following MK-0677 injection. Because the induction of Fos-LI is generally considered an indicator of increased neuronal activity, it is difficult to reconcile the immunocytochemical and electrophysiological results, although the electrophysiology was performed in anesthetized rats, which may mask some of the central consequences of systemic MK-0677 administration.

III. NONHYPOTHALAMIC ACTIONS OF MK-0667

The central nervous system (CNS) is enclosed within the blood–brain barrier (BBB), which prevents circulating moieties (e.g., pH, osmolarity, hormones) from potentially damaging direct interactions with the brain. However, the CNS is able to monitor the ambient levels of these circulating factors through regions that lack a complete BBB. These regions are termed circumventricular organs (CVOs) and consist of the subfornical organ (SFO), subcomissural organ (SCO), organosum of the lamina terminalis (OVLT), median eminence and the neurohypophysis in the forebrain, and the area postrema (AP) in the brain stem. Several CVOs are essential in relaying blood-borne information to hypothalamic nuclei involved in peptide release (11,12). Therefore, CVOs are a potential site at which GH secretagogues may exert their central actions.

With the exception of the arcuate nucleus, GH secretagogues do not induce Fos protein expression in any region of the forebrain (1), even though previous studies have shown that Fos induction can be used as a reliable indicator of cellular activation within these regions (11), thereby raising the question of how these secretagogues gain access to the CNS. One region of the brain not systemically investigated was the brain stem and the resident CVO, the AP. The AP is situated in the dorsal aspect of the brain stem overlying the nucleus tractus solitarius (NTS) and is involved in the emetic reflex. However, with the discovery of several classes of peptidergic receptor throughout this structure, the AP has also been ascribed roles in autonomic function, cardiovascular regulation, and food intake (13,14). Because the AP and NTS are involved in the regulation of several hypothalamic neurosecretory neurons we decided to examine whether these regions are involved in the central response to systemic GH secretagogue administration.

Our studies have revealed that systemic administration of GHRP-6, at the same dose known to activate arcuate nucleus neurons, induced Fos-LI within the AP and also in the dorsal aspect of the underlying NTS (Fig. 3). Similarly, systemic injection of MK-0677 also induced a similar number of cells to express Fos-LI in both these regions. At the level of obex Fos-LI was distributed evenly throughout the AP, but in the more rostral aspects of this CVO, Fos-LI was localized to the medial regions adjacent to the walls of the fourth ventricle. That this response was due to a direct secretagogue action, and not to the secondary effects of GH release, was confirmed when no Fos-LI was observed in the AP and NTS in response to systemic GHRH administration (Fig. 4). Furthermore, central injection of MK-0677, activated arcuate nucleus cells, but did not induce Fos-LI in the brain stem, suggesting that this response was a direct result of the blood-borne secretagogue and was not secondary to any of its central actions.

Release of GH is under the negative influence of somatostatin produced primarily in the periventricular nucleus (PeVN) of the hypothalamus. The PeVN has somatostatinergic projections to the arcuate nucleus, and electrical stimulation in this region decreases the neuronal activity of a population of GHRH neurosecretory arcuate nucleus neurons that are excited by GHRP-6 (15). Similarly, somatostatin decreases the activity of arcuate nucleus neurons excited by GHRP-6 in vitro and in vivo (see Chap. 12). Given that somatostatin seems to have inhibitory actions in the arcuate nucleus, we investigated whether stimulation of somatostain receptors would block MK-0677-induced activation of arcuate nucleus neurons. We found that in conscious male rats, systemic injection of a long-lasting somatostatin analog, Sandostatin, consistently attenuated the excitatory effects of MK-0677 in the arcuate nucleus (Fig. 5). Somatostatin receptors are distributed throughout the CNS, and the exact location of the receptors mediating this inhibitory effect remains to be established. To determine whether this effect occurs at the level of the brain stem, we examined the effect of Sandostatin administration on MK-0677-induced activation of the AP. MK-0677 consistently induced Fos-LI in the arcuate nucleus and the AP but, although prior administration of Sandostatin attenuated arcuate Fos-LI, it had

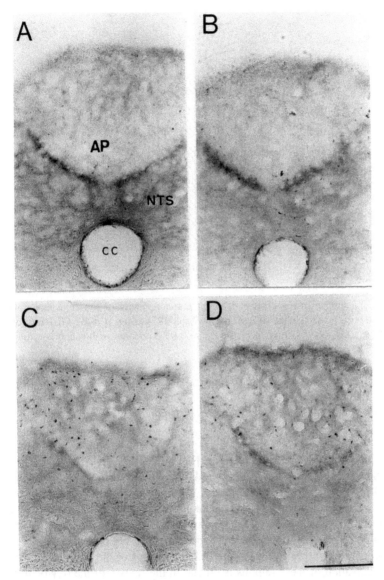

Figure 3 Photomicrographs showing Fos-like immunoreactivity in the area postrema (AP) and nucleus tractus solitarius (NTS) of conscious male rats injected systemically with (A) saline (0.2 mL), (B) GHRH (2 μg); (C) GHRP-6 (50 μg), (D) MK-0677 (50 μg). CC, central canal. Scale bar: 0.2 mm.

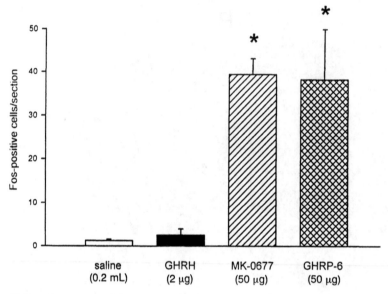

Figure 4 The number of Fos-positive cells in the area postrema of conscious male rats injected systemically with saline (0.2 mL, $n = 7$), GHRH (2 μg, $n = 6$), GHRP-6 (50 μg, $n = 7$), or MK-0677 (50 μg, $n = 8$). $*p < 0.05$ compared with saline controls.

no effect on the level of expression in the AP (see Fig. 5), suggesting that the inhibitory actions of Sandostatin on MK-0677-induced arcuate nucleus activation are not mediated at the level of the brain stem.

IV. A POSSIBLE ROLE FOR NOREPINEPHRINE IN THE CENTRAL RESPONSE TO GH SECRETAGOGUES

Studies of hypothalamic peptidergic systems have indicated that the NTS has a pivotal role in the integration of sensory and visceral information and subsequent peptide release (16). A major cell type that subserves these functions are the noradrenergic neurons in the A1 and A2 cell groups of the brain stem. For example, activation of these regions is crucial in the release of the oxytocin in response to various stimuli, such as parturition and cholecystokinin (CCK) administration (17,18). Furthermore, there are extensive noradrenergic projections from these brain stem regions to specific hypothalamic nuclei, including the arcuate nucleus (19). Noradrenergic tone directly influences GH release from the anterior pituitary gland with α-2-adrenoceptors mediating excitatory responses and α-1-and β-adrenoceptors mediating inhibitory actions (20). Addi-

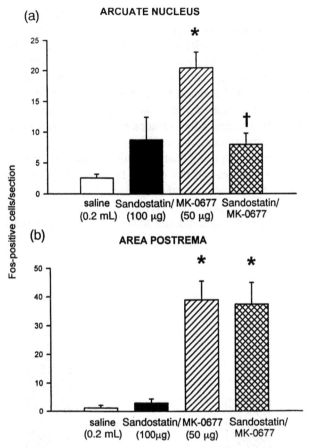

Figure 5 (a) The mean number of Fos-positive cells in the arcuate nucleus of conscious male rats injected with saline (0.2 mL, $n = 6$), Sandostatin (100 µg, $n = 6$), MK-0677 (50 µg, $n = 7$), or Sandostatin (100 µg) followed by MK-0677 (50 µg, $n = 7$). *$p < 0.05$ compared with saline-treated controls; †$p < 0.05$ compared with MK-0677-treated group. (b) The mean number of Fos-positive cells in the area postrema of conscious male rats injected with saline (0.2 mL, $n = 6$), Sandostatin (100 µg, $n = 6$), MK-0677 (50 µg, $n = 7$), or Sandostatin (100 µg) followed by MK-0677 (50 µg, $n = 7$). *$p < 0.05$ compared with saline-treated controls.

tionally, norepinephrine (NE) directly influences the activity of arcuate nucleus GHRH neurons and, thereby, GH release (21). Given the accumulating evidence for noradrenergic modulation of arcuate nucleus neuronal activity and the role of NE in regulating GH release, we examined the possibility that NE may be involved in GH secretagogue-induced activation of the arcuate nucleus.

Central NE levels were depleted by icv injection of the specific NE neurotoxin, 5-ADMP (Upjohn Ltd.), at a dose previously shown to significantly reduce hypothalamic NE concentrations (18); control animals were injected with vehicle. The extent of NE depletion was evaluated by immunocytochemical detection of hypothalamic dopamine β-hydroxylase (DBH). Five days after 5-ADMP injection, the extent of DBH staining within the arcuate nucleus was significantly reduced (by 65%) compared with control animals. Despite this marked reduction in NE levels, systemic administration of the GH secretagogue L-192,585, still induced the same amount of arcuate nucleus Fos-LI as seen in control animals (Fig. 6). Subsequently, using double-labeling techniques we have shown the population of brain stem neurons that display Fos-LI after GH secretagogue administration does not include cells that are immunoreactive for tyrosine hydroxylase. Therefore, it appears that NA does not have any direct modulatory influences on GH secretagogue-induced activation of arcuate nucleus neurons.

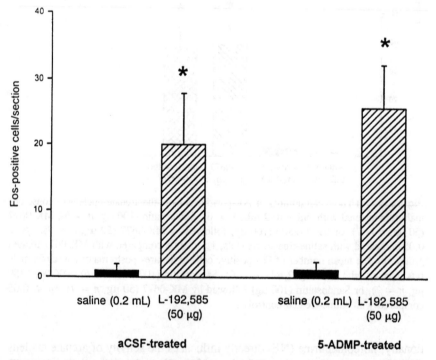

Figure 6 The mean number of Fos-positive cells in the arcuate nucleus of conscious male rats injected systemically with saline (0.2 mL, $n = 13$) or L-192,585 (50 µg, $n = 14$) 5 days after a central injection of aCSF (10 µL) or 5-ADMP (100 µg in 10 µL). *$p < 0.05$ compared with saline controls.

V. SUMMARY

The nonpeptide GH secretagogue MK-0677, can cross the blood–brain barrier and exert a specific action within the hypothalamic arcuate nucleus to activate neurons involved in the release of GHRH. In addition, MK-0677 also activates cells in the AP and NTS of the brain stem, possibly reflecting a mechanism by which this secretagogue gains access to the CNS to influence GH secretion. However, despite dense noradrenergic projections from the brain stem to the hypothalamus and the role NE plays in regulating GH release, GH secretagogues exert their central actions independently of this neurotransmitter. Similarly, the brain stem does not appear to be involved in the attenuation of MK-0677-induced activation of the arcuate nucleus by somatostatin receptor stimulation.

ACKNOWLEDGMENTS

This work is supported by Merck Research Laboratories. The authors would like to thank Dr. Colin H. Brown for reading and commenting on the manuscript.

REFERENCES

1. Dickson SL, Leng G, Robinson ICAF. Systemic administration of growth hormone-releasing peptide (GHRP-6) activates hypothalamic arcuate neurones. Neuroscience 1993; 53:303–306.
2. Dickson SL, Luckman SM. Induction of c-*fos* messenger ribonucleic acid in neuropeptide Y and growth hormone (GH)-releasing factor neurones in the rat arcuate nucleus following systemic injection of the GH secretagogue, GH-releasing peptide-6. Endocrinology 1997; 138:771–777.
3. Dickson SL, Leng G, Dyball REJ, Smith RG. Central actions of peptide and nonpeptide growth hormone secretagogues in the rat. Neuroendocrinology 1995; 61:36–43.
4. Clark RG, Carlsson LMS, Trojnar J, Robinson ICAF. The effects of a growth hormone-releasing peptide and growth hormone-releasing factor in conscious and anaesthetized rats. Neuroendocrinology 1989; 1:249–255.
5. Patchett AA, Nargund RP, Tata JR, Chen MH, Barakat KJ, Johnston DBR, Cheng K, Chan WWS, Butler B, Hickey G, Jacks T, Schleim K, Pong S-S, Chaung LYP, Chen HY, Frazier E, Leung KH, Chiu SHL, Smith RG. Design and biological-activities of L-163,191 (MK-0677)—a potent, orally-active growth-hormone secretagogue. Proc Natl Acad Sci USA 1995; 92:7001–7005.
6. Smith RG, Cheng K, Pong S-S, Leonard R, Cohen CJ, Arena JP, Hickey GJ, Chang CH, Jacks T, Drisko J, Robinson ICAF, Dickson SL, Leng G. Mechanism of action of GHRP-6 and non-peptidyl growth hormone secretagogues. In: Bercu

BB, Walker RE, eds. Growth Hormone Secretagogues. New York: Springer-Verlag, 1996:147–163.

7. Howard AD, Feighner SD, Cully DF, Arena JP, Liberator PA, Rosenblum CI, Hamelin M, Hrenluk DL, Palyha OC, Anderson J, Paress PS, Diaz C, Chou M, Liu KK, McKee KK, Pong S-S, Chaung L-Y, Elbrecht A, Dashkevicz M, Heavens R, Rigby M, Sirinathsinghji DJS, Dean DC, Melillo DG, Patchett AA, Nargund R, Griffin PR, DeMartino JA, Gupta SK, Schaeffer JM, Smith RG, Van der Ploeg LHT. A pituitary gland and hypothalamic receptor that functions in growth hormone release. Science 1996; 273:974–977.

8. Kamegai J, Hasegawa O, Minami S, Sugihara H, Wakabayashi I. The growth hormone-releasing peptide KP-102 induces c-*fos* expression in the arcuate nucleus. Mol Brain Res 1996; 39:153–159.

9. Locatelli V, Torsello A, Grilli R, Guidi M, Luoini M Deghenghi R, Muller EE. Growth hormone-releasing peptides stimulate feeding independently of their GH releasing activity. Endocrinol Metab 1997; 4:33.

10. Clark JT, Kalra PS, Crowley, WR, Kalra SP. Neuropeptide Y and human pancreatic polypeptide stimulate feeding behaviour in rats. Endocrinology 1984; 115:427–429.

11. Hamamura M, Nunez DJ, Leng G, Emson PC, Kiyama H. c-*fos* may code for a common transcription factor within the hypothalamic neural circuits involved in osmoregulation. Brain Res 1992; 572:42–51.

12. Richard D, Bourque CW. Synaptic activation of rat supraoptic neurons by osmotic stimulation of the organum vasculosum lamina terminalis. Neuroendocrinology 1992; 55:609–611.

13. Ferguson AV, Bains, JS. Electrophysiology of the circumventricular organs. Front Neuroendocrinol 1996; 17:440–475.

14. Johnson AK, Gross PM. Sensory circumventricular organs and brain homeostatic pathways. FASEB 1993; 7:678–686.

15. Dickson SL, Leng G, Robinson IC. Electrical stimulation of the rat periventricular nucleus influences the activity of hypothalamic arcuate neurones. J Neuroendocrinol 1994; 6:359–367.

16. Sawchenko PE, Swanson LW. Central noradrenergic pathways for the integration of hypothalamic neuroendocrine and autonomic responses. Science 1981; 214:685–687.

17. Luckman SM, Antonijevic I. Morphine blocks the induction of Fos immunoreactivity in hypothalamic magnocellular neurons at parturition, but not that induced in other brain regions. Ann NY Acad Sci 1993; 689:630–631.

18. Onaka T, Luckman SM, Antonijevic I, Palmer JR, Leng G. Involvement of the noradrenergic afferents from the nucleus tractus solitarius to the supraoptic nucleus in oxytocin release after peripheral cholecystokinin octapeptide in the rat. Neuroscience 1995; 66:403–412.

19. Sato A, Shioda S, Nakai Y. Catecholaminergic innervation of GRF-containing neurons in the rat hypothalamus revealed by electron-microscopic cytochemistry. Cell Tissue Res 1989; 258:31–34.

20. Willoughby JO, Chapman IM, Kapoor R. Local hypothalamic adrenoceptor activation in rat: alpha 1 inhibits and alpha 2 stimulates growth hormone secretion. Neuroendocrinology. 1993; 57:687–692.
21. Makara GB, Kiem DT, Vizi ES. Hypothalamic alpha-2A adrenoceptors stimulate growth hormone release in the rat. Eur J Pharmacol 1995; 287:43–48.

18

Neurochemical Identification of the Cells Expressing c-*fos* mRNA in the Rat Arcuate Nucleus Following Systemic Growth Hormone-Releasing Peptide-6 Injection

Suzanne L. Dickson
University of Cambridge, Cambridge, England

Simon M. Luckman
The Babraham Institute, Cambridge, England

I. GROWTH HORMONE SECRETAGOGUES ARE CENTRALLY ACTIVE COMPOUNDS

In the early 1980s, growth hormone (GH)-releasing peptide (GHRP-6) was emerging as the first identified number of a class of synthetic GH secretagogues (GHSs), the actions of which were selective and potent for GH secretion (1). GHRP-6 stimulates GH secretion from pituitary cells in vitro, providing evidence for a direct pituitary action (1) and the GH-releasing mechanism appears to differ from that of the endogenous GH-releasing factor (GRF; 2,3). By the late 1980s–early 1990s there was a great deal of circumstantial evidence that GHRP-6 may also act within the central nervous system (CNS) to stimulate GH secretion. For example, when administered together, the effects of GRF and GHRP-6 on GH secretion were merely additive in vitro (in pituitary perfusion experiments and in cell culture), but displayed an enormous synergy in vivo (4); these differences could not be explained by an interaction of these peptides at the pituitary level alone. Rather, it was suggested that the central actions of GHRP-6 may include the release of an unknown hypothalamic-releasing factor (a "U-factor") into the portal blood; according to this hypothesis, the U-factor would act together with GRF to stimulate GH secretion from the pituitary (4). That GHRP-6 may act centrally to alter the output of the hypothalamic GRF–somatostatin GH pulse generator was first suggested by Clark and colleagues (5). They argued that part of the GH-releasing mechanism of GHRP-6 is likely to include increased GRF release, because the GHRP-6–induced GH response was attenuated in rats passively immunized with GRF antiserum. Furthermore, they suggested that GHRP-6 may alter somatostatin secretion, because it disrupted the cyclic changes in GH release following regular injections of GRF (a response that has been attributed to cyclic changes in somatostatin secretion).

In 1993, we provided the first direct evidence that GHRP-6 activates a subpopulation of cells in the hypothalamic arcuate nucleus (6). In these experiments, we used two complementary approaches to detect activation of cells by GHRP-6. First, we observed increased electrical activity in a subpopulation of cells in this region, recorded in vivo with extracellular electrodes. Second, we demonstrated that a subpopulation of arcuate cells express Fos protein following an intravenous (iv) injection of GHRP-6. (Fos, the product of the immediate-early gene, c-fos, is expressed in many neuronal systems following activation, and it can be readily detected in cell nuclei using standard immunocytochemical techniques.) It seems likely that GHRP-6 acts directly within the central nervous system to induce expression of Fos protein in the arcuate nucleus because an intracerebroventricular (icv) injection of a low dose of GHRP-6 (0.1 µg) was as effective as an iv injection of 50 µg GHRP-6 for eliciting this response (7). More recently, we confirmed that GHRP-6 acts directly within the arcuate nucleus to activate cells in this region, for we observed changes in electrical activity of arcuate cells recorded in an in vitro hypothalamic slice preparation, from which all but closely adjacent inputs to the arcuate nucleus had been removed (8).

Although it now seems clear that GHRP-6 acts directly within the arcuate nucleus to activate cells in this region, it may not be the only site of its action within the central nervous system. The arcuate nucleus is the only forebrain structure to show an increase in expression of Fos protein following GHRP-6 injection; however, we have also observed a small degree of activation of cells in the brain stem, in the area postrema, and in cells in close proximity to the nucleus tractus solitarii (9; see Chap. 17). It remains to be determined whether these cells influence GH secretion and whether they form part of an afferent projection to the hypothalamus or, indeed, to higher brain structures.

One of the key questions about the central GH-releasing mechanism of GHRP-6 is whether the cells activated include the GRF-containing neurons. Certainly, the distribution of the arcuate cells expressing Fos protein following GHRP-6 injection is consistent with this hypothesis, for they display a degree of overlap in distribution with the GRF-containing neurons, notably in the ventrolateral portion (6). Furthermore, many of the arcuate cells excited by

GHRP-6 in our our electrophysiological studies also fulfilled multiple criteria for identification as GRF neurons (6,10). More direct evidence is suggested by studies in sheep, demonstrating increased GRF concentrations in portal blood following systemic injection of the hexarelin,* another peptidyl GH secretagogue (11).

It is unlikely that the sole action of GHRP-6 within the CNS is the activation of GRF neurons in view of the large synergistic release of GH when GHRP-6 and GRF are administered concomitantly (4). Both our electrophysiological studies and the functional mapping studies with Fos immunocytochemistry have led us to speculate that the cells activated by GHRP-6 may be a heterogeneous population. Notably, in GHRP-6-injected rats, there is a large cluster of Fos-positive cells in the ventromedial aspect of the arcuate nucleus (6), where few GRF cells have been described (12,13).

II. GHRP-6 ACTIVATES NEUROSECRETORY CELLS IN THE ARCUATE NUCLEUS

In our electrophysiological studies (in vivo) we demonstrated that the arcuate cells excited by GHRP-6 include cells that were antidromically identified as projecting to the median eminence—and, therefore, are likely to be neurosecretory cells—(6,7; see Chap. 12). Following GHRP-6 injection, the predominant response recorded at the cell bodies of these antidromically identified arcuate cell groups was excitatory, whereas the predominant response of cells not fulfilling criteria for antidromic activation (most of which are unlikely to be neurosecretory cells) was inhibitory (7). The diversity of electrophysiological responses recorded in these studies provided evidence that the arcuate cells activated by GHRP-6 include both neurosecretory and non-neurosecretory cells. This hypothesis was supported by a study in which we determined whether arcuate cells expressing Fos protein following GHRP-6 injection were retrogradely labeled following systemic injection of the retrograde tracer Fluorogold (14). When injected by this route, Fluorogold does not cross the blood–brain barrier and is not transported synaptically, rather it is taken up by neurones that project to areas supplied by fenestrated capillaries or to the periphery (15). Thus, retrogradely labeled cells were identified in regions of the hypothalamus containing neurohypophysiotropic neurons (the paraventricular nucleus and the supraoptic nucleus) and in regions containing adenohypophysiotropic neurons (including the arcuate nucleus, the periventricular nucleus, and anterior hypothalamic areas). In GHRP-6 injected rats, which had been pretreated with

---

*Hexarelin is a trademark for examorelin (I.N.N.).

Fluorogold, we were able to identify single cells in the arcuate nucleus that were activated by GHRP-6 (from the detection of Fos protein in the cell nucleus) and that were likely to be neurosecretory cells (because the cell cytoplasm contained Fluorogold; Fig. 1). Most cells expressing Fos protein following GHRP-6 injection (68–82%) were identified as projecting outside the blood–brain barrier (and, therefore, are likely to be neurosecretory cells), leaving a further 18–32% for which the absence of Fluorogold suggests that these cells are unlikely to be neurosecretory cells.

Given that most arcuate cells activated by GHRP-6 appear to be neurosecretory, it is likely that these cells will include either or both of the two principal neurosecretory populations in this region, the GRF or the dopaminergic (tyrosine hydroxylase-containing) groups (see 16, 17, and references therein). These neurons overlap in distribution, particularly in the ventrolateral portion of the arcuate nucleus (18). Most other neurosecretory peptides and transmitters synthesized in the arcuate nucleus are colocalized in GRF or tyrosine hydroxylase-containing cells. Of the possible non-neurosecretory populations of cells in the arcuate nucleus that may be a target for GHRP-6 action are notably the pro-opiomelanocorticotropin (POMC)- and the somatostatin-containing cells (16,17).

Another major population of cells present in the arcuate nucleus is the neuropeptide Y (NPY)-containing cells: these cells are scattered throughout the arcuate nucleus, notably in the ventromedial arcuate nucleus (16,17). Although NPY is present in portal blood in concentrations that are higher than in the peripheral circulation (19), it is not clear whether the arcuate NPY cells are true neurosecretory adenohypophysiotropic cells that project to the external zone of the median eminence (20,21). In rats treated parenterally with monosodium glutamate (MSG; this causes cytotoxic lesion of 80–90% of arcuate cells that project outside of the blood–brain barrier), there was a depletion of NPY-immunoreactive cell bodies in this region, consistent with the idea that NPY cells project outside the blood–brain barrier. However, MSG treatment did not alter the NPY fibers in the external layer of the median eminence, suggesting that these fibers may originate from elsewhere, perhaps as ascending noradrenergic fibers from the brain stem that colocalize NPY (16).

III. NEUROCHEMICAL IDENTIFICATION OF THE ARCUATE CELLS ACTIVATED BY GHRP-6

It may seem relatively straightforward to determine the neurochemical identity of the arcuate cells activated by GHRP-6 by combining immunocytochemistry for Fos protein (detected in the nucleus) with immunocytochemistry for other neurochemical markers (for peptides present in the cell cytoplasm including, for

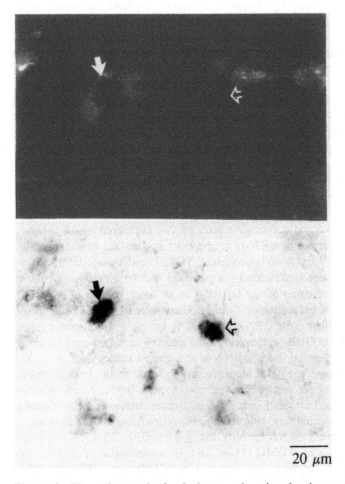

Figure 1 Photomicrograph of a single coronal section showing neurons in the arcuate nucleus of a rat injected with 50 μg GHRP-6. Cells with nuclear staining of Fos protein were detected under bright-field microscopy, and cells projecting outside of the blood–brain barrier were identified by the presence of retrogradely transported Fluorogold, detected under ultraviolet microscopy. Single Fos-positive cells were identified as either labeled (solid arrows) or unlabeled (open arrows) with Fluorogold. Between 68 and 82% of the cells expressing Fos protein were retrogradely labeled with Fluorogold, indicating that they project outside of the blood–brain barrier and, therefore, are likely to be neurosecretory cells. Section thickness = 40 μm. (From Ref. 14.)

example, GRF or tyrosine hydroxylase). However, for GRF, it would appear that there are insufficient levels of GRF present in the cell body for detection using the immunocytochemical approach. Most studies detecting GRF peptide have used colchicine-injected animals; the colchicine arrests axonal transport and facilitates immunocytochemical detection of peptide in the cell body. However, colchicine is also a powerful stimulus for the induction of Fos protein in the CNS and could not be incorporated into studies investigating whether the arcuate cells that express Fos protein following GHRP-6 injection include the GRF-containing neurons. Our attempts to double-label cells for Fos protein and tyrosine hydroxylase in non–colchicine-injected rats were more successful: approximately 7% of the cells expressing Fos protein following GHRP-6 injection were identified as tyrosine hydroxylase-containing and, therefore, are likely to be dopaminergic cells in this region (14).

More recently, we used an alternative strategy to identify the arcuate cells activated by GHRP-6: on consecutive sections, we compared the distribution of c-*fos* mRNA with that of mRNAs for peptides synthesized in arcuate cells, including GRF, NPY, tyrosine hydroxylase, POMC, and somatostatin (22). In these studies, rats were anesthetized and perfused with fixative 30 min following an iv injection of either 50 μg GHRP-6 or an equal volume of saline vehicle. Paraffin-embedded section of 7 μm thickness were processed using in situ hybridization for either c-*fos* mRNA or mRNAs for the neurochemical markers (Figs. 2 and 4). In GHRP-6 treated rats the mean ($\pm$ standard error of mean) number of cells expressing c-*fos* mRNA in the arcuate nucleus (23 ± 2 cells per section per rat; $n = 5$) was significantly higher than for vehicle-treated controls (2 ± 1 cells per section per rat; $n = 5$; $p < 0.001$, Mann–Whitney U-test). Superimposed camera lucida maps of arcuate nucleus (Fig. 3) indicated that, in GHRP-6–injected rats, neurochemically identifiable cells expressing c-*fos* mRNA also express mRNA for NPY ($51 \pm 4\%$), GRF ($23 \pm 1\%$), tyrosine hydroxylase ($11 \pm 3\%$), POMC ($11 \pm 2\%$), or somatostatin ($4 \pm 1\%$). We conclude that most of arcuate cells activated by GHRP-6 are NPY- and GRF-containing cells.

Although using this technique it is impossible to determine the absolute number of each cell type that is activated by GHRP-6, we can provide figures that allow the relative proportions of each population activated to be compared. The mean number of cells expressing GRF mRNA was 14 ± 1 per section; of these 38% were identified as also containing c-*fos* mRNA in the consecutive section (Fig. 4A). Despite the similar distribution and greater number per section (38 ± 3 cells per section) of POMC mRNA-containing neurons, only $6 \pm 1\%$ of these also expressed c-*fos* mRNA (see Fig. 4C). The largest population of neurochemically identified neurons in this study were those containing NPY mRNA (41 ± 3 cells per section). Of these, almost one-third ($30 \pm 3\%$)

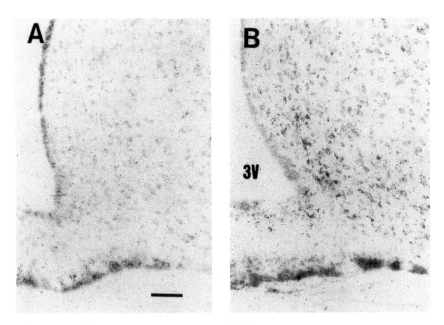

Figure 2 The c-*fos* mRNA in arcuate nucleus of conscious male rats injected iv with (A) isotonic saline or (B) 50 μg GHRP-6. 3V, third ventricle. Bar represents 70 μm. (From Ref. 22.)

also contained c-*fos* mRNA in the consecutive section (Fig. 4B). Somatostatin (14 ± 1 cells per section) and tyrosine hydroxylase (14 ± 2 cells per section) mRNA-containing neurons only colocalized c-*fos* in 7 ± 2% and 14 ± 3% of the cells, respectively.

One of the limitations of using a consecutive section approach, comparing cellular distributions of two different mRNAs is the confidence with it is possible to be certain that we are observing the same cell on the neighboring section: any cell showing an overlap in the distribution of its mRNA was scored positive. In our experience, it is easier to be confident of coexpression of, for example, GRF and c-*fos* mRNA, because cells expressing these mRNAs were not densely clustered in the ventrolateral arcuate nucleus. In contrast, the cells expressing c-*fos* mRNA and NPY mRNA were densely clustered in the more ventromedial regions of the arcuate nucleus, and so it could be argued that we have overestimated the actual number of NPY cells expressing c-*fos* mRNA following GHRP-6 injection. However, for POMC-containing cells, another group of cells that are present in abundance in the arcuate nucleus, only a small proportion of these cells overlap in distribution with the cells expressing c-*fos*

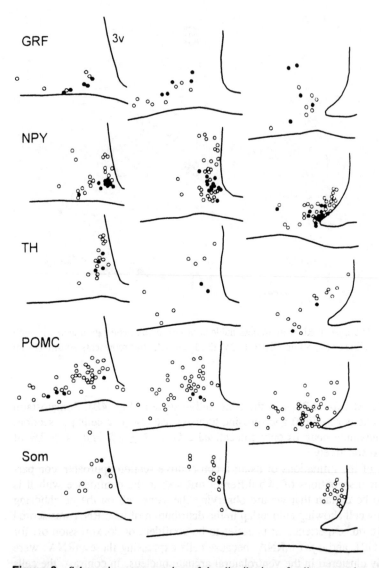

Figure 3 Schematic representation of the distribution of cells expressing mRNAs for GRF, NPY, TH, POMC, and somatostatin (Som) with cells expressing c-*fos* mRNA in the arcuate nucleus of a conscious male rat injected with 50 µg GHRP-6. Cells expressing these neuropeptide markers that could be identified as coexpressing c-*fos* mRNA on the consecutive section are represented by closed circles and those not coexpressing c-*fos* mRNA are represented by open circles. Three sets of consecutive sections are illustrated that correspond to three different levels of the arcuate nucleus: bregma –2.12 mm (left), –2.56 mm (middle), and –3.14 mm (right). 3V, third ventricle. (From Ref. 22.)

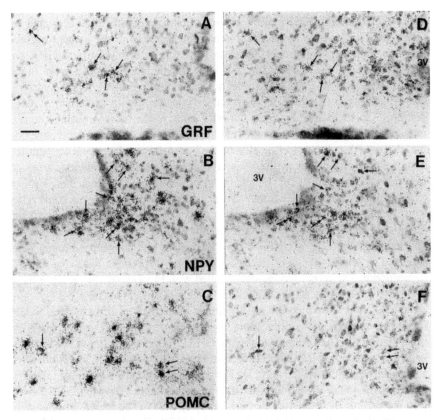

Figure 4 (A–C) Representative sections through the arcuate nucleus of a single rat showing the expression of mRNAs for the neurochemical markers, GRF, NPY, and POMC; (D–F) consecutive sections to the former showing the expression of c-*fos* mRNA induced by a single injection of GHRP-6. Arrows indicate the neurones expressing both c-*fos* and the corresponding neurochemical marker. 3V, third ventricle. Bar represents 40 μm. (From Ref. 22.)

mRNA, even in the most ventromedial region where cells expressing c-*fos* mRNA are densely clustered. Thus, although possible that we are overestimating the actual proportion of cells expressing c-*fos* in response to GHRP-6 injection that are NPY-containing, it seems unlikely that this is a major problem.

Most of the cells expressing GRF mRNA were located in the more ventrolateral regions of the arcuate nucleus. These GRF neurons have established neuroendocrine status, and they project to the median eminence, where GRF is released into the portal system of capillaries for transport to the anterior pituitary. Thus, activation of neurosecretory GRF-containing neurons by GHRP-

6 may account for at least some of the neurosecretory neurons activated by GHRP-6. The tyrosine hydroxylase-containing neurons in the arcuate nucleus are also neurosecretory cells. Because tyrosine hydroxylase and GRF are coexpressed in a small proportion of cells in the arcuate nucleus, notably in the ventrolateral portion (18), a small degree of activation of the tyrosine hydroxy-lase-containing cells by GHRP-6 would be consistent with activation of GRF neurons. Our demonstration that the NPY cells in the arcuate nucleus consti-tute a major population of cells activated by GHRP-6 would lead us to conclude that these NPY cells project outside of the blood–brain barrier and account for a substantial proportion of the cells retrogradely labeled following systemic injection of Fluorogold (14). It remains to be determined whether these project to the median eminence or whether they are open to the circulation in some other way. Interestingly, arcuate NPY neurons may detect circulating levels of the *obese* gene protein, leptin (23). Activation of only a small proportion of POMC and somatostatin cells by GHRP-6 may account for the nonneurosecretory cells activated by GHRP-6 (that is, the cells expressing Fos protein following GHRP-6 injection that were not retrogradely labeled with Fluorogold; 14) because these two cells groups appear not to project to the median eminence (16,17).

The extent to which the NPY neurons participate in neuroendocrine GH regulation is unknown. Furthermore, it is difficult to identify a role for NPY in mediating the GH response following GHRP-6 injection because NPY stimu-lates GH secretion when administered iv (24), but inhibits GH secretion when administered by the icv route (25). However, activation of NPY by GHRP-6 may explain some of the other physiological responses following GHRP-6 in-jection, such as the increased feeding behavior (26) or the increased release of corticotropin (adrenocorticotropic hormone; ACTH; 27). The arcuate NPY cells give rise to a major hypothalamic pathway that sends afferent fibers to the anterior hypothalamus, the preoptic area, the paraventricular nucleus (PVN), as well as the ventromedial and dorsomedial hypothalamus (28,29). Multiple daily injections of NPY into the PVN increases daily food intake and body weight (30). Activation of NPY neurons by GHRP-6 administration, therefore, may provide a mechanism to explain the increase in feeding behavior in rats following GHRP-6 administration (31). NPY infused directly into the PVN also increases the secretion of both corticosterone and ACTH (32–35). In the PVN, there is ultrastructural evidence for NPY immunoreactive contacts with corti-cotropin-releasing hormone (CRH)-immunoreactive cell bodies (36), suggest-ing that NPY cells make direct contact with CRF cells within the PVN. GHRP-6 (and other peptide and nonpeptide mimetics) also increase ACTH and cortisol secretion in many species (37–39).

In summary, our data suggest that GHRP-6 acts within the central ner-vous system (including a direct action on cells in the arcuate nucleus) to acti-vate neurons in this region, including the NPY and GRF populations. The recent cloning of the GH secretagogue receptor (shown to be present at both pituitary and hypothalamic sites; 40) has paved the way toward establishing whether the arcuate cells expressing the GH secretagogue receptor include the GRF and NPY populations.

ACKNOWLEDGMENTS

We thank colleagues who participated in the research described in this chapter: Professor Gareth Leng, Dr. Odile Viltart, and Dr. Richard E. J. Dyball. SML is a BBSRC Postdoctoral Fellow.

REFERENCES

1. Bowers CY, Momany FA, Reynolds GA, Hong A. On the in vitro and in vivo activity of a new synthetic hexapeptide that acts on the pituitary to specifically release growth hormone. Endocrinology 1984; 114:1537–1545.
2. McCormick GF, Millard WJ, Badger TM, Bowers CY, Martin JB. Dose–response characteristics of various peptides with growth hormone-releasing activity in the unanaesthetized male rat. Endocrinology 1985; 117:97–105.
3. Sartor O, Bowers CY, Reynolds GA, Momany FA. Variables determining the growth hormone response of His-D-Trp-D-Phe-Lys-NH$_2$ in the rat. Endocrinology 1985; 117:1441–1447.
4. Bowers CY, Sartor AO, Reynolds GA, Badger TM. On the actions of the growth hormone-releasing hexapeptide, GHRP. Endocrinology 1991; 128:2027–2035.
5. Clark RG, Carlsson LMS, Trojnar J, Robinson ICAF. The effects of a growth hormone-releasing peptide and growth hormone-releasing factor in conscious and anaesthetized rats. J Neuroendocrinol 1989; 1:289–255.
6. Dickson SL, Leng G, Robinson ICAF. Systemic administration of growth hormone-releasing peptide (GHRP-6) activates hypothalamic arcuate neurones. Neuroscience 1993; 53:303–306.
7. Dickson SL, Leng G, Dyball REJ, Smith RG. Central actions of peptide and non-peptide growth hormone secretagogues in the rat. Neuroendocrinology 1995; 61:36–43.
8. Dickson SL, Doutrelant-Viltart O, McKenzie DN, Dyball REJ. Growth hormone-releasing peptide activates rat arcuate neurones recorded in vitro. Program of the 10th International Congress of Endocrinology, San Francisco, 1996:680 (Abstr 1).
9. Bailey ART, Dickson SL, Leng G. Increased expression of Fos protein in the area postrema of the conscious male-rat following systemic injection of growth-hormone releasing peptide (GHRP-6). J Physiol 1996; 489P:172.
10. Dickson SL, Leng G, Robinson ICAF. Electrical stimulation of the rat periventricular nucleus influences the activity of hypothalamic neurones. J Neuroendocrinol 1994; 6:359–367.

11. Guillaume V, Magnan E, Cataldi M, Dutour A, Sauze N, Renard M, Razafindraibe H, Conte-Devolx B, Deghenghi R, Lenaerts V, Oliver C. Growth hormone (GH)-releasing hormone secretion is stimulated by a new GH-releasing hexapeptide in sheep. Endocrinology 1994; 135:1073–1076.

12. Jacobowitz DM, Schulte H, Chrousos GP, Loriaux DL. Localization of GRF-like immunoreactive neurones in the rat brain. Peptides 1983; 4:521–524.

13. Sawchenko PE, Swanson LW, Rivier J, Vale WW. The distribution of growth hormone-releasing factor (GRF) immunoreactivity in the CNS of the rat: an immunohistochemical study using antisera directed against rat hypothalamic GRF. J Comp Neurol 1985; 237:100–115.

14. Dickson SL, Doutrelant-Viltart O, Dyball REJ, Leng G. Retrogradely labeled neurosecretory neurones of the rat hypothalamic arcuate nucleus express Fos protein following systemic injection of growth hormone (GH)-releasing peptide (GHRP-6) J Endocrinol 1996; 151:323–331.

15. Merchenthaler I. Neurons with access to the general circulation in the rat central nervous system: a retrograde tracing study with Fluorogold. Neuroscience 1991; 44:655–662.

16. Meister B. Gene expression and chemical diversity in hypothalamic neurosecretory neurones. Mol Neurobiol 1993; 7:87–110.

17. Everitt BJ, Meister B, Hökfelt T, Melander T, Terenius L, Rökaeus Å, Theodorsson-Norheim E, Dockray G, Edwardson J, Cuello C, Elde R, Goldstein M, Hemmings H, Ouimet C, Wallas I, Greengard P, Vale W, Weber E, Wu J-Y, Chang K-J. The hypothalamic arcuate nucleus–median eminence complex: immunohistochemistry of transmitters, peptides and DARPP-32 with special reference to co-existence in dopamine neurones. Brain Res Rev 1986; 11:97–155.

18. Niimi M, Takahara J, Sato M, Kawanishi K. Identification of dopamine and growth hormone-releasing factor-containing neurones projecting to the median eminence of the rat by combined retrograde tracing and immunohistochemistry. Neuroendocrinology 1992; 55:92–96.

19. McDonald JK, Lumpkin MD, Samson WK, McCann SMM. Neuropeptide YY affects secretion of luteinizing hormone and growth hormone in ovariectomized rats. Proc Natl Acad Sci USA 1985; 82:561–564.

20. Li S, Hisano S, Daikoku S. Mutual synaptic associations between neurones containing neuropeptide Y and neurones containing enkephalin in the arcuate nucleus of the rat hypothalamus. Neuroendocrinology 1993; 57:306–313.

21. Clarke I, Jessop D, Millar R, Bloom S, Lightman S, Coen CW, Lew R, Smith I. Many peptides that are present in the external zone of the median eminence are not secreted into the hypophysial portal blood of sheep. Neuroendocrinology 1993; 57:765–775.

22. Dickson SL, Luckman SM. Induction of c-*fos* messenger ribonucleic acid in neuropeptide Y and growth hormone (GH)-releasing factor neurones in the rat arcuate nucleus following systemic injection of the GH secretagogue, GH-releasing peptide-6. Endocrinology 1997; 138:771–777.

23. Stephens TW, Basinski M, Bristow PK, Buevalleskey JM, Burgett SG, Craft L, Hale J, Hoffmann J, Hsiung HM, Kriauciunas A, Mackellar W, Rostek PR, Schoner B, Smith D, Tinsley FC, Zhang XY, Heiman M. The role of neuropep-

tide Y in the antiobesity action of the obese gene product. Nature 1995; 377:530–552.

24. Chabot J-G, Enjalbert A, Pelletier G, Dubois PM, Morel G. Evidence for a direct action of neuropeptide Y in the rat pituitary gland. Neuroendocrinology 1988; 47:511–517.

25. Catzeflis C, Pierroz DD, Rohner-Jeanrenaud F, River JE, Sizonenko PC, Aubert ML. Neuropeptide Y administered chronically into the lateral ventricle profoundly inhibits both the gonadotropic and the somatotropic axis in adult female rats. Endocrinology 1993; 132:224–234.

26. Clark JT, Kalra PS, Crowley WR, Kalra SP. Neuropeptide Y and human pancreatic polypeptide stimulate feeding in rats. Endocrinology 1984; 115:427–429.

27. Akabayashi A, Watanabe Y, Wahlestedt C, McEwen BS, Paez X, Leibowitz SF. Hypothalamic neuropeptide Y, its gene expression and receptor activity: relation to circulating corticosterone in adrenalectomized rats. Brain Res 1993; 665:201–212.

28. Bai FL, Yamano M, Emson PC, Smith AD, Powell JF, Tohyama M. An arcuatoparaventricular and dorsomedial hypothalamic neuropeptide Y-containing system which lacks noradrenaline in the rat. Brain Res 1985; 331:172–175.

29. Baker RA, Herkenham M. Arcuate nucleus neurones that project to the hypothalamic paraventricular nucleus: neuropeptidergic identity and consequences in adrenalectomy on mRNA level in the rat. J Comp Neurol 1985; 358:518–530.

30. Stanley BG, Kyskouli SE, Lampert S, Leibowitz SF. Neuropeptide Y chronically injected into the hypothalamus: a powerful neurochemical inducer of hyperphagia and obesity. Peptides 1986; 7:1189–1192.

31. Locke W, Kirgis HD, Bowers CY, Abdoh AA. Intracerebroventricular growth hormone-releasing peptide-6 stimulates eating without affecting plasma growth hormone responses in rats. Life Sci 1995; 56:1347–1352.

32. Wahlstedt C, Skagerberg G, Ekman R, Heilig M, Sundler F, Hakanson R. Neuropeptide Y (NPY) in the area of the hypothalamic paraventricular nucleus activates the pituitary–adrenocortical axis in the rat. Brain Res 1987; 417:33–38.

33. Haas DA, George SR. Neuropeptide Y-induced effects on hypothalamic corticotropin-releasing factor content and release are dependent upon noradrenergic/adrenergic neurotransmission. Brain Res 1989; 498:333–338.

34. Saphier D, Feldman S. Adrenoreceptor specificity in the central regulation of adrenocortical secretion. Neuropharmacology 1989; 28:1231–1237.

35. Albers HE, Ottenweller JE, Liou SY, Lumpkin MD, Anderson ER. Neuropeptide Y in the hypothalamus: effect on corticosterone and single-unit activity. Am J Physiol 1990; 258:R376–R382.

36. Liposits Z, Seivers L, Paull WK. Neuropeptide-Y and ACTH-immunoreactive innervation of corticotropin releasing factor (CRF)-synthesizing neurones in the hypothalamus of the rat: an immunocytochemical analysis at the light and electron microscopic levels. Histochemistry 1988; 88:227–234.

37. Hickey GJ, Jacks T, Judith F, Taylor J, Schoen WR, Krupa D, Cunningham P, Clark J, Smith RG. Efficacy and specificity of L-692,429, a novel nonpeptidyl growth hormone secretagogue in beagles. Endocrinology 1994; 134:695–701.

38. Thomas GB, Fairhall KM, Robinson ICAF. GH, ACTH and corticosterone responses to GHRP-6, CRF and AVP in conscious rats. Program of the 10th International Congress of Endocrinology, San Francisco, 1996:592 (Abstr 1).
39. Gertz BJ, Barrett JS, Eisenhandler R, Krupa DA, Wittreich JM, Seibold JR, Schneider SH. Growth hormone response in man to L-692,429, a novel nonpeptide mimic of growth hormone-releasing peptide. J Clin Endocrinol Metab 1993; 77:1393–1397.
40. Howard AD, Feighner SD, Cully DF, Arena JP, Liberator PA, Rosenblum CI, Hamelin M, Hrenluk DL, Palyha OC, Anderson J, Paress PS, Diaz C, Chou M, Liu KK, McKee KK, Pong S-S, Chaung L-Y, Elbrecht A, Dashkevicz M, Heavens R, Rigby M, Sirinathsinghji DJS, Dean DC, Melillo DG, Patchett AA, Nargund R, Griffin PR, DeMartino JA, Gupta SK, Schaeffer JM, Smith RG, Van der Ploeg LHT. A pituitary gland and hypothalamic receptor that functions in growth hormone release. Science 1996; 273:974–977.

19
Do Growth Hormone-Releasing Peptides Play a Role in the Neurohypophysis?

Bjarne Fjalland and Thomas Anderson
The Royal Danish School of Pharmacy, Copenhagen, Denmark

Peter B. Johansen
Novo Nordisk, Copenhagen, Denmark

I. INTRODUCTION

It is well established that release of posterior pituitary hormones is also regulated at the level of the neurohypophysis (1,2). Many endogenous substances are involved in this regulation. The release of oxytocin is under inhibitory control of endogenous opioids (3). Furthermore, the opioid-binding sites in the nerve terminals are predominantly κ, and substances interfering with κ-receptors (i.e., dynorphin and some selective κ-receptor agonists) inhibit the evoked release of oxytocin and vasopressin from the isolated neurohypophysis and from isolated nerve terminals (4,5), indicating the presence of functional κ-receptors on terminals of oxytocin and vasopressin neurons.

Many other neurotransmitters or modulators are present in the posterior pituitary. Immunohistochemical techniques have shown that growth hormone-releasing hormone (GHRH) and somatostatin-secreting neurons terminate in the neurohypophysis (6,7), and that the posterior pituitary contains relatively high amounts of somatostatin (8). Furthermore, vasopressin is involved in the action of some somatostatin analogs (9). Recently, specific-binding sites for GH-releasing peptides (GHRPs) have been shown, and the existence of an endogenous ligand for these receptors is suggested (10,11). Therefore, we found it of interest to examine whether small, synthetic molecules, with stimulating effects on growth hormone release, could interfere with the release of hormones from the posterior pituitary.

II. METHOD

Female Sprague-Dawley rats (200–250 g) in spontaneous estrus were used. The animals were decapitated and the skin and skull overlying the brain removed. After carefully teasing the pituitary free of the surrounding membranes, the whole brain with the intact hypophysis was transferred to a petri dish containing the medium bubbled with a mixture of 95% O_2 and 5% CO_2, pH 7.4, of the following composition: 115 mM NaCl, 6.0 mM KCl, 0.8 mM Na_2HPO_4, 1.5 mM $CaCl_2$, 0.9 mM $MgCl_2$, 22 mM $NaHCO_3$, and 10 mM D-glucose. Under a stereomicroscope, the neurointermediate lobe was separated from the anterior lobe and was cut at the level of the infundibulum. The pituitary stalk was tied, with a fine cotton thread, to a 0.5-mm platinum wire electrode and immersed in 2 mL of the medium at 37°C. The preparation was preincubated for 1 h. Each neurointermediate lobe was electrically stimulated three times with 60-min–rest periods interposed. Drugs were present 30 min before, and during the second stimulation period. Trains of rectangular pulses (1 V, 2 ms, 30 Hz for 10 s, with an interval of 10 s, for a total of 2.5 min) were applied between the platinum wire (cathode) carrying the neurohypophysis and a platinum wire dipped into the incubation medium. During the stimulation period the gland was raised to the surface of the incubation medium and the gas supply interrupted. The preparation was again immersed into the medium immediately after stimulation and the gas supply restored. The medium was collected for radioimmunoassay (12,13) after a 10-min period following the stimulation. The response was quantified as the ratio (S_2/S_1) between the amounts of hormone released on the second and the first stimulation, respectively.

To obtain an immediate preliminary measure of the response to the stimulation, a semiquantitative bioassay was used. The uteri were dissected from the rats and freed from fat. A thread was attached at each end and the organ was superfused with the medium. Contractions were recorded isotonically and compared with standard dose–response curves obtained by superfusion of the muscle with oxytocin.

III. RESULTS

Phasic, submaximal stimulation of the pituitary stalk led to an evoked release of neurohypophysial hormones that was strongly dependent on the pattern of electrical stimulation (14). In the present experiments, the release of vasopressin and oxytocin decreased about 15% from the first to the second stimulation, giving a control ratio S_2/S_1 of about 0.85 (see Figs. 1 and 2). GHRH [rGRF(1–29)NH_2] significantly inhibited the release of both vasopressin and oxytocin (about 25%) at a concentration of 10^{-7} M (results not shown).

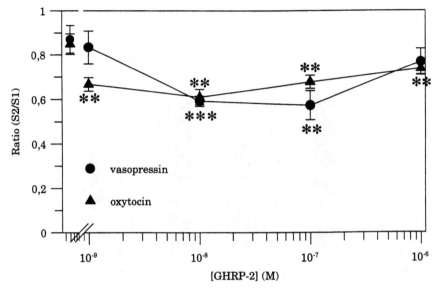

Figure 1 Effect of GHRP-2 on electrically evoked release of vasopressin and oxytocin from rat neurointermediate lobes. The results are expressed as the ratio $S_2/S_1 \pm$ SEM of hormone release during the second stimulation, in the presence of test substance, to the release during the first stimulation in control medium. $n = 6-9$. Different from control: $**p < 0.01$; $***p < 0.001$.

Growth hormone-releasing peptide-2 (GHRP-2 [D-Ala^1-D-βNal^2-Ala^3-Trp^4-D-Phe^5-Lys^6-NH$_2$]) was examined in the micromolar to nanomolar range (Fig. 1). It inhibited the release of both hormones in a biphasic way, with a maximal effect at about 10^{-8} M. Vasopressin release was significantly reduced in the range of 10^{-8} to 10^{-7} M, whereas the output of oxytocin was inhibited from 10^{-9} to 10^{-6} M.

Growth hormone-releasing peptide-2 induces a dose-dependent increase in growth hormone release in ovine pituitary cells in the concentration range of 10^{-10} to 10^{-7} M (15), an effect probably mediated by influx of Ca^{+2} through voltage-gated Ca^{+2} channels (16). The effect of GHRP-2 on GH release was also biphasic; a concentration of 10^{-6} M induced less stimulation (15). Furthermore, the dose–response relation for release of corticotropin (adrenocorticotropic hormones ACTH) is biphasic (bell-shaped) in conscious pigs following intravenous (iv) bolus injections of GHRP-2, ranging from 0.2 to 200 nmol/kg (Raun K, personal communication).

The substance P antagonist (D-Arg^1,D-Phe^5,D-Trp^{7,9},Leu^{11})-substance P is an antagonist of GHRP in the adenohypophysis and displaces GHRP-2 in EP-

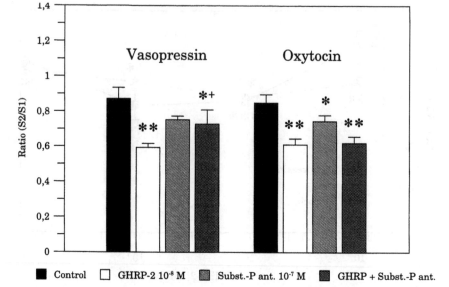

Figure 2 Influence of a substance P antagonist (D-Arg^1,D-Phe^5,D-Trp^{7,9},Leu^{11}-substance P) on the GHRP-2-induced inhibition of release of vasopressin and oxytocin from electrically stimulated isolated rat neurointermediate lobes. The results are expressed as the ratio S_2/S_1 $\pm$ SEM of hormone release during the second stimulation, in the presence or absence of test substance, to the release during the first stimulation in control medium. $n = 6$–9. Different from control: *$p < 0.05$, **$p < 0.01$. Different from GHRP-2: $^+p < 0.05$. Ant, antagonist.

1 cells (17). This antagonist did not significantly influence the release of vasopressin (Fig. 2), but reduced the oxytocin release by about 10%. The inhibitory effect of GHRP-2 (10^{-8} M) on vasopressin release was partly reversed by the antagonist (10^{-7} M), whereas the substance had no influence on the reduced secretion of oxytocin induced by GHRP-2.

IV. CONCLUSION

The experiments have demonstrated, that GH secretagogues exert inhibitory effects, in the nanomolar range, on release of vasopressin and oxytocin from the rat neurohypophysis and that the effect of GHRP-2 was biphasic. Furthermore, the effect of GHRP-2 on vasopressin secretion was partly reversed by a substance known to displace GHRP-2 in other cells, whereas the effect on oxytocin was not influenced. The results suggest, that vasopressin release from the posterior pituitary may be under the influence of GHRPs.

REFERENCES

1. Boersma CJC, Van Leeuwen FW. Neuron–glia interactions in the release of oxytocin and vasopressin from the rat neural lobe: the role of opioids, other neuropeptides and their receptors. Neuroscience 1994; 62:1003–1020.
2. Falke N. Modulation of oxytocin and vasopressin release at the level of the neurohypophysis. Prog Neurobiol 1991; 36:465–484.
3. Bicknell RJ, Chapman C, Leng G. Effects of opioid agonists and antagonists on oxytocin and vasopressin release in vitro. Neuroendocrinology 1985; 41:142–148.
4. Herkenham M, Rice KC, Jacobsen AE, Rothman RB. Opiate receptors in rat pituitary are confined to the neural lobe and are exclusively kappa. Brain Res 1986; 382:365–371.
5. Zhao BG, Chapman C, Bicknell RJ. Functional κ-opioid receptors on oxytocin and vasopressin nerve terminals isolated from the rat neurohypophysis. Brain Res 1988; 462:62–66.
6. Anthony ELP, Bruhn TO. Do some traditional hypophysiotropic hormones play nontraditional roles in the neurohypophysis? Implications of immunocytochemical studies. Ann NY Acad Sci 1993; 689:469–472.
7. Mikkelsen JD, Schmidt P, Sheikh SP, Larsen PJ. Non-vasopressinergic, non-oxytocinergic neuropeptides in the rat hypothalamo–neurohypophyseal tract: experimental immunohistochemical studies. Prog Brain Res 1992; 92:367–371.
8. Patel YC, Reichlin S. Somatostatin. In: Methods of Hormone Radioimmunoassay, 2d ed. 1979:77–99.
9. Fan NY, Powers CA, Stier CT Jr. Lack of antidiuretic activity of lanreotide in the diabetes insipidus rat. J Pharmacol Exp Ther 1996; 276:875–881.
10. Howard AD, Feighner SD, Cully DF, Arena JP, Liberator PA, Rosenblum CI, Hamelin M, Hreniuk DL, Palyha OC, Anderson J, Paress PS, Diaz C, Chou M, Liu KK, McKee KK, Pong SS, Chaung LY, Elbrecht A, Dashkevicz M, Heavens R, Rigby M, Sirinathsinghji DJS, Dean DC, Melillo DG, Patchett AA, Nargund R, Griffin PR, DeMartino JA, Gupta SK, Schaeffer JM, Smith RG, Van der Ploeg LHT. A receptor in pituitary and hypothalamus that functions in growth hormone release. Science 1996; 273:974–977.
11. Pong SS, Chaung LYP, Dean DC, Nargund RP, Patchett AA, Smith RG. Identification of a new G-protein-linked receptor for growth hormone secretagogues. Mol Endocrinol 1996; 10:57–61.
12. Christensen JD, Fjalland B. Lack of effect of opiates on release of vasopressin from isolated rat neurohypophysis. Acta Pharmacol Toxicol 1982; 50:113–116.
13. Fjalland B, Christensen JD, Grell S. GABA receptor stimulation increases the release of vasopressin and oxytocin in vitro. Eur J Pharmacol 1987; 142:155–158.
14. Jørgensen A, Fjalland B, Christensen JD, Treiman M. Dihydropyridine ligands influence the evoked release of oxytocin and vasopressin dependent on stimulation conditions. Eur J Pharmacol 1994; 259:157–163.
15. Chen DWC, Zhang J, Bowres CY, Clarke IJ. The effects of GH-releasing peptide-6 (GHRP-6) and GHRP-2 on intracellular adenosine 3'5'-monophosphate (cAMP) levels and GH secretion in ovine and rat somatotrophs. J Endocrinol 1996; 148:197–205.

16. Chen C, Clarke IJ. Effects of growth hormone-releasing peptide-2 (GHRP-2) on membrane Ca^{+2} permeability in cultured ovine somatotrophs. J Neuroendocrinol 1996; 7:179–186.
17. Tai S, Kaji H, Okimura Y, Abe H, Chihara K. Neurokinin receptor antagonists inhibit the binding of growth hormone-releasing peptide to EP-1 human neuroblastoma cells. Biochem Biophys Res Commun 1995; 214:454–460.

20

Evaluation of Pituitary Function Using Growth Hormone Secretagogues

Barry B. Bercu
*University of South Florida College of Medicine, Tampa,
and All Children's Hospital, St. Petersburg, Florida*

Richard F. Walker
University of South Florida, Tampa, Florida

I. INTRODUCTION

Reduced growth hormone (GH) secretion is a common cause of delayed physical development in children. Deficits in GH secretion during development can be attributed to hypophyseal or hypothalamic dysfunctions, the most common cause being insufficient or inadequate presentation of neurosecretory regulatory hormones (1). It is, however, possible that intrinsic pituitary defects can reduce production, secretion, or both, of GH. Despite these different etiologies, growth deficiencies are almost universally treated with recombinant GH, essentially foregoing the alternative treatment: pharmacological enhancement of endogenous GH secretion. Although human GHRH is easily synthesized and readily available for treatment of short stature, variable and unpredictable responses have delayed its use as a therapeutic agent until very recently.

The progressive decrease in GH secretion during aging could result from several factors. Two of the most obvious are inadequate stimulation of the pituitary gland by hypophysiotropic factors, or degeneration of pituitary-based mechanisms for GH production or secretion. Of the two, it is most likely that decremental changes in the relation of the pituitary with GH regulatory hormones of hypothalamic origin are the primary etiologic factors in age-related decline in GH secretion. However, this view is seemingly inconsistent with many reports that state the pituitary becomes refractory to stimulation during aging. For example, except for two reports (2,3), there is general consensus that GH se-cretion in response to GH-releasing hormone (GHRH) administration in vivo declines with advancing aging (4–12). Assuming that GHRH is the only stimu-latory agent controlling GH secretion, this progressive decrement could result from several factors. One factor could be that prolonged reduction in pituitary stimulation by GHRH causes desensitization to the GH secretagogue (GSH) because hormones often induce their own receptors. Support for reduced stimu-lation of the pituitary by GHRH derives from the fact that available or appro-priate GH secretagogues seem to decline during aging (13–15). GHRH respon-siveness in aging rats, however, was immediately restored by coadministration with GH-releasing peptide (GHRP; 16), suggesting that age-related, reduced efficacy of GHRH may be due to the absence or reduced concentrations of other, as yet undefined, endogenous cosecretagogues (for a detailed overview of this subject see Ref. 17).

About two decades ago, a family of xenobiotic GH secretagogues were discovered that released GH by a mechanism separate from GHRH (18). The prototype for the xenobiotic secretagogues was His-D-Trp-Ala-Trp-D-Phe-Lys-NH$_2$ or GH-releasing hexapeptide (GHRP-6). GHRP potentiated the action of GHRH when both compounds were administered together, and the potency of GHRP was dramatically reduced when endogenous GHRH was reduced by passive immunization (19). The reciprocal test was not possible because an endogenous ligand for GHRP has not been identified. The recent report of the identification of the receptor should aid in this quest (20). Nonetheless, because functional dependence of GHRP can be demonstrated by passive immunization against GHRH, the reverse may also be true (i.e., GHRH activity may be af-fected by an endogenous GHRP analog). This possibility that GH secretion may be controlled by more factors than GHRH and somatostatin has been previously suggested (21).

The relevance of these findings to the present discussion is that they support the hypothesis that GHRH and an endogenous analog of GHRP are complementary GH secretagogues that together provide appropriate stimulation of the pituitary gland to sustain normal GH production and release. By this logic, GHRH and GHRP could be used alone and in combination to diagnose pitu-itary-based causes on inadequate GH secretion (22; Fig. 1). The model assumes that a "normal" response to administration of GHRP (or GHRP-mimic) or GHRH requires the presence of its endogenous analog (i.e., GHRH or GHRP, respectively). A blunted response to either exogenous GH secretagogues is interpreted as indicating a deficiency of its endogenous complement (see Fig. 1B and C). Blunted responses to both exogenous GH secretagogues adminis-tered sequentially, implies deficiencies of both endogenous complements (see Fig. 1D). This condition can be differentiated from inherent pituitary problems, such as those involving receptor or second-messenger deficits by a "normal"

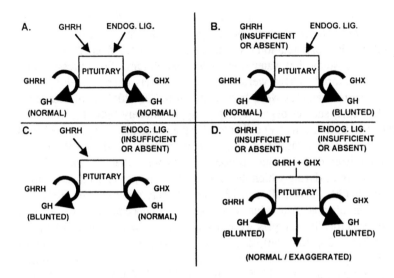

Figure 1 Schematic representation of the hypothetical model in support of the complementary relation of GHRH and the endogenous ligand (endog. lig) for GHRP or GHRP nonpeptide mimic. See text for discussion of the hypothesis. GHX, peptide and nonpeptide forms of GHRP. (From Ref. 33.)

response to GHRH and GHRP coadministration. A blunted response following coadministration of both secretagogues would indicate inherent pituitary dysfunction, rather than inadequate endogenous stimuli. Here we present preliminary data in children and middle-aged adults to test this hypothesis.

II. MATERIALS AND METHODS

The GHRP-2 was synthesized by Kaken Pharmaceutical Co., Ltd (Japan). GHRP-2 was provided by Wyeth–Ayerst (King of Prussia, PA), GRF 1-29 by Serono Laboratories, Inc. (Norwell, MA), and GHRH (1-44) NH$_2$ by ICN Pharmaceuticals (Costa Mesa, CA). GHRH(1-44)NH$_2$ is equivalent to GRF 1-29 in potency; it was substituted for GRF 1-29 when GRF 1-29 was no longer commercially available.

Subjects were recruited from the pediatric endocrine clinics of All Children's Hospital, St. Petersburg, Florida, and the University of South Florida College of Medicine, Tampa, Florida. Adult subjects were recruited from the local community. The protocol was approved by both the All Children's Hospital and University of South Florida IRBs. Informed consent was obtained from the subjects, parents (and legal guardians), and children when appropriate.

Children were from age 3½ and up, including 17 boys and 9 girls. Adult age range was from 29 to 68; 5 men and 2 women. Each subject had a medical history and physical examination. All subjects were fasted from 2200 h until completion of the study, but water was allowed ad libitum. Blood pressure and pulse were monitored. There were no adverse effects except flushing with GHRH. The serum samples were frozen and stored at –20°C until assayed.

Briefly, each subject began testing at approximately 0900 h, ½ h after placement of an intravenous (iv) catheter. For the sequential study, three blood samples for basal concentrations of serum GH were drawn, and GHRP-2 (1 μg/kg) was administered as a bolus. Blood samples were then drawn at 5-min intervals for 20 min, during which the GH secretory response occurred, and at longer intervals (30, 45, 60, 90, and 120 min). Ninety minutes later GHRH (1 μg/kg) was administered as a bolus, when basal GH concentrations were again established. Specifically, samples were drawn at the new zero time (210 min) and then at 215, 220, 225, 240, 255, 270, 300, and 330 min. For the combined GHRP-2 plus GHRH (1.0 μg/kg for each secretagogue) study blood samples were drawn at –30, –15, 0, 5, 10, 15, 30, 45, 60, 90, 120, 180, and 210 min. GH concentrations in each serum sample were determined by polyclonal radioimmunoassay (23). Interassay coefficient of variation was 13%, and intra-assay coefficient of variation was 9%. The study population comprised children with normal GH secretion and those with GH secretory dysfunction of various clinical diagnoses. The adults were normal volunteers. The children who were normal volunteers were siblings of children being evaluated for growth disturbances. The purpose of evaluating different etiologies of GH secretory dysfunction was an attempt to validate the hypothesis that complementary endogenous GH secretagogues must be present for optimal expression of GHRP or GHRH stimulatory potentials.

A total of 45 tests were performed in 38 subjects. Three GH-deficient and one control child had both sequential and combined GHRP-2 plus GHRH testing. Two adult volunteers had the GHRP-2 study alone repeated. One non–GH-deficient, slow-growing, short-stature child had sequential GHRP-2 plus GHRH testing while on and off clonidine therapy for attention deficit disorder.

All values are expressed as the mean ± SEM. The area under curve (AUC) was calculated for 90 min after GHRP-2 and GHRH administration in the sequential study. Statistical analyses were done using ANOVA and Scheffe tests.

III. RESULTS

Table 1 summarizes the peak concentrations after GHRP-2 and GHRH during sequential testing. The most profound and consistent results were in the adult volunteers. Note the blunted GH response to GHRH in normal volunteer adults

Table 1 Peak GH Concentrations Following GHRP-2 and GHRH in the Various Groups

Group	Peak GH after GHRP-2 (μg/L)	Peak GH after GHRH (μg/L)	Peak GHRP-2/ GHRH ratio	Area under curve (μg/L · 90 min) GHRP-2	Area under curve (μg/L · 90 min) GHRH
GH-deficient children (n = 15)	20.1 ± 5.5 (< 0.5-> 80)	19.6 ± 5.1 (0.3-70.2)	1.3 ± 0.4 (0.3-5.6)	995 ± 371	924 ± 232
Control children (n = 8)	42.2 ± 4.3 (31.7-62.0)	39.8 ± 7.8 (19.2-88.9)	1.4 ± 0.3 (0.4-2.9)	1598 ± 274	2201 ± 437
Slowly growing, non-GH-deficient children (n = 8)	63.6 ± 24.9 (4.9-190.3)	31.4 ± 8.4 (1.5-78.8)	2.9 ± 1.3 (0.4-3.3)	2460 ± 953	1544 ± 449
Adult volunteers (n = 7)	52.0 ± 15.1 (14.6-123.2)	6.8 ± 2.4 (2.5-16.3)	15.0 ± 4.1 (0.9-36.2)	2785 ± 692	285 ± 93
Men (n = 5)	59.3	3.1	20.5		
Women (n = 2)	19.2	16.1	1.2		
Control vs. GH-deficient children	$p < 0.02$	$p < 0.05$			$p < 0.01$
Control children vs. adult volunteers		$p < 0.01$	$p < 0.01$		$p < 0.02$
GH-deficient children vs. adult volunteers	$p < 0.05$		$p < 0.001$	$p < 0.02$	
Slow-growing, non-GH-deficient children vs. adult volunteers		$p < 0.02$	$p < 0.02$		$p < 0.02$
GH-deficient vs. slow-growing, non-GH-deficient children	$p < 0.05$				

Adult peak GH: GHRP-2 vs. GHRH, $p < 0.02$.
Adult area under curve: GHRP-2 vs. GHRH, $p < 0.02$.

versus adolescent children (Fig. 2). Both groups had exuberant GH responses to GHRP-2. There was a marked sex difference in the limited number of adult volunteers studied (Fig. 3). In all adult subjects except for the two women, aged 29 and 58 years, peak GH concentration after GHRP-2 administration was significantly greater when compared with GHRH-stimulated GH release. One adult woman had a ratio of 1.5 (peak GH after GHRP-2/peak GH after GHRH), with a peak GH of 23.7 µg/L after GHRP-2 and 15.9 µg/L after GHRH (see Fig. 3), and the other had a peak GH of 14.6 and 16.3 µgL, respectively (ratio = 0.9). For the remainder of the group (all men), the mean GH concentration after GHRP-2 and GHRH administration was 55.0 and 5.2 µg/L, respectively (ratio = 9.0). Representative examples are shown in Figure 4. The most exaggerated ratio was in a 33-year-old male volunteer (peak GH after GHRP-2 and GHRH was 123.2 and 3.4 µg/L, respectively; ratio = 36.2; data not shown). In this limited study, both adult women had greater GH responses to GHRH than did the men.

Table 1 summarizes the data in GH-deficient, slow-growing, non–GH-deficient, and control children. Classic GH deficiency is based on two or more blunted GH stimulation tests of less than 10 µg/L, delayed bone age, and poor

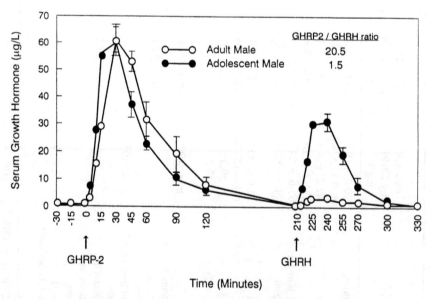

Figure 2 The GH secretory responses to sequential GHRP-2 and GHRH administration in healthy male volunteers and adolescent children. GHRP-2/GHRH ratio refers here and other figure legends to peak GH concentration after each individual stimulus in the sequential provocative test. Data expressed as mean ± SEM.

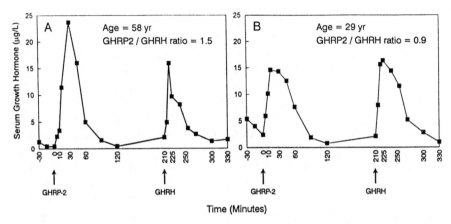

Figure 3   The GH secretory responses to sequential GHRP-2 and GHRH in two adult female volunteers.

growth velocity. Non–GH-deficient children have a growth velocity less than the 25th percentile for age, normal GH provocative tests, and delayed bone age. The ratio of peak GH response to GHRP-2 plus GHRH is similar in all three groups with the highest ratio in the slow-growing, non–GH-deficient group (1.3 and 1.4 for GH-deficient and control groups, respectively; 2.9 in the slow-growing, non–GH-deficient group; p = NS). As a group, peak GH concentrations after GHRP-2 were lowest in the GH-deficient group compared with the control children and slow-growing, non–GH-deficient group (20.1, 42.2, and 63.6 µg/L, respectively; control vs. GH-deficient, p < 0.02, GH-deficient

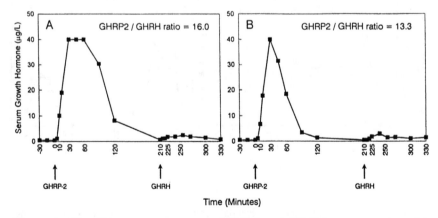

Figure 4   The GH secretory response to GHRP-2 and GHRH in representative adult male volunteers.

versus slow-growing, non–GH-deficient children, $p < 0.05$); AUC: 995 ± 371, 1598 ± 274, and 2460 ± 953 µg/L · 90 min, respectively; p = NS for all groups. Also, as a group peak GH after GHRH was lowest in the GH-deficient patients (19.6 µg/L in GH-deficient, 39.8 µg/L in control, and 31.4 µg/L in slow-growing, non–GH-deficient groups; control vs. GH-deficient, $p < 0.05$; AUC: 924 ± 232, 2201 ± 437, 1544 ± 449 µg/L · 90 min, respectively; control vs. GH-deficient children, $p < 0.01$, GH-deficient vs. slow-growing, non–GH-deficient children, $p < 0.02$). On the other hand, there was marked within-group variation, demonstrating individual differences. Figure 5 demonstrates some extreme individual examples of children who were slow-growing, non–GH-deficient; however, the peak GH after GHRP-2 plus GHRH ratio was preserved (3.3 vs. 2.4). Of interest, GH-deficient children exposed to radiation causing central nervous system damage also had a variable response (Fig. 6). GH responses to GHRP-2 and GHRH were somewhat muted except for two patients, one with an exaggerated response to GHRP-2 and the other to GHRH. The child with the most exaggerated GH response to GHRH (peak GH 70.2 µg/L) had central nervous system damage causing precocious puberty; however, the only other hypothalamic–pituitary deficiency was GH deficiency. Children with direct damage to the pituitary gland had markedly decreased responses to both stimuli as well as decreased response to combined stimuli. This is clearly demonstrated in a patient with hypophysitis and essentially no visible pituitary tissue on magnetic resonance imaging (Fig. 7). In general, the patients with multiple hormone deficiency had the lowest GH responses to the secretagogues (Figs. 7 and 8).

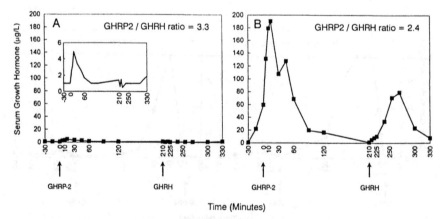

Figure 5 The GH secretory to sequential GHRP-2 and GHRH in two slow-growing, non–GH-deficient children. Although the secretory patterns were markedly different, the peak GH GHRP-2/GHRH ratios were preserved.

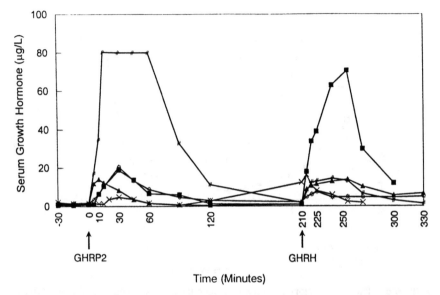

Figure 6 The GH secretory response to sequential GHRP-2 and GHRH in six children previously exposed to radiation therapy affecting the central nervous system.

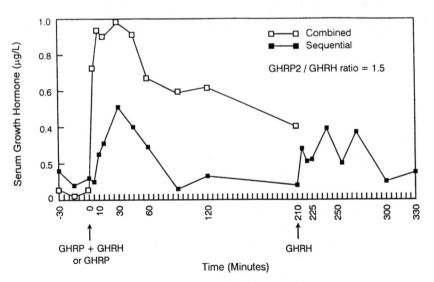

Figure 7 The GH secretory response to sequential and combined GHRP-2 and GHRH in a patient with hypophysitis. (From Ref. 22.)

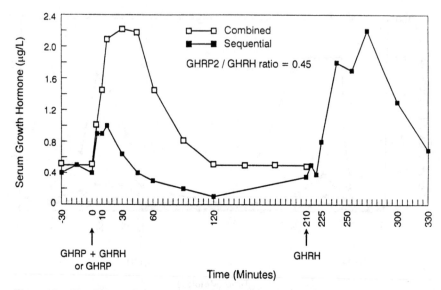

Figure 8 The GH secretory response to sequential and combined GHRP-2 and GHRH in a child with septo-optic dysplasia and multiple hormone deficiency.

There was a statistically significant positive correlation between peak GH after GHRP-2 and mean peak GH after standard provocative testing (Fig. 9; $r = 0.62$, $p < 0.02$). On the other hand, in the same sequential study, there was no correlation between peak GH after GHRH and mean peak GH after standard provocative testing (Fig. 10; $r = 0.24$, $p = $ NS).

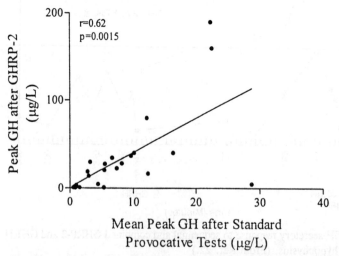

Figure 9 Correlation of peak GH after GHRP-2 versus mean peak GH after standard provocative stimuli in GH-deficient and slow-growing, non–GH-deficient children.

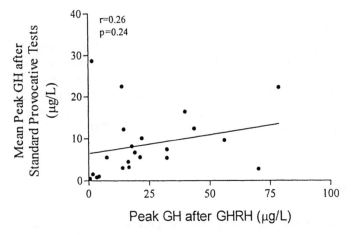

Figure 10   Correlation of peak GH after GHRH versus mean peak GH after standard provocative stimuli in GH-deficient and slow-growing, non–GH-deficient children.

Four children had sequential and combined GHRP-2 plus GHRH studies. Two patients with hypopituitarism (multiple hormone deficiency) had an increased response to the combined stimulus (see Figs. 7 and 8). The peak GH (highest peak of GH response to GHRP-2 or GHRH) of the two secretagogues in sequence versus in combination increased 1.2-, 2.3-, and 2.3-fold. Figure 11 shows the response of both tests in a control child. His peak GH after GHRP-

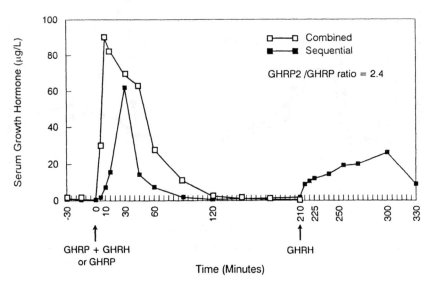

Figure 11   The GH secretory response to sequential and combined GHRP-2 and GHRH in a normal-statured child.

2 was 62.0 µg/L and 26.1 µg/L after GHRH; following the combined stimulus peak GH was 89.6 µg/L demonstrating a 1.4-fold increase.

In an additional child, the sequential GHRP-2 plus GHRH study was performed on a regimen of on and off clonidine. He was treated with clonidine for attention deficit disorder. While taking clonidine, the peak GH value to GHRP-2 administration was 81.0 µg/L and 70.0 µg/L to GHRH (ratio 1:2); on the other hand, after not taking clonidine for 2 weeks, peak GH value to GHRP-2 administration was 34.4 µg/L, and 19.0 µg/L to GHRH (ratio 1.8; Fig. 12).

IV. DISCUSSION

Here we provide preliminary data to support our previously reported hypothesis and rationale for a three-step provocative testing analysis to determine the neuroendocrine bases of GH secretory function (Fig. 13; 22). The results of our study support the hypothesis that GHRH and an endogenous ligand or analog of GHRP (and nonpeptidyl mimics) are complementary GH secretagogues that together provide appropriate stimulation of the pituitary gland to sustain nor-

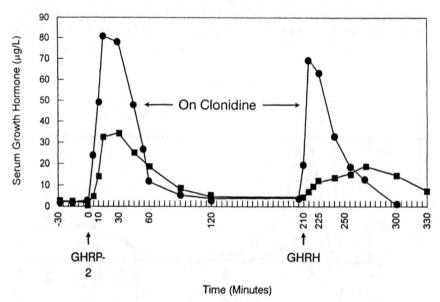

Figure 12 The GH secretory response to GHRP-2 and GHRH in a child with attention deficit disorder while taking clonidine and after not taking clonidine for 2 weeks. (From Ref. 22.)

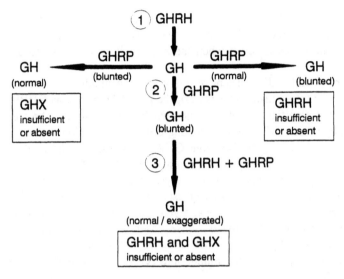

Figure 13 A three-step provocative pituitary function test using responses to sequentially administered and coadministered GH secretagogues as diagnostic parameters for determining the neuroendocrine basis of GH secretory dysfunction. GHX, peptidyl and nonpeptidyl forms of GHRP. (From Ref. 22.)

mal GH production and release. By this logic, GHRH and GHRP could be used alone and in combination to diagnose pituitary-based inadequate GH secretion in slow-growing children and "aging" adults. We devised this model to directly test pituitary GH secretory capability (see Fig. 1). The model assumes that a "normal" response to exogenous GHRP or GHRH requires the presence of its endogenous analog (i.e., GHRH or GHRP, respectively). A blunted response to either exogenous GH secretagogue is interpreted as indicating a deficiency of its endogenous complement. Blunted responses to both exogenous GH secretagogues administered sequentially, implies deficiencies of both endogenous complements. This condition can be differentiated from inherent pituitary problems, such as those involving receptor or second-messenger deficits by a "normal" response to GHRH and GHRP coadministration. A blunted response following coadministration of both GH secretagogues would indicate inherent pituitary dysfunction, rather than inadequate endogenous stimuli. More "definitive" support for this type of provocative testing would be provided by long-term therapy demonstrating improved growth in children and "metabolic" improvement in adults associated with the prolonged recombinant GH therapy.

Other investigators have compared GHRP alone and in combination with GHRH in various experimental settings (24–40). The observation of synergy was quite consistent. Pombo et al. have also demonstrated an absence of GH secretion to either one or both stimuli in children with neonatal stalk section (34). On the other hand, our studies have shown preservation of the relative synergy in a qualitative sense, even when there is significant damage to the pituitary.

The most profound and consistent results were seen in adult volunteers. In all the older adults (more than 30 years old), the peak GH valve was GHRP-2 administration was significantly greater than that following GHRH. As previously reported by others, GH release following GHRH provocation decreases with aging and to a lesser degree following GHRP provocation. The relative GH output, as expressed by the peak GH ratio (GHRP/GHRH), was most exaggerated in a 33-year-old man. For the 29-year-old woman the ratio was reversed. These preliminary observations suggest a dichotomous reversal of GH secretory response to these secretagogues at a younger adult age than might be expected. According to our hypothesis, these observations would imply a diminution in endogenous ligand for GHRP in aging. We have previously reported experimental evidence from animals and relevant clinical studies in children (17,28,29).

The use of GHRH to evaluate the extent to which GH can be released is invalid, because, after many years of laboratory and clinical studies, it is now recognized that responses to provocative challenges are not dose-dependent (31). In fact, maximal GHRH-mediated GH secretion can be amplified by an ancillary factor, such as prostaglandins, arginine, or other such GHRP (32,33). Even when GHRH is administered to subjects with normally functioning GH neuroendocrine axes, it stimulated different degrees of GH secretion (30,31). This was attributed to interaction with endogenous GH-regulatory peptides, including somatostatin or even GHRH itself (25). Blunted responses to GHRH were attributed to interaction of the GHRH stimulus with somatostatin. Alternatively, when GHRH was given during "troughs" of episodic secretion or during the early-rising stages of each episode, the stimulated response was greater than it was when GHRH was administered at the peak of the endogenous GH secretory episode or during its descending phase (26). Given this normal variation in the response to GHRH provocative stimuli, the diagnostic value of this peptide became limited.

Interpretation of the results must be in the context of the timing of the exogenous bolus secretagogue relative to the phase of the endogenous GH secretory pulse (26), provocative testing reproducibility, and the known difficulties in the diagnosis of growth disorders (1,35–40). The preliminary observations described here may, partly explain some of the confusion related to diagnostic testing of children with GH secretory abnormalities. With these considerations in mind we tested children with various causes of GH secretory dysfunction, ranging from hypopituitarism to that resulting from radiation therapy.

The GH-deficient children had less GH secretion following GHRP-2, but there was overlap with normal and slow-growing, non–GH-deficient children. The most profoundly affected were children with multiple hypothalamic–pituitary deficiency, suggesting that, for most children, the deficiency in GHRH is relative. The same is true for the GHRH provocative responses, but to a smaller degree, suggesting a smaller deficiency in endogenous ligand for GHRP. It is of interest that the relation of synergy between GHRP-2 and GHRH was preserved in the children who had profound GH deficiency, and even in the child who had a hypophysitis, with very little residual pituitary tissue (no pituitary discernible by magnetic resonance imaging). Interestingly, although there was variability in GH secretory patterns of two slow-growing, non–GH-deficient children, the ratio of peak GH to GHRP-2 plus GHRH remained intact (Fig. 5). This suggests an important functional qualitative component that these two children might have in common. In our study, there was no correlation of peak GH value after GHRH administration with mean peak GH levels to standard GH provocative testing, whereas there was a positive correlation between peak GH values after GHRP-2 administration and mean peak GH response to standard provocative stimuli.

Interpretation of the GH response must be in the clinical context. For example, a child with short stature might be thought to release GH well within the normal range with a provocative dose of GHRH until it was recognized that the child was receiving clonidine for attention deficit disorder. As noted in our study, after clonidine was removed (and hypothetically its influence on endogenous GHRH secretion), the GH response was diminished to both GHRP and GHRH in a sequential provocative study.

Radiation of the head for treatment of cancer in children retards growth, presumably by reducing the secretion of endogenous GHRH from hypothalamic neurons. We have suggested a hierarchy of dysfunction (extrahypothalamic neurotransmitters > hypothalamic GHRH or somatostatin secretion > pituitary GH secretion (41). This hypothesis should also include damage and dysfunction of the neurons producing the endogenous ligand for GHRP. In our study, we demonstrate a variable response consistent with the complexity described in the foregoing as well as the other potential reasons for variability discussed earlier.

V. SUMMARY

We have presented data in support of a novel diagnostic test for evaluating pituitary function in slowly growing children and aging adults. The objective of the test is to determine whether it is feasible to use GHRP and GHRH as diagnostic tools to investigate the etiology of GH deficiency. Additional diagnostic studies are necessary to corroborate these preliminary observations. We hope that the data resulting from the further application of the principles on which the diagnostic test is based will allow appropriate selection of therapeutic entities, ranging from GHRH or GHRP given separately or in combination, or alternatively recombinant GH, the latter for subjects lacking a pituitary mechanism for GH production or secretion.

ACKNOWLEDGMENTS

We thank Dr. C. Y. Bowers, Tulane University and Dr. B. Gonen for providing GHRP-2 and Wyeth–Ayerst for grant support. We thank Dr. B. Rubalcava of ICN Pharmaceuticals for providing GHRH 1–44.

REFERENCES

1. Spiliotis BE, August GP, Hung W, Sonis W, Mendelson W, Bercu BB. Growth hormone neurosecretory dysfunction: a treatable cause of short stature. JAMA 1984; 251:2223–2230.
2. Wehrenberg WB, Ling N. The absence of an age-related change in the pituitary response to growth hormone-releasing factor in rats. Neuroendocrinology 1983; 37:463–466.
3. Pavlov EP, Harman SM, Merriam GR, Gelato MC, Blackman MR. Responses of growth hormone (GH) and somatomedine-C to GH-releasing hormone in healthy aging men. J Clin Endocrinol Metab 1986; 62:595–600.
4. Ceda GP, Valenti G, Butturini U, Hoffman AR. Diminished pituitary responsiveness to growth hormone-releasing factor in aging male rats. Endocrinology 1986; 1818:2109–2114.
5. Ghigo E, Goffi S, Arvat E, Nicolosi M, Procopio M, Bellone J, Imperiale E, Mazza E, Baracchi G, Camanni F. Pyridostigmine partially restores the GH responsiveness to GHRH in normal aging. Acta Endocrinol 1990; 123:169–174.
6. Iovino M, Monteleone P, Steardo L. Repetitive growth hormone-releasing hormone administration restores the attenuated growth hormone (GH) response to GH-releasing hormone testing in normal aging. J Clin Endocrinol Metab 1989; 69:910–913.
7. Lang I, Schernthaner G, Pietschmann P, Kurz R, Stephenson JM, Templ H. Effects of sex and age on growth hormone response to growth hormone-releasing hormone administration restores the attenuated growth hormone (GH) response to GH-releasing hormone testing in normal aging. J Clin Endocrinol Metab 1989; 69:910–913.
8. Lang I, Kruz R, Geyer G, Tragl KH. The influence of age on human pancreatic growth hormone releasing individuals. J Clin Endocrinol Metab 1987; 65:535–540.

9. Shibaski T, Shizume K, Nakahara M, Masuda A, Jibiki K, Demura H, Wakabayashi I, Ling N. Age-related changes in plasma growth hormone response to growth hormone-releasing factor in man. J Clin Endocrinol Metab 1984; 58:212–214.

10. Sonntag WE, Gough MA. Growth hormone releasing hormone induced release of growth hormone in aging rats: dependence on pharmacological manipulation of endogenous somatostatin release. Neuroendocrinology 1988; 47:482–488.

11. Sonntag WE, Steger RW, Forman LJ, Meites J. Decreased pulsatile release of growth hormone in old male rats. Endocrinology 1980; 107:1875–1879.

12. Sonntag WE, Hylka VW, Meites J. Impaired ability of old male rats to secrete growth hormone in vivo but not in vitro in response to hpGRF (1–44). Endocrinology 1983; 113:2305–2307.

13. De Gennaro Colonna V, Zoli M, Cocchi D, Maggi A, Marrama P, Agnati LF, Muller EE. Reduced growth hormone releasing factor (GHRF)-like immunoreactivity and GHRF gene expression in the hypothalamus of age rats. Peptides 1989; 10:705–708.

14. Morimoto N, Kawakami F, Makin S, Chihara K, Hasegawa M, Ibata Y. Age-related changes in growth hormone releasing factor and somatostatin in the rat hypothalamus. Neuroendocrinology 1988; 47:459–464.

15. Ono M, Miki N, Shizume K. Release of immunoreactive growth hormone-releasing factor (GRF) and somatostatin from incubated hypothalamus in young and old male rats (abstr). Neuroendocrinology 1986; 43:111.

16. Walker RF, Yang S-W, Bercu BB. Robust growth hormone (GH) secretion in aged female rats co-administered GH-releasing hexapeptide (GHRP-6) or GH releasing hormone (GHRH). Life Sci 1991; 49:1499–1504.

17. Bercu BB, Walker RF. A diagnostic test employing growth hormone secretagogues for evaluating pituitary function in the elderly. In: Bercu BB, Walker RF, eds. Growth Hormone Secretagogues. New York: Springer-Verlag, 1996:289–305.

18. Bowers CY, Sartor AO, Reynolds GA, Badger TM. On the actions of growth hormone-releasing hexapeptide, GHRP. Endocrinology 1991; 128:2027–2035.

19. Bercu BB, Yang S-W, Masuda R, Walker RF. Role of selected endogenous peptides in growth hormone releasing hexapeptide (GHRP-6) activity: analysis of GHRH, TRH, and GnRH. Endocrinology 1992; 130:2579–2586.

20. Howard AD, Feighner SD, Cully DF, Arena JP, Liberator PA, Rosenblum CI. A receptor in pituitary and hypothalamus that functions in growth hormone release. Science 1996; 273:974–977.

21. Goth MI, Lyons CE, Canny BJ, Thorner MO. Pituitary adenylate cyclase activating polypeptide, growth hormone (GH)-releasing peptide and GH-releasing hormone stimulate GH release through distinct pituitary receptors. Endocrinology 1992; 130:939–944.

22. Bercu BB, Walker RF. Evaluation of pituitary function in children using growth hormone secretagogues. J Pediatr Endocrinol Metab 1996; (suppl 3):325–332.

23. Bercu BB, Weideman CA, Walker RF. Sex differences in growth hormone secretion by rats administered GH-releasing hexapeptide (GHRP-6). Endocrinology 1991; 129:2592–2598.

24. Walker RF, Yang S-W, Bercu BB. Robust growth hormone (GH) secretion in aged female rats co-administered GH-releasing hexapeptide (GHRP-6) and GH-releasing hormone (GHRH). Life Sci 1991; 49:499–1504.

25. Bercu BB, Yang S-W, Masuda K, Walker RF. Role of selected endogenous peptides in growth hormone releasing hexapeptide (GHRP-6) activity: analysis of GHRH, TRH and GnRH. Endocrinology 1992; 130:2579–2586.

26. Cho KH, Yang SW, Hu C-S, Bercu BB. Growth hormone (GH) response to growth hormone-releasing hormone (GHRH) varies with intrinsic growth hormone secretory rhythm in children: Reduced variability using somatostatin pretreatment. J Pediatr Endocrinol 1992; 5:155–165.

27. Bercu BB, Yang S-W, Masuda R, Hu C-S, Walker RF. Effects of co-administered growth hormone (GH) releasing hormone and GH-releasing hexapeptide on maladaptive aspects of obesity in Zucker rats. Endocrinology 1992; 131:2800–2804.

28. Walker RF, Yang S-W, Masuda R, Bercu BB. Effects of GH releasing peptides on stimulated GH secretion in old rats. In: Bercu BB, Walker RF, eds. Basic and Clinical Aspects of Growth Hormone II. New York: Springer-Verlag, 1994:167–192.

29. Walker RF, Ness GC, Zhao Z, Bercu BB. Effects of stimulated GH secretion of age-related changes in plasma cholesterol and hepatic low density lipoprotein messenger RNA concentrations. Mech Aging Dev 1994; 2:857–862.

30. Bercu BB, Walker RF. A diagnostic test employing growth hormone secretagogues for evaluating pituitary function in the elderly. In: Bercu BB, Walker RF, eds. Growth Hormone Secretagogues. New York: Springer-Verlag, 1996:289–305.

31. Walker RF, Bercu BB. An animal model for evaluating growth hormone secretagogues. In: Bercu BB, Walker RF, eds. Growth Hormone Secretagogues. New York: Springer-Verlag, 1996:253–287.

32. Bercu BB, Walker RF. Evaluation of pituitary function in children using frowth hormone secretagogues. J Pediatr Endocrinol Metab 1996; 9(suppl 3):325–332.

33. Bercu BB, Walker RF. Novel growth hormone secretagogues: clinical applications. Endocrinologist 1997; 7:51–64.

34. Pombo M, Barreiro J, Peñalva A, Peino R, Dieguez C, Casanueva FF. Absence of growth hormone (GH) secretion after the administration of either GH-releasing hormone (GHRH), GH-releasing peptide (GHRP-6), or GHRH plus GHRP-6 in children with neonatal pituitary stalk transection. Trends Endocrinol Metab 1995; 80:3180–3184.

35. Tuilpakov AN, Bulatov AA, Peterkova VA, Elizarova GP, Volevodz NN, Bowers CY. Growth hormone (GH)-releasing effects of synthetic peptide GH-releasing peptide-2 and GH-releasing hormone $(1-29NH_2)$ in children with GH insufficiency and idiopathic short stature. Metabolism 1995; 9:1199–1204.

36. Patchett AA, Nargund RP, Tata JR, Chen M-H, Barakat KJ, Johnston DBR, Cheng K, Chan WW-S, Butler B, Hickey G, Jacks T, Schleim K, Pong S-S, Chaung L-YP, Chen HY, Frazier E, Leung KH, Chiu S-HL, Smith RG. Design and biological activities of L-163,191 (MK-0677): a potent, orally active growth hormone secretagogue. Proc Natl Acad Sci USA 1995; 92:7001–7005.

37. Popovic V, Micic D, Damjanovic S, Djurovic M, Simic M, Gligorovic M, Dieguez

C, Casanueva FF. Evaluation of pituitary GH reserve with GHRP-6. J Pediatr Endocrinol Metab 1996; 9:289–298.

38. Bowers CY, Granda-Ayala R. GHRP-2, GHRH, and SRIF interrelationships during chronic administration of GHRP-2 to humans. J Pediatr Endocrinol Metab 1996; 9:261–270.
39. Pombo M, Leal-Cerro A, Barreiro J, Peñalva A, Reino R, Mallo F, Diequez C, Casanueva FF. Growth hormone releasing hexapeptide-6 (GHRP-6) test in the diagnosis of GH-deficiency. J Pediatr Endocrinol Metab 1996; 9:333–338.
40. Bercu BB, Walker RF. Growth hormone secretagogues in children with altered growth. Acta Paediatr Scand. In press.
41. Jorgensen EV, Schwartz ID, Hvizdala E, Barbosa J, Phuphanich S, Shulman DI, Root AN, Estrada J, Hu C-S, Bercu BB. Neurotransmitter control of GH secretion in children after cranial radiation therapy. J Pediatr Endocrinol 1993; 6:131–142.

21

Clinical Use of Growth Hormone Secretagogues

Mark A. Bach
Merck Research Laboratories, Merck and Co., Inc., Rahway, New Jersey

I. INTRODUCTION

The growth of interest in clinical uses for growth hormone (GH) has largely been fueled by the unlimited supplies made available through recombinant technology since 1985. The seminal observation by Bowers et al. that met-enkephalin analogs could stimulate secretion of growth hormone from pituitary cells (1), followed by the continuing refinement of GH secretagogues (GHSs) to improve specificity of hormonal stimulation (2) has, in itself, led to a growing field of study that may improve our understanding of the control of endogenous GH secretion. Although it has been demonstrated that the GH-releasing peptides (GHRPs) elicit GH release in humans when delivered intravenously (3), intranasally (4), or orally (5), their bioavailability is limited. Consequently, much effort has been put toward the development of low molecular weight compounds that mimic the action of the GHRPs through binding at the same receptor (6–8). These compounds allow improved oral bioavailability compared with peptides and may expand the potential clinical usefulness of GH secretagogues. The first of these compounds L-629,429, is a nonpeptidyl benzolactam (9). In 1995, Patchett et al. described L-163,191 (MK-0677), a spiropiperidine GHRP mimetic with good oral activity (10). These two compounds, as well as GHRP-6, GHRP-2, and hexarelin* have been studied in humans in several patient populations. Although the differences among these compounds will be highlighted, for the purpose of this review, GH secretagogues will refer to the GHRPs and GHRP mimetics that act through the same recently described receptor (11,12).

---

*Hexarelin is a trademark for examorelin (I.N.N.).

The pharmacology of the GH secretagogues, particularly relative to duration of action, will be briefly discussed. In addition, the rationales for clinical target selection and strategies for clinical development will be discussed.

II. PHARMACOLOGY

The mechanism of GH release by the GH secretagogues is complex and not completely understood. Both animal and human data demonstrate that the secretagogues bind to pituitary somatotrophs and cause direct stimulation of GH secretion (11,13–23). The secretagogues also bind to cells within the hypothalamus (24) where the GH secretagogue receptor has been identified (11). Most studies suggest that the physiological action of the secretagogues occurs both at the pituitary and at the level of the hypothalamus; therefore, an intact hypothalamic–pituitary axis is required for a vigorous GH response (see Chap. 3). Consistent with this, animal studies have shown stimulation of hypothalamic GHRH secretion in response to hexarelin, but no change in hypophysial portal somatostatin levels (25). A review of the mechanisms of action of the GH secretagogues can be found in this text (see Chap. 11).

This chapter will focus primarily on the characteristics of the GH secretagogues that are relevant to their development as therapeutic agents. There have been few direct comparisons of specific agents, making conclusions difficult. However, although all of the GH secretagogues apparently act through the same receptor, there appear to be fundamental differences in physiological responses to these agents.

The hormonal responses to a single bolus dose of a GH secretagogue have been remarkably consistent for each of the peptidyl compounds. Following an intravenous, oral, or intranasal dose of a secretagogue, a single broad GH peak is detected (26–34). Where measured, this GH peak is accompanied by smaller peaks of prolactin and corticotropin (adrenocorticotropic hormone; ACTH)-mediated cortisol secretion (33,35). It is possible that multiple doses of a secretagogue would induce feedback mechanisms, such that GH responses would not be observed after multiple doses. Down-regulation, or desensitization, after multiple doses of hexarelin has been reported (36). To be useful as a therapeutic agent, some degree of hormonal response must presumably be sustained. In contrast with the study just noted, other authors have reported that the GH response to multiple doses of hexarelin is sustained. Ghigo et al. (37) dosed healthy elderly subjects with hexarelin intranasally or orally for 8 or 15 days and found that the immediate GH secretory response was sustained (Fig. 1), although there was no significant increase in serum insulin-like growth factor-I (IGF-I) levels. Mericq et al. (38) and Pihoker et al. (39) treated GH-deficient

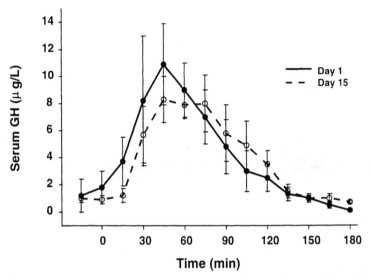

Figure 1 Seven healthy elderly women were treated with hexarelin orally (20 mg tid) for 15 days. Blood samples for GH were obtained every 15 min from −15 min to 180 min postdose on the days of the first and last dose. The GH AUC was not significantly different between the first and last doses. (Adapted from Ref. 37.)

children with GHRP-2 for 6 months, with no significant increase in serum IGF-I, but reported an increase in growth velocity. Laron et al. (40) treated non–GH-deficient, short children with hexarelin, 60 μg/kg/tid intranasally, for up to 8 months and reported a significant increase in serum IGF-I. The apparent variability in serum IGF-I response to multiple doses of peptide secretagogues may be related to differences in response to GHRP-2 and hexarelin, differences in responsiveness in the patient populations studied, or intersubject variability in response-making detection of a small response difficult when the sample size tested is limited.

The GH response to a single dose of GHRP-6 is similar to other peptidyl secretagogues. Bowers et al. (41) infused GHRP-6 (0.1, 0.3, or 1 μg/kg) into 18 healthy young adult men over 1 min. Approximately 40 min after infusion of the highest dose, they demonstrated a peak GH response of approximately 60 μg/L (Fig. 2a). This broad GH peak gradually returned to baseline levels approximately 150 min after the dose. In a revealing study, Huhn et al. (42) infused GHRP-6 into 8 healthy young adult men. In this study, in a double-blind random crossover design, subjects received saline or GHRP-6 (1 μg/kg per hour) over a 24-h period. Serum was sampled for GH levels at 10 min intervals. Rather than the single GH peak observed with a bolus of GHRP-6,

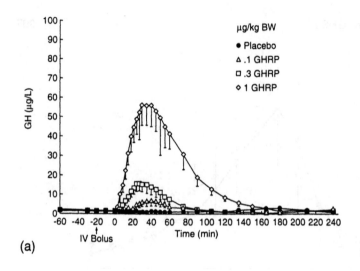

(a)

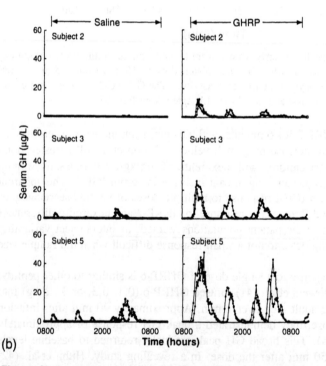

(b)

Figure 2 (a) Healthy men (23–36 years old) were treated, in a partial crossover design, with a single iv bolus dose of GHRP-6 to generate this dose response curve. Mean (± SEM) GH levels (n = 6–9 per group) are shown. (b) Serum samples were collected every 10 min from each of three normal subjects during 24-h saline or GHRP-6 (1 μg/kg · h, iv) infusions. Samples were analyzed for GH by immunoradiometric assay. (Adapted from: (a), Ref. 41; (b), Ref. 42.)

a sustained, upregulated pulsatile pattern of GH secretion (compared with saline) was detected (see Fig. 2b). A similar observation was reported by Jaffe et al. (43). Thus, the GH response to sustained serum levels of secretagogue was fundamentally different from that of a bolus dose. This observation leads to speculation that, although acting at the same receptor, dosing with longer-acting secretagogues may result in serum hormonal profiles that are very different from those seen with the relatively short-acting peptides.

MK-0677 is a GHRP mimetic (41) with a half-life of 5–6 h in dogs (44). The GH secretory response to a single oral dose in healthy young men was very similar to that of the peptidyl secretagogues (45). Similarly, a single oral dose of MK-0677 (25 mg) in healthy elderly men and women resulted in a single, broad peak of GH secretion over approximately 8 h (Fig. 3). In another study, healthy elderly men and women were treated with MK-0677 25 mg orally (46). GH, cortisol, and prolactin were sampled every 20 min for 24 h before and after 14 daily doses. On day 14, there was an increase in GH peak amplitude and area under the curve (AUC), but no detectable change in peak number and no postdose peak in GH secretion. The GH secretory pattern from a representative subject is shown in Figure 4. Serum cortisol AUC was unchanged from the baseline. With MK-0677 treatment there was an approximately 24% increase in mean serum prolactin levels on day 14; these levels remained within the

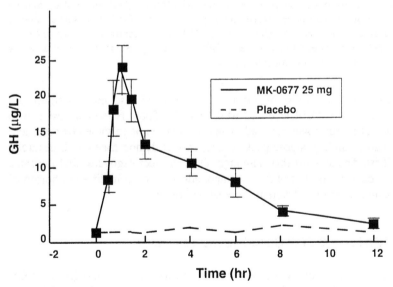

Figure 3   Healthy elderly subjects were treated with a single oral dose of MK-0677 (25 mg or placebo). Subjects underwent serial sampling every 20 min for 12 h for GH levels. (Adapted from Ref. 45.)

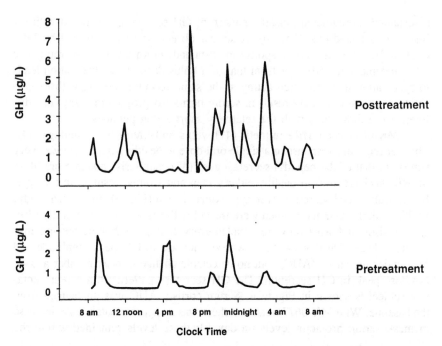

Figure 4   Healthy elderly subjects were treated with MK-0677 (25 mg or placebo) daily at 10–11 PM. Subjects underwent serial sampling every 20 min for 24 h for GH at baseline and after 14 days of treatment. Shown are the 24-h GH concentrations from a 72-year-old man before and after 14 daily doses of MK-0677 25 mg. Blood was collected every 20 min from 8 AM until the following 8 AM.

normal range. Two weeks of dosing with MK-0677 resulted in a significant increase in serum IGF-I levels as well (Fig. 5). Thus, repetitive dosing with this oral GH secretagogue resulted in an upregulation of pulsatile GH secretion, rather than a single postdose peak of GH, as was demonstrated with hexarelin (37). These data suggest that, although all of the secretagogues bind the same receptor, other properties of these compounds may result in different hormonal profiles and, possibly, different therapeutic potential.

III.   STRATEGIES FOR CLINICAL DEVELOPMENT

The approved clinical use of recombinant human GH (rhGH) includes GH deficiency in adults and children, short stature in association with renal insufficiency, AIDs-related wasting, and short stature associated with Turner's syndrome. In addition, the clinical use of GH is being explored for several other

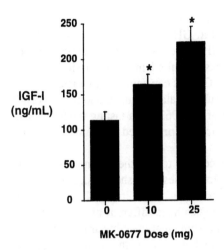

Figure 5 Healthy elderly men and women were treated with MK-0677 (10 mg, $n = 12$; 25 mg, $n = 6$; or placebo, $n = 14$) daily at 10–11 PM. Serum IGF-I levels were obtained after 14 days of treatment. (Adapted from Ref. 46.)

conditions, including age-related muscle loss (sarcopenia; 47–56), improvement in perisurgical or parenteral nutrition and other catabolic states (57–66), obesity (67–69), and non–GH-deficient, short stature (70–82; Table 1). GH secretagogues could provide an attractive alternative to GH for some of these conditions. Potential advantages of secretagogues include an oral route of administration as well as the more physiological GH secretory pattern attained. However, it is also possible that secretagogues will have limited efficacy in many GH-deficient adults and a subset of GH-deficient children in whom the hypothalamic–pituitary axis is not intact. In addition, secretagogues may be unable to completely overcome the GH resistance associated with catabolic states (83,84). Therefore, in some conditions in which GH therapy is currently being explored (e.g., wound healing, postsurgical rehabilitation; 57,85–87) or used (e.g., AIDS wasting; 88), the secretagogues may have limited efficacy. However, Van den Berghe et al. (12) performed a study in which an iv infusion of

Table 1 Clinical Uses of Recombinant Human Growth Hormone

Approved	Exploratory
GH-deficient children	Elderly
Growth in renal failure	Perisurgical
Turner's syndrome	Obesity

GHRP-2 (1 µg/kg per hour) was administered for a 9-h period to critically ill adults who met specific inclusion criteria. GHRP-2 administration resulted in a four- to six-fold increase in mean GH secretion, with an increase in peak amplitude and basal level. This increase resulted in an approximately 60% increase in serum IGF-I, suggesting that GHRPs can stimulate biologically active increases in GH in the presence of presumed catabolism. In addition, recently, Clemmons et al. (89) reported the use of MK-0677 in a fasting model for catabolic states. In this study, MK-067 partially prevented the nitrogen wasting associated with fasting.

Controlled clinical studies are necessary to test the therapeutic potential of GH secretagogues. However, the pilot studies published to date suggest the potential for salutary effects in several conditions, including catabolic states. Because short- and longer-acting secretagogues have different properties relative to hormonal responses, one cannot also assume that the clinical response to one compound is generalizable to others. In addition, as with any new drug, there may be unknown and unanticipated side effects. Planning the clinical development of growth hormone secretagogues requires careful consideration of the relative advantages and disadvantages of these compounds in each potential patient population. A review of the clinical literature on GH secretagogues in key patient populations follows.

A. Short Stature: Diagnosis and Therapy

Growth hormone-deficient short stature may result from a defect at the pituitary, within the central nervous system (CNS), or as a result of hypothalamic–pituitary disconnection. Idiopathic GH deficiency probably is caused by hypothalamic or neurosecretory defects, and accounts for most of the GH deficiency in children (90). The GH secretagogues may be useful in identifying the etiology of GH deficiency (see Chap. 20). Patients with low or absent GH response to secretagogues could be tested in combination with GHRH to differentiate between pituitary and hypothalamic defects. In short, normal children, the GH secretory response to hexarelin varied with pubertal status (91). A clinical role for a GH secretatogue diagnostic test will require establishment of normal ranges, probably adjusted for pubertal status.

1. Growth Hormone-Deficient Children

Growth hormone-deficient children are an obvious clinical target for treatment with GH secretagogues. However, it seems intuitive that only a subset of these children will respond to secretagogues with GH secretion, for an intact hypothalamic–pituitary connection and a functioning pituitary are required for an optimal response. Several groups have performed studies that begin to define the characteristics of secretagogue responders among GH-deficient children. Loche et al. (92) studied 15 children and 4 adults who met the classic criteria for GH deficiency, compared with 45 short, normal children. GH therapy was discontinued for 2–4 weeks before testing with a single dose (2 µg/kg iv) of hexarelin. Among the GH-deficient children, those with structural defects detected on magnetic resonance imaging (MRI) had a lower mean peak GH response (5.5 ± 2.3 ng/mL) than those with normal MRI (mean peak GH 63.0 ± 6.5 ng/mL). The latter response was indistinguishable from that of short, normal children (mean peak GH responses of 51.7 ± 3.7 ng/mL). Tiulpakov et al. (93) performed a similar experiment in which nine GH-deficient children and five children with idiopathic short stature received a single dose of GHRP-2 (1 µg/kg iv). In this study, one GH-deficient children responded with a GH peak response higher than 30 ng/mL; the others had peak responses less than 1.5 ng/mL. In contrast, the children with idiopathic short stature demonstrated peak GH responses from 8.7 to more than 100 ng/mL (Fig. 6). Mericq et al. (94) tested 22 prepubertal GH deficient children for a GH response to GHRP-1 (1 µg/kg iv), approximately 60% of the subjects had a significant response, defined as four times the standard deviation of the GH assay. However, the mean response was 7.5 ± 8.0 ng/mL, substantially less than the response reported in normal subjects. Subsequently, patients who responded to a single dose were treated daily with GHRP-2 (0.3 µg/kg sc). Doses were increased to 1 and 3 µg/kg at 2-month intervals. After 6 months of treatment, mean growth veloc-

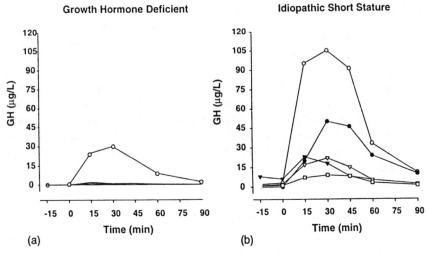

Figure 6 The GH secretory response to a single bolus dose of GHRP-2 (1 µg/kg, iv) in (a) children with GH deficiency ($n = 9$) and (b) children with idiopathic short stature ($n = 5$). (Adapted from Ref. 97.)

ity approximately doubled (2.5 ± 0.5 cm/yr to 5.6 ± 1.5 cm/yr). Interestingly, there was no significant change in the serum IGF-I level, perhaps implying a suboptimal dose or dosing regimen was used (38).

Results of the efficacy trials of GH secretagogues in GH-deficient children described to date (38,92,93) are intriguing. However, additional studies are necessary to determine the profile of responsiveness and degree of efficacy in this population. In selecting children who demonstrated a hormonal response to a single dose of secretagogue, some potential responders may have been excluded. A priming effect with repeated exposure to a secretagogue may be necessary to accurately ascertain the potential for response in an individual. It is also possible that the initial response may not be sustained after multiple doses, for downregulation of the GH response to growth hormone secretagogues has been described after prolonged dosing (95), presumably mediated by rising IGF-I levels. Thus, it may not be possible to sustain the supraphysiological GH levels seen with exogenous GH treatment which, in turn, may limit the ability to achieve "catch-up" growth. However, it is also possible that the GH levels achieved with recombinant GH may result in GH receptor downregulation (96), leading to the declining growth velocities seen with increasing duration of GH treatment. Thus, because of the differing serum GH profile, the first-year growth comparison of GH secretagogues to GH may not be representative of the long-term results. An additional consideration is that all of the clinical studies in GH-deficient children employed short-acting peptide secretagogues. It is possible that different pharmacodynamic responses with longer-acting compounds (e.g., MK-0677) may result in a different efficacy profile.

Finally, the cohorts of GH-deficient children tested by Mericq (38) and Tiulpakov (97) may not be representative of most children currently diagnosed as GH deficient and receiving replacement therapy. Growth hormone deficiency is heterogeneous, representing a spectrum, ranging from absolute deficiency to the upper limit of arbitrarily defined "deficient" range, with a peak GH response of 10 μg/L to classic provocative stimuli. As the criteria for diagnosis of deficiency have gradually become less stringent, a smaller proportion of these patients would be classified as absolutely deficient and, hence, least likely to respond to secretagogues. Therefore, in spite of the modest clinical efficacy described in Chilean children (38), additional efficacy trials of growth hormone secretagogues are needed to determine efficacy in other populations.

2.   Non–Growth Hormone-Deficient Short Stature

Several groups have investigated the use of GH secretagogues in children with non–GH-deficient short stature. These studies include the group reported by Tiulpakov (97), Laron (98; also see Chap. 22), and Bellone (99). The GH response to secretagogues was higher than that seen in GH-deficient children and was similar to historical reports of the response in children of normal stature. Laron et al. (100) have reported increases in growth velocity in these children treated with intranasal hexarelin. In an uncontrolled trial, eight prepubertal children were treated with hexarelin (60 µg/kg tid) intranasally for up to 8 months. Growth velocity increased significantly (mean ± SD) from 5.3 ± 0.8 to 8.3 ± 1.7 cm/yr. Whether this degree of benefit would translate into increased adult height after long-term therapy is unknown. Therefore, use of GH secretagogues in non–GH-deficient short stature should be limited to controlled clinical studies in children with extreme short stature for whom clinical use may be more justified and efficacy can be defined.

3. Other Growth Disorders

Recombinant human GH is approved for treatment of girls with Turner's syndrome and growth failure in children with renal failure. Both of these groups of patients would be expected to have intact hypothalamic–pituitary axes and, therefore, to secrete GH in response to secretagogue stimulation. Adults with chronic renal failure secrete GH in response to hexarelin (101). However, it is unknown whether sufficiently high levels of GH can be stimulated in these patients over a long enough time to demonstrate clinical efficacy. Before pursuing efficacy trials in this population, pilot studies to demonstrate hormonal responsiveness to GH secretagogues would be needed.

B. Potential Indications in Adults

Classic GH deficiency as well as normal aging are associated with decreased GH secretion when compared with young healthy adults. In addition, catabolic conditions, including type I diabetes, postoperative recovery, and malnutrition from a variety of causes, are characterized by relative GH resistance, resulting in elevated levels of GH and decreasing circulating IGF-I (84). The rationale for treatment and the therapeutic potential of growth hormone secretagogues in these conditions is discussed in the following.

1. Aging

Growth hormone secretion decreases after puberty, and by the seventh decade, many elderly subjects have endogenous GH and IGF-I levels that fall in the range of classically GH-deficient patients (102,103; Fig. 7). Muscle strength and mass also decrease with increasing age, and these losses are further exacerbated by acute illness or immobilization. Because GH is an anabolic hormone, these observations led to the hypothesis by Rudman et al. (104) that replacing GH in elderly subjects may result in improvements in physical functioning. In 1990, these authors presented data from a study in which 12 healthy elderly men, with

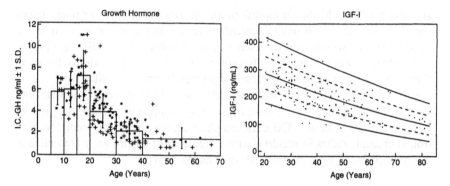

Figure 7 Decrease in GH and IGF-I secretion with advancing age: Integrated concentration of GH was determined every 20 min for 24 h in 173 nonobese subjects from 7 to 65 years of age. (Adapted from Refs. 103 and 102.)

low serum IGF-I, were treated with rhGH for 6 months, and were compared with untreated controls. The treated group demonstrated an increase in IGF-I and lean mass and a decrease in fat mass. Subsequently, several other groups have performed similar studies in healthy elderly men or women and have also demonstrated salutary effects on body composition. However, most investigators have not yet been able to demonstrate an improvement in strength or function with GH treatment (38,51,97,105,106). However, in a well-designed trial, Welle et al. (56) recently reported a 10–12% increase in muscle strength in healthy elderly after 3 months of treatment with relatively low doses (0.03 mg/ kg three times weekly) of recombinant human GH. Several additional studies are in progress, including a series of studies funded by the National Institute on Aging as part the "Trophic Factors Initiative." Data from these studies will become available over the next few years.

Difficulty in demonstrating functional efficacy in the studies of GH in the elderly may result from several factors. All studies yet published have recruited healthy elderly subjects, in whom demonstration of functional improvement may be most difficult. In addition, poor tolerability of exogenous GH resulted in subject attrition or in the need to reduce doses in most studies. Because poor tolerability may be related to the nonphysiological pattern and supraphysiological levels of GH attained with exogenous GH, it is possible that GH secretagogues may have an advantage in tolerability, particularly in the elderly population.

Secretagogues with oral bioavailability, such as hexarelin, have been demonstrated to stimulate GH secretion in elderly subjects (26,107). MK-0677 has been tested in elderly subjects, with good general tolerability and sustained increases in pulsatile GH secretion and serum IGF-I levels in elderly subjects (see Figs. 4 and 5; 34). MK-0677 increases serum IGF-I levels in functionally impaired elderly subjects (108), and data will soon be available on the effects on body composition and strength in this population.

Development of GH secretagogues for treatment of the "frail" elderly presents additional challenges. Age-related musculoskeletal impairment as a result of muscle wasting (sarcopenia) is not well recognized as a treatable clinical syndrome. Thus, selection of an appropriate patient population remains difficult, and selection of endpoints for efficacy trials remains a challenge. In addition, given the inherent day-to-day variability in function in the frail target population, as well as the presence of a host of concomitant conditions, demonstration of clinically meaningful efficacy will be difficult.

2. Prolonged Corticosteroid Exposure

Prolonged exposure to corticosteroids, as a result of Cushing's syndrome or through iatrogenic causes, results in nitrogen loss and muscle wasting. The anabolic action of GH may ameliorate some of this wasting. Several studies have attempted to address this, with mixed results. Gertz et al. demonstrated that L-692,429, a nonpeptidyl GHRP analog, can partially overcome suppression of GH secretion associated with corticosteroid exposure (109). In a study in patients with untreated Cushing's syndrome, ten patients (nine with pituitary adenoma and one with adrenal adrenal adenoma) and five normal adults controls received GHRP-6, 100 µg iv (110). The postdose GH AUC was suppressed approximately 77% in Cushing's syndrome patients, compared with normal controls. However, another group (83) has report lack of a GH response to GHRP-6 (1 µg/kg) in four patients with Cushing's syndrome (cause of hypercortisolism not published). This disparity in results may be due to individual variation in responsiveness to GHRP-6 or, possibly, to different causes of Cushing's syndrome in the two reports.

The presence of a significant GH response to secretagogues in at least a subset of patients suggests the possibility of anabolic benefit if the response is sustained with multiple doses over time. However, as the GH secretagogues may exhibit modest lack of hormonal specificity, stimulation of small quantities of ACTH may complicate assessment of treatment status and could affect eventual clinical benefit. Thus, any long-term study in this population would require close monitoring.

IV. SUMMARY

The Second International Symposium on Growth Hormone Secretagogues presented an opportunity to observe the progress in this field since the previous meeting 2 years ago. At that time, a great deal of chemical discovery work was in progress, and clinical work was limited to relatively short-term experience with short-acting peptides. In the ensuing 2 years since this symposium, discovery work has continued, and we have gained a great deal of information on responsiveness to secretagogues in many pathological states. We have also begun to accumulate data on longer-term dosing, and to delineate the potential differences in hormonal profiles between short-acting and longer-acting compounds. The preliminary information available will guide us to rational selection of clinical targets for drug development. In addition, in light of the identification of the unique receptor for these compounds (11), perhaps by the time of the next international symposium, we will have isolated the presumed endogenous ligand for the secretagogues, further contributing to our understanding of the mechanism of regulation of GH secretion.

As is discussed elsewhere in this volume (see Chap. 20), the secretagogues have potential to play a diagnostic role in assessing defects in hypothalamic–pituitary axis function. In therapeutic applications, the GH secretagogues may hold specific advantages compared with rhGH. In particular, oral dosing resulting in a physiological GH profile and sustained elevations in serum IGF-I (46) may be better tolerated than GH, particularly in elderly subjects. However, little long-term safety and efficacy data are yet available for any GH secretagogues. Concerns remain over the modest lack of hormonal specificity of these compounds, although these effects may diminish with repeated dosing. In addition, as with any chemical entity, there is the possibility of other non–GH-mediated side effects that have not yet been identified.

The GH secretagogues may be useful in the treatment of GH deficiency, Turner's syndrome, and other growth disorders, a variety of catabolic conditions, and age or corticosteroid-related musculoskeletal wasting. Clinical studies are currently under way to evaluate the usefulness of secretagogues in several of these conditions. It seems likely that secretagogues will prove to have efficacy in a subset of GH-deficient patients with an intact hypothalamic–pituitary axis. In frail elderly subjects, early evidence suggests an acceptable tolerability profile, but long-term clinical benefit remains exclusive. However, the potential to affect the inevitable functional decline of aging remains an attractive goal. The next few years will bring answers to questions on the clinical potential for GH secretagogues.

REFERENCES

1. Bowers CY, Chang J, Momany F, Folkers K. Effect of the enkephalins and enkephalin analogs on release of pituitary hormones in vitro. In: McIntyre I, ed. Molecular Endocrinology. Amsterdam: Elsevier/North Holland, 1977:287.
2. Bowers CY, Reynolds GA, Chang D, Hong A, Chang K, Momany F. A study

on the regulation of growth hormone release from the pituitaries of rats in vitro. Endocrinology 1981; 108:1071–1080.

3. Ilson BE, Jorkasky DK, Curnow RT, Stote RM. Effect of a new synthetic hexapeptide to selectively stimulate growth hormone release in healthy human subjects. J Clin Endocrinol Metab 1989; 69:212–214.

4. Hayashi S, Okimura Y, Yagi H, Uchiyama T, Takeshima Y, Shakutsui S, Oohashi S, Bowers CY, Chihara K. Intranasal administration of His-D-Trp-Ala-Trp-D-Phe-LysNH₂ (growth hormone releasing peptide) increased plasma growth hormone and insulin-like growth factor-I levels in normal men. Endocrinol Jpn 1991; 38:15–21.

5. Bowers CY, Alster DK, Frenz JM. The growth hormone-releasing activity of a synthetic hexapepetide in normal men and short stature children after oral administration. J Clin Endocrinol Metab 1992; 74:292–298.

6. McDowell RS, Elias KA, Stanley MS, et al. Growth hormone secretagogues: characterization, efficacy, and minimal bioactive conformation. Proc Natl Acad Sci USA 1995; 92:11165–11169.

7. Elias KA, Ingle GS, Burnier JP, Hammonds RG, McDowell RS, Rawson TE, Somers TC, Stanley MS, Cronin MJ. In vitro characterization of four novel classes of growth hormone-releasing peptide. Endocrinology 1995; 136:5694–5699.

8. Langeland Johansen N, Madsen K, Sehested Hansen B, Ankersen M, Lau J, Thogersen H, Andersen KE. Structure activity relationship of two novel sets of GHRP receptor agonists (abstr). Endocrinol Metab 1997; 4:34.

9. Smith RG, Cheng K, Schoen WR, Pong SS, Hickey G, Jacks T, Butler B, Chan WWS, Chaung LYP, Judith F, Taylor J, Wyvratt MJ, Fisher MH. A nonpeptidyl growth-hormone secretagogue. Science 1993; 260:1640–1643.

10. Patchett AA, Nargund RP, Tata JR, et al. Design and biological activities of L-163,191 (MK-0677): a potent, orally active growth hormone secretagogue. Proc Natl Acad Sci USA 1995; 92:7001–7005.

11. Howard AD, Feighner SD, Cully DF, Arena JP, Liberator PA, Rosenblum CI, Hamelin M, Hreniuk DL, Palyha OC, Anderson J, Paress PS, Diaz C, Chou M, Liu KK, McKee KK, Pong S, Chaung I, Elbrecht A, Dashkevicz M, Heavens R, Rigby M, Sirinathsinghji DJS, Dean DC, Melilo DG, Patchett AA, Nargund R, Griffin PR, DeMartino JA, Gupta SK, Schaeffer JM, Smith RG, Van der Ploeg LHT. A receptor in pituitary and hypothalamus that functions in growth hormone release. Science 1996; 273:974–977.

12. Van den Berghe G, de Zegher F, Veldhuis JD, Wouters P, Awouters M, Verbruggen W, Schetz M, Verwaest C, Lauwers P, Bouillon R, Bowers CY. The somatotropic axis in critical illness: effect of continuous growth hormone (GH)-releasing hormone and GH-releasing peptide-2 infusion. J Clin Endocrinol Metab 1997; 82:590–599.

13. Veeraragavan K, Sethumadhavan K, Bowers CY. Growth hormone-releasing peptide (GHRP) binding to porcine anterior pituitary and hypothalamic membranes. Life Sci 1992; 50:1149–1155.

14. Sethumadhavan K, Veeraragavan K, Bowers CY. Demonstration and characterization of the specific binding of growth hormone-releasing peptide to rat ante-

rior pituitary and hypothalamic membranes. Biochem Biophys Res Commun 1991; 178:31–37.

15. Badger TM, Millard WJ, McCormick GF, Bowers CY, Martin JB. The effects of growth hormone (GH)-releasing peptides on GH secretion in perifused pituitary cells of adult male rats. Endocrinology 1984; 115:1432–1438.

16. Bowers CY, Momany FA, Reynolds GA, Hong A. On the in vitro and in vivo activity of a new synthetic hexapeptide that acts on the pituitary to specifically release growth hormone. Endocrinology 1984; 114:1537–1545.

17. Soliman EB, Hashizume T, Kanematsu S. Effect of growth hormone (GH)-releasing peptide (GHRP) on the release of GH from cultured anterior pituitary cells in cattle. Endocr J 1994; 41:585–591.

18. Fletcher TP, Thomas GB, Willoughby JO, Clarke IJ. Constitutive growth hormone secretion in sheep after hypothalamopituitary disconnection and the direct in vivo pituitary effect of growth hormone releasing peptide 6. Neuroendocrinology 1994; 60:76–86.

19. Wu D, Chen C, Zhang J, Katoh K, Clarke I. Effects in vitro of new growth hormone releasing peptide (GHRP-1) on growth hormone secretion from ovine pituitary cells in primary culture. J Neuroendocrinol 1994; 6:185–190.

20. Bresson Bepoldin L, Dufy Barbe L GHRP-6 induces a biphasic calcium response in rat pituitary somatotrophs. Cell Calcium 1994; 15:247–258.

21. Wu D, Chen C, Katoh K, Zhang J, Clarke IJ. The effect of GH-releasing peptide-2 (GHRP-2 or KP 102) on GH secretion from primary cultured ovine pituitary cells can be abolished by a specific GH-releasing factor (GRF) receptor antagonist. J Endocrinol 1994; 140:R9–13.

22. Cheng K, Chan WW, Butler B, Wei L, Schoen WR, Wyvratt MJ JR, Fisher MH, Smith RG. Stimulation of growth hormone release from rat primary pituitary cells by L-692,429, a novel nonpeptidyl GH secretagogue. Endocrinology 1993; 132:2729–2731.

23. Blake AD, Smith RG. Desensitization studies using perifused rat pituitary cells show that growth hormone-releasing hormone and His-D-Trp-Ala-Trp-D-Phe-Lys-NH$_2$ stimulate growth hormone release through distinct receptor sites. J Endocrinol 1991; 129:11–19.

24. Smith RG, Pong S, Hickey G, Jacks T, Cheng K, Leonard R, Cohen CJ, Arena JP, Chang CH, Drisko J, Wyvratt MJ, Fisher M, Nargund R, Patchett AA. Modulation of pulsatile GH release through a novel receptor in hypothalamus and pituitary gland. Recent Prog Horm Res 1996; 51:261–286.

25. Guillaume V, Magnan E, Cataldi M, et al. Growth hormone (GH)-releasing hormone secretion is stimulated by a new GH-releasing hexapeptide in sheep. Endocrinology 1994; 135:1073–1076.

26. Arvat E, Gianotti L, Grottoli S, Imbimbo BP, Lenaerts V, Deghenghi R, Camanni F, Ghigo E. 1994; Effects of hexarelin, a synthetic hexapeptide, alone and combined with GHRH or arginine on GH secretion in normal elderly subjects (abstr). Neuroendocrinology 60:32.

27. Casati G, Palmieri E, Biella O, Guglielmino L, Arosio M, Faglia G. Effects of hexarelin (EP 23905–MF 6003) on circulating GH, PRL, TSH, gonadotropins and

alpha-subunit of glycoprotein hormone levels in acromegalic patients (abstr). Eur J Endocrinol 1994; 130:189.

28. Ghigo E, Arvat E, Gianotti L, Imbimbo BP, Lenaerts V, Deghenghi R, Camanni F. Growth hormone-releasing activity of hexarelin, a new synthetic hexapeptide, after intravenous, subcutaneous, intranasal, and oral administration in man. J Clin Endocrinol Metab 1994; 78:693–698.

29. Giustina A, Bussi AR, Deghenghi R, Imbimbo B, Licini M, Poiesi C, Wehrenberg WB. Comparison of the effects of growth hormone-releasing hormone and hexarelin, a novel growth hormone-releasing peptide-6 analog, on growth hormone secretion in humans with or without glucocorticoid excess. J Endocrinol 1995; 146:227–232.

30. Imbimbo BP, Mant T, Edwards M, Amin D, Dalton N, Boutignon F, Lenaerts V, Wuthrich P, Deghenghi R. Growth hormone-releasing activity of hexarelin in humans. a dose–response. Eur J Clin Pharmacol 1994; 46:421–425.

31. Kearns G, Pihoker C, Bowers C. Pharmacokinetics (PK) and pharmacodynamics (PD) of growth-hormone releasing peptide-2 (GHRP) in children. Clin Pharmacol Ther 1996; 59:P1 71.

32. Laron Z, Frenkel J, Gil-Ad I, Klinger B, Lubin E, Wuthrich P, Boutignon F, Lengerts V, Deghenghi R. Growth hormone releasing activity by intranasal administration of a synthetic hexapeptide (hexarelin). Clin Endocrinol 1994; 41:539–541.

33. Di Vito L, Arvat E, Broglio F, Gianotti L, Ramunni J, Maccagno B, Boghen MF, Deghenghi R, Ghigo E. Comparison between the activity of hexarelin and GHRP-2 on GH, PRL, ACTH and cortisol secretion in man (abstr). Endocrinol Metab 1997; 4:P-081.

34. Arvat E, Di Vito L, Ramunni J, Gianotti L, Maccagno B, Broglio F, Deghenghi R, Camanni F. Age-related variations in the GH- PRL- and ACTH-releasing activities of hexarelin in man (abstr). Endocrinol Metab 1997; 4:0-064.

35. Frieboes RM, Murch H, Maier P, Schier T, Holsboer F, Steiger A. Growth hormone-releasing peptide-6 stimulates sleep, growth hormone, ACTH and cortisol release in normal man. Neuroendocrinology 1995; 61:584–589.

36. Klinger B, Silbergeld A, Deghenghi R, Frenkel J, Laron Z. Desensitization from long-term intranasal treatment with hexarelin does not interfere with the biological effects of this growth hormone-releasing peptide in short children. Eur J Endocrinol 1996; 134:716–719.

37. Ghigo E, Arvat E, Gianotti L, Grottoli S, Rizzi G, Ceda GP, Boghen MF, Deghenghi R, Camanni F. Short-term administration of intranasal or oral hexarelin, a synthetic hexapeptide, does not desensitize the growth hormone responsiveness in human aging. Eur J Endocrinol 1996; 135:407–412.

38. Mericq V, Cassorla F, Salazar T, Avila A, Iniguez G, Bowers CY, Merriam GR. Treatment with GHRP-2 accelerates the growth of GH deficient children. J Invest Med 1996; 44:A103.

39. Pihoker C, Kearns GL, Bowers C. Pharmacokinetics (PK) and pharmacodynamics (PD) of growth-hormone releasing peptide-2 (GHRP-2)—a phase-1 study in children. J Invest Med 1996; 44:A60.

40. Laron Z, Frenkel J, Deghenghi R, Anin S, Klinger B, Silbergeld A. Intranasal administration of the GHRP hexarelin accelerates growth in short children. Clin Endocrinol 1995; 43:631-635.

41. Bowers CY, Reynolds GA, Durham D, Barrera CM, Pezzoli SS, Thorner MO. Growth hormone (GH)-releasing peptide stimulates GH release in normal men and acts synergistically with GH-releasing hormone. J Clin Endocrinol Metab 1990; 70:975-982.

42. Huhn WC, Hartman ML, Pezzoli SS, Thorner MO. Twenty-four-hour growth hormone (GH)-releasing peptide (GHRP) infusion enhances pulsatile GH secretion and specifically attenuates the response to a subsequent GHRP bolus. J Clin Endocrinol Metab 1993; 76:1202-1208.

43. Jaffe CA, Ho PJ, Demottfriberg R, Bowers CY, Barkan AL. Effects of a prolonged growth-hormone (GH)-releasing peptide infusion on pulsatile GH secretion in norman men. J Clin Endocrinol Metab 1993; 77:1641-1647.

44. Leung KH, Miller RR, Cohn DA, Colletti C, McGowan E, Feeney WP, Nargund R, Rosegay A, Wallace MA, Chiu SHL. Pharmacokinetics and disposition of MK-677, a novel growth hormone secretagogue, in rats and dogs (abstr). ISSX Proc 1996; 10:277.

45. Murphy MG, 1997 personal communication.

46. Chapman IM, Bach MA, Van Cauter E, Farmer M, Krupa D, Taylor AM, Schilling LM, Cole KY, Skiles EH, Pezzoli SS, Hartman ML, Veldhuis JD, Gormley GJ, Thorner MO. Stimulation of the growth hormone (GH)/IGF-I axis by daily oral administration of a GH secretagogue (MK-0677) in healthy elderly subjects. J Clin Endocrinol Metab 1996; 81:4249-4257.

47. Shetty KR, Duthie EH Jr. Anterior pituitary function and growth hormone use in the elderly. Endocrinol Metab Clin North Am 1995; 24:213-231.

48. Marcus R, Butterfield G, Holloway L, Gilliland L, Baylink D, Hintz R, Sherman B. Effects of short-term administration of recombinant human growth hormone to elderly people. J Clin Endocrinol Metab 1990; 70:519.

49. Taaffe DR, Jin IH, Vu TH, Hoffman AR, Marcus R. Lack of effect of recombinant human growth hormone (GH) on muscle morphology and GH–insulin-like growth factor expression in resistance-trained elderly men. J Clin Endocrinol Metab 1996; 81:421-425.

50. Thompsin JL, Butterfield GE, Marcus R, Hintz RL, Van Loan M, Ghiron L, Hoffman AR. The effects of recombinant human insulin-like growth factor-I and growth hormone on body composition in elderly women. J Clin Endocrinol Metab 1995; 80:1845-1852.

51. Taaffe DR, Pruitt L, Reim J, Hintz RL, Butterfield G, Hoffman AR, Marcus R. Effect of recombinant human growth hormone on the muscle strength response to resistance exercise in elderly men. J Clin Endocrinol Metab 1994; 79:1361-1366.

52. Holloway L, Butterfield G, Hintz RL, Gesundheit N, Marcus R. Effects of recombinant human growth hormone on metabolic indices, body composition, and bone turnover in healthy elderly women. J Clin Endocrinol Metab 1994; 79:470-479.

53. Yaraheski KE, Zachwieja JJ. Growth hormone therapy for the elderly: the fountain of youth proves toxic [letter]. JAMA 1993; 270:1694.
54. Growth hormone therapy in elderly people [editorial]. Lancet 1991; 337:1131–1132.
55. Rudman D, Feller AG, Cohn L, Shetty KR, Rudman IW, Draper MW. Effects of human growth hormone on body composition in elderly men. Horm Res 1991; 36(suppl 1):73–81.
56. Welle S, Thornton C, Statt M, McHenry B. Growth hormone increases muscle mass and strength but does not rejuvenate myofibrillar protein synthesis in healthy subjects over 60 years old. J Clin Endocrinol Metab 1996; 81:3239–3243.
57. Rasmussen LH, Steenfos HH. [Growth hormone and surgery.] Ugeskr Laeger 1992; 154:1019–1023.
58. Wong WK, Soo KC, Nambiar R, Tan YS, Yo SL, Tan IK. The effect of recombinant growth hormone on nitrogen balance in malnourished patients after major abdominal trauma. Aust NZ J Surg 1995; 65:109–113.
59. Byrne TA, Morrissey TB, Gatzen C, Benfell K, Nattakom TV, Scheltinga MR, LeBoff MS, Ziegler TR, Wilmore DW. Anabolic therapy with growth hormone accelerates protein gain in surgical patients requiring nutritional rehabilitation. Ann Surg 1993; 218:400–418.
60. Voerman BJ, Strack van Schijndel RJM, de Boer H, Groeneveld J, Nauta JP, van der Veen EA, Thijs LG. Effects of human growth hormone on fuel utilization and mineral balance in critically ill patients on full intravenous nutritional support. J Crit Care 1994; 9:143–150.
61. Roe CF, Kinney JM. The influence of human growth hormone on energy sources in convalescence. Metab Renal Physiol 1962; 13:369–371.
62. Koea JB, Douglas RG, Shaw JHF, Gluckman PD. Growth hormone therapy initiated before starvation ameliorates the catabolic state and enhances the protein-sparing effect of total parenteral nutrition. Br J Surg 1993; 80:740–744.
63. Wilmore DW. Catabolic illness: strategies for enhancing recovery. N Engl J Med 1991; 325:695–700.
64. Revhaug A, Mjaaland M. Growth hormone and surgery. Horm Res 1997; 40:99–101.
65. Vara-Thorbeck R, Ruiz-Requena E, Guerrero-Fernandez JA. Effects of human growth hormone on the catabolic state after surgical trauma. Horm Res 1996; 45:55–60.
66. Katakami H, Arimura A, Frohman LA. Growth hormone (GH)-releasing factor stimulates hypothalamic somatostatin release: an inhibitory feedback effect on GH secretion. Endocrinology 1986; 118:1872–1877.
67. Richelsen B, Pedersen SB, Borglum JD, Moller Pedersen T, Jorgensen J, Jorgensen JO. Growth hormone treatment of obese women for 5 wk: effect on body composition and adipose tissue LPL activity. Am J Physiol 1994; 266:E211–E216.
68. Drent ML, Wever LD, Ader HJ, van der Veen EA. Growth hormone administration in addition to a very low calorie diet and an exercise program in obese subjects. Eur J Endocrinol 1995; 132:565–572.

69. Skaggs SR, Crist DM. Exogenous human growth hormone reduces body fat in obese women. Horm Res 1991; 35:19–24.

70. Barton JS, Gardineri HM, Cullen S, Hindmarsh PC, Brook CG, Preece MA. The growth and cardiovascular effects of high dose growth hormone therapy in idiopathic short stature. Crit Endocrinol (Oxf) 1995; 42:619–626.

71. Cuttler L, Silvers JB, Singh J, Marrero U, Finkelstein B, Tannin G, Neuhauser D. Short stature and growth hormone therapy. A national study of physician recommendation patterns [see comments]. JAMA 1996; 276:531–537.

72. Spagnoli A, Spadoni GL, Cianfarani S, Pasquino AM, Troiani S, Boscherini B. Prediction of the outcome of growth hormone therapy in children with idiopathic short stature. A multivariate discriminant analysis. J Pediatr 1995; 126:905–909.

73. Wit JM, Boersma B, de Muinck Keizer Schrama SM, et al. Long-term results of growth hormone therapy in children with short stature, subnormal growth rate and normal growth hormone response to secretagogues. Dutch Growth Hormone Working Group. Clin Endocrinol (Oxf) 1995; 42:365–372.

74. de Zegher F, Maes M, Gargosky SE, Heinrichs C, Du Caju MV, Thiry G, De Schepper J, Craen M, Breysem L, Lofstrom A, Jonsson P, Bourguignon JP, Malvaux P, Rosenfeld RG. High-dose growth hormone treatment of short children born small for gestational age. J Clin Endocrinol Metab 1996; 81:1887–1892.

75. Rekers Mombarg LT, Rijkers GT, Massa GG, Wit JM. Immunologic studies in children with idiopathic short stature before and during growth hormone therapy. Dutch Growth Hormone Working Group. Horm Res 1995; 44:203–207.

76. Ranke MB, Lindberg A. Growth hormone treatment of idiopathic short stature: analysis of the database from KIGS, the Kabi Pharmacia International Growth Study. Acta Paediatr Suppl 1994; 406:18–23.

77. Farello G, De Simone M, Gentile T, De Matteis F. [The treatment with biosynthetic growth hormone (GH) of familial short stature.] Minerva Pediatr 1994; 46:347–350.

78. Loche S, Cambiaso P, Setzu S, Carta D, Marini R, Borrelli P, Cappa M. Final height after growth hormone therapy in non–growth-hormone-deficient children with short stature. J Pediatr 1994; 125:196–200.

79. Zadik Z, Chalew SA, Zung A, Lieberman E, Kowarski AA. Short stature: new challenges in growth hormone therapy. J Pediatr Endocrinol 1993; 6:303–310.

80. Albertsson Wikland K. Characteristics of children with idiopathic short stature in the Kabi Pharmacia International Growth Study, and their response to growth hormone treatment. International Board of the Kabi Pharmacia International Growth Study. Acta Paediatr Suppl 1993; 391:75–78.

81. Hopwood NJ, Hintz RL, Gertner JM, et al. Growth response of children with non–growth-hormone deficiency and marked short stature during three years of growth hormone therapy. J Pediatr 1993; 123:215–222.

82. Moore WV, Moore KC, Gifford R, Hollowell JG, Donaldson DL. Long-term treatment with growth hormone of children with short stature and normal growth hormone secretion. J Pediatr 1992; 120:702–708.

83. Rodgers BD. Catabolic hormones and growth hormone resistance in acquired immunodeficiency syndrome and other catabolic states. Proc Soc Exp Biol Med 1996; 212:324–331.

84. Bentham J, Rodriguez-Arnao J, Ross RJM. Acquired growth hormone resistance in patients with hypercatabolism. Horm Res 1993; 40:87–91.

85. Jorgensen PH, Bang C, Andreassen TT, Flyvbjerg A, Orskov H. Dose–response study of the effect of growth hormone on mechanical properties of skin graft wounds. J Surg Res 1995; 58:295–301.

86. Welsh KM, Lamit M, Morhenn VB. The effect of recombinant human growth hormone on wound healing in normal individuals. J Dermatol Surg Oncol 1991; 17:942–945.

87. Gore DC, Honeycutt D, Jahoor F, Wolfe RR, Herndon DN. Effect of exogenous growth hormone on whole-body and isolated-limb protein kinetics in burned patients. Arch Surg 1991; 126:38–43.

88. Mulligan K, Grunfeld C, Hellerstein MK, Neese RA, Schambelan M. Anabolic effects of recombinant human growth hormone in patients with wasting associated with human immunodeficiency virus infection. J Clin Endocrinol Metab 1993; 77:956–962.

89. Clemmons DR, Plunkett LJ, Polvino WM. Administration of the GH secretagogue L-163,191 stimulates an anabolic response in calorically restricted normal volunteers (abstr). Endocrinol Metab 1997; 4:35.

90. August GP, Lippe BM, Blethen SL, Rosenfeld RG, Seelig SA, Johanson AJ, Compton PG, Frane JW, McClellan BH, Sherman BM. Growth hormone treatment in the United States: demographic and diagnostic features of 2331 children. J Pediatr 190; 116:899–903.

91. Loche S, Cambiaso P, Carta D, Setzu S, Imbimbo BP, Borrelli P, Pintor C, Cappa M. The growth hormone-releasing activity of hexarelin, a new synthetic hexapeptide, in short normal and obese children and in hypopituitary subjects. J Clin Endocrinol Metab 1995; 80:674–678.

92. Loche S, Cambiaso P, Merola B, Colao A, Faedda A, Imbimbo BP, Deghenghi R, Lombardi G, Cappa M. The effect of hexarelin on growth hormone (GH) secretion in patients with GH deficiency. J Clin Endocrinol Metab 1995; 80:2692–2696.

93. Tiulpakov AN, Brook CGD, Pringle PJ, Peterkova VA, Volevodz NN, Bowers CY. GH responses to intravenous bolus infusions of GH releasing hormone and GH releasing peptide 2 separately and in combination in adult volunteers. Clin Endocrinol 1995; 43:347–350.

94. Mericq V, Cassorla F, Garcia H, Avila A, Bowers CY, Merriam GR. Growth hormone (GH) responses to GH-releasing peptide and to GH-releasing hormone in GH-deficient children. J Clin Endocrinol Metab 1995; 80:1681–1684.

95. Copinschi G, Van Onderbergen A, L'Hermite-Baleriaux M, Mendel C, Caufriez A, Leproult R, Bolognese J, De Smet M, Thorner MO, Van Cauter E. Effects of a 7-day treatment with a novel, orally active, growth hormone secretagogue, MK-677, on 24 hour GH profiles, insulin-like growth factor I, and adrenocortical function in normal young men. J Clin Endocrinol Metab 1996; 81:2776–2782.

96. Hochberg Z, Even L, Peleg I, Youdim MBH, Amit T. The effect of human growth hormone therapy on GH binding protein in GH-deficient children. Acta Endocrinol 1991; 125:23–27.

97. Tuilpakov AN, Bulatov AA, Peterkova VA, Elizarova GP, Volevodz NN, Bow-

ers CY. Growth-hormone (GH)-releasing effects of synthetic peptide GH-releasing peptide-2 and GH-releasing hormone (1–29NH(2)) in children with GH insufficiency and idiopathic short stature. Metab Clin Exp 1995; 44:1199–1204.

98. Laron Z, Bowers CY, Hirsch D, Almonte AS, Pelz M, Keret R, Gilad I. Growth hormone-releasing activity of growth hormone-releasing peptide-1 (a synthetic heptapeptide) in children and adolescents. Acta Endocrinol 1993; 129:424–426.

99. Bellone J, Ghizzoni L, Aimaretti G, Volta C, Boghen MF, Bernasconi S, Ghigo E. Growth hormone-releasing effect of oral growth-hormone-releasing-peptide-6 (GHRP-6) administration in children with short stature. Eur J Endocrinol 1995; 133:425–429.

100. Moon HD, Simpson ME, Li CH, Evans HM. Neoplasms in rats treated with pituitary growth hormone. I. Pulmonary and lymphatic tissues. Cancer Res 1950; 10:297–308.

101. Kyrgialanis A, Kakavas I, Voudiclari S, Deghenghi R, Moschogianni H, Tolis G. Pituitary somatotrope responsiveness to two growth hormone secretagogues: GRF (1–29) and the GHRP—hexarelin in hemodialysed patients with chronic renal failure (CRF) (abstr). Nephrol Dial Transplant 1995; 10:985.

102. Zadik Z, Chalew SA, McCarter RJ Jr, Meistas M, Kowarski AA. The influence of age on the 24-hour integrated concentration of growth hormone in normal individuals. J Clin Endocrinol Metab 1985; 60:513–516.

103. Juul A, Bang P, Hertel NT, Main K, Dalgaard P, Jorgensen K, Muller J, Hall K. Skakkeback NE. Serum insulin-like growth factor-I in 1030 healthy children, adolescents, and adults: relation to age, sex, stage of puberty, testicular size, and body mass index. J Clin Endocrinol Metab 1994; 78:744–752.

104. Rudman D, Feller A, Nagraj H, Gergans G, Lalitha P, Goldberg A, Schlenker R, Cohn L, Rudman I, Mattson D. Effects of human growth hormone in men over 60 years old. N Engl J Med 1990; 323:1–6.

105. Papadakis MA, Grady D, Black D, Tierney MJ, Gooding GA, Schambelan M, Grunfeld C. Growth hormone replacement in healthy older men improves body composition but not functional ability. Ann Intern Med 1996; 124:708–716.

106. Yarasheski KE, Zachwieja JJ, Campbell JA, Bier DM. Effect of growth hormone and resistance exercise on muscle growth and strength in older men. Am J Physiol 1995; 268:E268–76.

107. Arvat E, Gianotti L, Grottoli S, Imbimbo BP, Lenaerts V, Deghenghi R, Camanni F, Ghigo E. Arginine and growth hormone-releasing hormone restore the blunted growth hormone-releasing activity of hexarelin in elderly subjects. J Clin Endocrinol Metab 1994; 79:1440–1443.

108. Plotkin D, Ng J, Farmer M, Gelato M, Kaiser F, Kiel D, Korenman S, McKeever C, Munoz D, Schwartz R, Krupa D, Gormley G, Bach MA. Use of MK-677, an oral GH secretagogue, in frail elderly subjects (abstr). Endocrinol Metab 1997; 4:35.

109. Gertz BJ, Sciberras DG, Yogendran L, Christie K, Bador K, Krupa D, Wittreich JM, James L. L-694,429, a nonpeptide growth hormone (GH) secretagogue, reverses glucocorticoid suppression of GH secretion. J Clin Endocrinol Metab 1994; 79:745–749.

110. Leal-Cerro A, Pumar A, Garcia-Garcia E, Dieguez C, Casanueva FF. Inhibition of growth hormone release after the combined administration of GHRH and GHRP-6 in patients with Cushing's syndrome. Clin Endocrinol (Oxf) 1994; 41:649–654.

22

Short- and Long-Term Growth Hormone Secretagogue Administration in Children

Zvi Laron
Schneider Children's Medical Center, Tel Aviv, Israel

I. INTRODUCTION

The newly described peptidergic (1–4) and nonpeptidergic growth hormone (GH) secretagogues (GHSs; 5) that enhance endogenous GH secretion are of special interest to the pediatricians and pediatric endocrinologists who, in their practice, encounter many diseases and conditions that are benefited by GH administration. The GHSs act through a receptor that is independent from that of growth hormone-releasing hormone (GHRH; 6,7). The GHSs act synergetically with GHRH (8,9) amplifying the pulsatile GH release and increasing the insulin-like growth factor-I (IGF-I) secretion. These actions are most probably induced by suppressing somatostatin secretion (10). A major advantage of these substances over GHRH, human GH (hGH) and IGF-I is that they are active not only when injected, but also by the oral and intranasal routes. The number of reports on the use of the GHSs in the pediatric clinic is limited, involving almost exclusively the peptidergic substances. Most reports deal with bolus effects. This chapter presents a review of published data on the use of this new class of hormone-like drugs in children and adolescents, with a discussion on their future possible role in pediatric endocrinology.

II. IMMEDIATE RESPONSE TO GHS

A. Experience of Others

Bowers et al. (11) tested nine short children with an oral bolus of 300 µg/kg growth hormone-releasing peptide-6 (GHRP-6; His-D-Trp-Ala-Trp-D-Phe-Lys-NH_2). Four children who had increased serum GH previously to routine pharmacological tests (hypoglycemia, arginine, clonidine, or glucagon) responded to GHRP-6, with a peak of 20–31.8 µg/L hGH. Of five children who had no hGH rise to the pharmacological stimuli, only one responded to GHRP-6 with a peak of 8 µg/L.

Bellone et al. (12) tested the intravenous (iv) response of a bolus injection of hexarelin* (His-D-2-methyl-Trp-D-Phe-Lys-NH_2; Europeptides, Argenteuil, France) 2 µg/kg in 12 children (8 boys and 4 girls with familial short stature and registered peak hGH levels of 19–92 µg/L.

Penalva et al. (13) tested 12 prepubertal children (aged 7–11 years) by an iv bolus of 1 µg/kg GHRP-6. The mean GH peak was 57 $\pm$ 11.2 mU/L in the boys and 44.2 $\pm$ 5.4 mU/L in the girls (1 mU/L = 0.5 ng/mL; 1 ng/mL = 1 µg/L).

Hayashi et al. (14) tested the effect of an iv bolus of GHRP-6 (1 µg/kg) in 41 short patients, which included an undefined number of children. They reported no response to the GHRP in patients with pituitary stalk resection.

Tuilpakov et al. (15) investigated the immediate iv response to 1 µg/kg of GHRP-2 (D-Ala-D-β-Nal-Ala-Trp-D-Phe-Lys-NH_2) compared with GHRH(1–29) in two groups of children and adolescents: group 1, nine patients with GH deficiency (age range 6.2–15.6 years); and group 2, five patients (aged 6.9–13.6 years) with idiopathic short stature (ISS). In group 1, five of nine patients did not respond to GHRP-2 and three showed only a slight increase in GH levels. In one patient, the peak GH reached 30 µg/L. It was not stated how the same patient responded to stimuli such as hypoglycemia, clonidine, or arginine. In the five children with ISS the GH response to GHRP-2 varied from low to 50 µg/L. The response to a combined administration of GHRP-2 and GHRH(1–29) was higher than with each drug alone. In the combined test, five of eight children of group 1 had a peak concentration between 2.5 and 6.8 µg/L; two had peaks of 15 and 19 µg/L (denoting either hypothalamic or partial GH deficiency, personal observation), and one patient had a peak of 125 µg/L (not previously reported in a true GH-deficient patient, personal observation).

B. Personal Experience

Our group has had clinical experience with GHSs in children since 1992. In the first stages, we investigated the comparative potency of the heptapeptide

---

*Hexarelin is a trademark for examorelin (I.N.N.).

GHRP-1 (Ala-His-D-β-Nal-Ala-Trp-D-Phe-Lys-NH$_2$) and GHRH(1–29) by iv bolus, using the same dose, 1 µg/kg (16). We investigated 15 healthy children and adolescents with ISS (6 prepubertal, 9 pubertal) and 8 juvenile patients with pituitary insufficiency: 4 with isolated GH deficiency (IGHD), 2 with multiple pituitary hormone deficiencies (MPHD), 1 with partial hGH deficiency, and one with GHRH deficiency (Table 1). Eleven of 23 subjects also underwent an iv GHRH(1–29) test (1 µg/kg). All the healthy children responded with a progressive rise in plasma hGH. peaking at 15–30 min, with a significantly higher rise (p < 0.05) in the pubertal than in the prepubertal group. The hGH response to GHRH(1–29) in these children was similar or slightly higher. Six hypopituitary patients had no response to either GHRP-1 or GHRH; the patient with partial GH deficiency had a peak of 6.5 µg/L (at 5 min) to GHRP-1 and 9.2 µg/L (at 15 min) to GHRH. One patient had no response to hGH to hypoglycemia, clonidine, and GHRP-1, but plasma hGH rose to 10 µg/L after GHRH administration.

The aim of another study was to compare the effect of one iv bolus (0.1 µg/kg) of the hexapeptide hexarelin with an intranasal (20 µg/kg) dose of the same drug. We tested 12 children with ISS, aged from 5 to 15 years. The two tests were performed with a 1-week interval between them. The mean peak response between the two tests for the whole group was similar: in the iv test, it was 79.2 ± 49.3 mU/L (mean ± SD) and in the intranasal test 69 ± 39.4 mU/L (i.e., 39.6 ± 24.5 µg/L and 34.5 ± 17.7 µg/L, respectively). However, the peak response after the iv administration occurred earlier (15–30 min) than that after intranasal puff administration (30–60 min) (17).

After we had observed no undesirable effects and proved the effectiveness of this GHS, we proceeded to perform 1-week studies (18) looking for an appropriate dose for long-term administration. We found that twice or thrice daily intranasal administration of hexarelin evoked a significant rise of serum

Table 1 Effect of 1 µg/kg Intravenous Bolus of GHRP-1 on Serum hGH Levels (µg/L; Mean ± SD)

Group	n	Basal	Peak
Idiopathic short stature			
Prepubertal	6	2.2 ± 1.0	20.2 ± 5.0
Pubertal	9	3.3 ± 1.8	34.5 ± 8.9
GH deficiency			
Pituitary lesion	6	0.6 ± 0.2	0.6 ± 0.1
Hypothalamic lesion	1	< 0.5	< 0.5
Partial GH deficiency	1	< 0.5	6.5

hGH 1 µg/L = ng/mL

IGF-I, the growth mediator of GH (19), and of alkaline phosphate, a sensitive marker of the growth process in children.

For the long-term studies, we decided to use a dose of 60 μg/kg tid by nasal spray, a dose that was well tolerated by the children. We thus continued the study up to a year, when the supply of the drug was stopped.

This study comprised eight prepubertal children (seven boys and one girl) aged 5.4–11.7 years, referred to our clinic because of psychosocial problems related to their short stature; we followed them without treatment for periods of 1 or more years. All were short (> -1.9 SDS height), but had a normal hGH response to clonidine (20) or insulin hypoglycemia tests (> 20 mU/L; 21). An additional inclusion criteria was a response of hGH of 40–60 mU/L to one intranasal dose (20 μg/kg) of hexarelin (22).

Their pertinent clinical data are summarized in Table 2. The study was approved by the local Ethics Committee and by the Ministry of Health, and written informed consent was obtained from all parents.

The patients were examined weekly during the first month of treatment; biweekly during the second month, and monthly thereafter. Harpenden stadiometers were used by the same person for recumbent or standing anthropometric measurements. A Harpenden caliper served to measure suprailiac, triceps, and subscapular skinfolds. We estimated bone age using the Greulich and Pyle atlas (23) from a hand and wrist radiograph, with separate readings for carpal and phalangeal bones. The height was plotted on Tanner, Whitehouse, and Takaishi growth charts (24).

Blood for blood count, routine chemistry including liver tests, serum IGF-I, and insulin were drawn after an overnight fast before and every 2–3 months during hexarelin administration. Serum hGH, insulin, thyroid-stimulating hormone (TSH), and free thyroxine (fT4), were measured by radioimmunoassay (RIA) as described previously (25). Serum IGF-I was measured by RIA after acid ethanol extraction, followed by cryoprecipitation, as previously described (26). The sensitivity of the assay was 25 pmol/L, the within-assay coefficient of variation for a concentration of 20 nM/L of IGF-I was 4.7%. All sera were determined in a single assay; blood chemistry was determined by autoanalyzer (Hitachi, Japan). Statistical analysis was performed using Student's paired t test.

III. RESULTS OF LONG-TERM TREATMENT

Hexarelin administration was associated with a significant rise in serum IGF-I (Fig. 1), and a growth-promoting effect (Fig. 2) was observed as early as 3 months after the start of treatment (17). The short children, who before treatment had grown 5.4 ± 0.8 cm/yr (mean ± SD), increased growth velocity to 8.3 ± 1.7 cm/yr during 7–10 months of hexarelin treatment (p < 0.0001).

Table 2 Clinical Data from Eight Prepubertal Children with Idiopathic Short Stature (ISS)

Patient Sex	CA (yr)[a]	BA (yr)[a]	Before Treatment				
			Height (cm)	SDS	Growth velocity (cm/yr)	Weight (kg)	Skinfold subscapular (mm)
1 M	5^4	4	99.5	−2.3	6.5	14.2	6
2 M	6^2	4^6	97.8	−3.7	4.4	11.6	6
3 M	6^2	4^9	107.2	−1.9	5.5	14.9	5
4 F	6^4	3^6	105.5	−2.2	6.0	15.5	6
5 M	7^9	6	114.9	−1.9	5.9	20.9	8
6 M	8^5	6	116.5	−2.1	5.7	18.0	5
7 M	10^2	6	121.8	−2.5	4.8	21.9	5
8 M	11^7	9	133.5	−2.3	4.1	28.6	7
Mean			112.1		5.4	18.2	6.0
± SD			± 12.1		± 0.8	± 5.4	± 1.1

[a]Superscript refers to months.

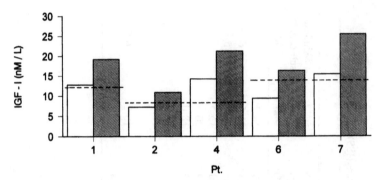

Figure 1 Serum IGF-I rise during 3 months of intranasal hexarelin administration (60 µg/kg tid) to boys with idiopathic short stature (ISS). Open box, before treatment; hatched box, hexarelen treatment; broken lines, lower normal levels of IGF-I for age: $p < 0.003$. (From Ref. 17.)

During the same periods the children increased their weight by 1.3 ± 0.5 kg (mean $\pm$ SD); concomitantly the subscapular skinfolds, which are reduced in this type of short stature (27), decreased by 1.1 ± 0.8 mm ($p < 0.007$) when comparing the individual measurements. An increase in the head circumference of 0.6 ± 0.3 cm/yr was also observed ($p < 0.002$).

After 7 days of hexarelin administration (18) serum IGF-I rose from a mean of 10.4 ± 3.9 (SD) nM/L to 14.5 ± 5.4 nM/L ($p < 0.02$) and remained

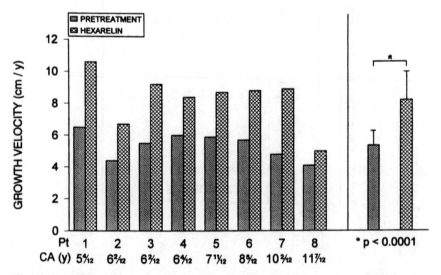

Figure 2 Change of growth velocity (cm/yr) during long-term hexarelin treatment of eight boys with ISS.

Table 3 Serum IGF-I (nM/L) Levels Before, During, and After Intranasal
Hexarelin Treatment

Patient No.	Basal	7 d	1 mo	3 mo	End of treatment	3 mo off treatment
1	5.5	9.6	8.8	9.0	9.8	—
2	7.2	7.5	6.2	7.1	9.6	7.1
3	5.8	12.0	9.8	8.3	8.7	—
4	12.7	22.2	17.3	16.3	13.9	16.6
5	14.2	15.1	16.3	15.6	22.3	18.6
6	13.0	12.3	12.9	18.7	—	—
7	9.4	15.2	16.3	13.5	21.0	—
8	15.4	22.4	23.3	24.3	32.2	22.8
Mean	10.0	14.5	13.9	14.1	16.8	16.3
± SD	4.1	5.4	5.5	5.9	8.8	6.6
			$p < 0.02$			

essentially constant throughout the whole treatment periods ($p < 0.02$). three months after stopping hexarelin, they were still in the same range (Table 3). Serum inorganic phosphorus and alkaline phosphatase levels increased significantly from pretreatment levels of 1.5 ± 0.1–1.8 ± 0.1 mmol/L ($p < 0.004$) and from 219 ± 74 to 261 ± 75 U/L ($p < 0.05$), respectively. These changes could already be observed in the first month of treatment. Blood glucose, electrolytes, creatinine, and liver tests were unchanged. There was no significant effect on thyroid function, as measured by plasma T_4, T_3, and TSH. Serum prolactin and cortisol varied in their response to hexarelin, remaining within the limits of normal (Table 4).

Determinations of serum procollagen levels revealed that the basal serum procollagen I, III, and ICTP (pyridinoline cross-linked carboxy-terminal telopeptide) were higher than those reported for IGF-I deficiency (28) and were within the normal limits for age. Hexarelin treatment caused doubling of the serum ICTP, reflecting type-I collagen degradation. These collagens account for over 90% of the noncalcified matrix. There was no significant change in type-III procollagen, found mainly in the skin, blood vessels, and intestine. Bone age advanced less or parallel with age.

The drug was well tolerated, there was no local irritation, and no undesirable effects were observed nor reported by the parents.

Forthwith three illustrations of the effect and aftereffect of long-term hexarelin treatment in boys with ISS. As can be seen, all boys had a pretreatment follow-up of 2 or more years. Figure 3 shows that one boy, the product of normal-sized parents, grew below the third percentile. Hexarelin treatment

Table 4 Serum Cortisol and Prolactin (Mean ± SD) Before and During Intranasal Hexarelin Treatment

	Months			
	0	3	6	8
Cortisol (nM/L)	444 ± 213	458 ± 273	462 ± 265	334 ± 216
Prolactin (μg/L)	14.4 ± 13.3	11.8 ± 8.9	12.6 ± 9.7	6.4 ± 2.3

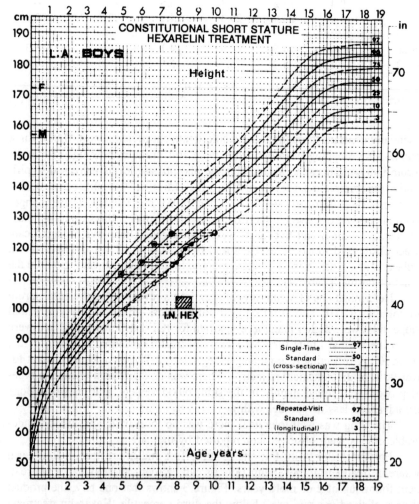

Figure 3 Growth chart of a boy treated by intranasal hexarelin (60 μg/kg tid).

elevated him into the seventh percentile without advancing the bone age. After stopping treatment, he continued to grow for a few months along the same percentile, but 1 year later his height corresponded to the third percentile, still taller than before treatment. Figure 4 illustrates a boy who for years also grew under the lower limit of the norm. Hexarelin administration at a more-advanced age raised his height to the fifth percentile without advancing the bone age. The concomitantly developing puberty led to continuing growth along the same percentile, but induced an advancement in the bone age. Figure 5 illustrates the growth of another boy: hexarelin administration normalized his growth up to the tenth percentile. Half a year after stopping hexarelin, he was started on GH

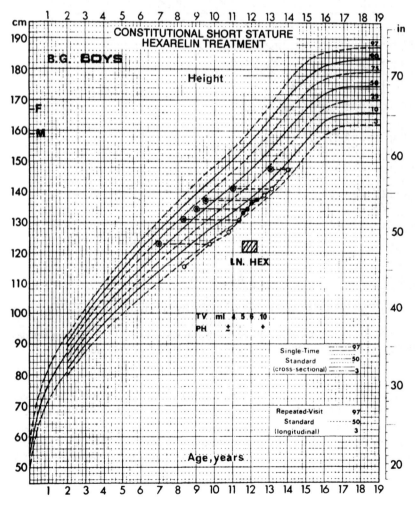

Figure 4 Growth chart of a boy treated by intranasal hexarelin (60 μg/kg tid).

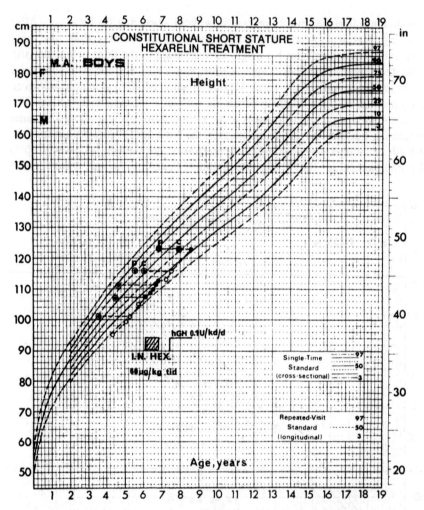

Figure 5 Growth chart of a boy treated by intranasal hexarelin (60 µg/kg per day) followed, after an interruption, by treatment with hGH (0.1 U/kg tid sc)

treatment (in another country, in a dose of 0.1 U/kg per day) and 1 year thereafter had crossed the tenth percentile without showing (by report) pubertal signs. The only other report of long-term treatment of children with a GHS was a poster by Pihoker et al. (29), who treated six children with GH deficiency (probably partial, personal observation) aged 7–9, for 15–24 months by GHRP-2 (10–15 µg/kg) administered intranasaly tid. As a result, the mean growth velocity increased from 3.5 to 6 cm/yr.

IV. DESENSITIZATION

Because desensitization has been reported during treatment of children with GHRH (30), it was only natural that we investigated whether long-term hexarelin treatment of the boys with ISS would induce down-regulation of the GH-stimulating effect of this secretagogue.

The study was performed by comparing the effect of iv as well as intranasal boluses of hexarelin to simulate GH secretion at various stages of the study (31). Figure 6 illustrates that already after 7 days of intranasal hexarelin administration, the GH-stimulating effect was halved and was maintained during 6 months continued treatment. Figure 7 shows that measurements of the GH response to hexarelin by an iv bolus at the end of the trial was one-fourth of the basal response, but 3 months after stopping the drug, a catchup was clearly registered. We concluded that the partial down-regulation observed did not affect the growth-promoting effect because adequate levels of serum IGF-I were maintained.

V. DISCUSSION

In recent years, when height below accepted norms presents as a stigma in society (32,33), the desire to become taller has increased with the availability of growth-stimulating drugs, such as GHRH, hGH, or IGF-I (30,34,35). Many clinical trials are being performed with the aforementioned drugs, all of which

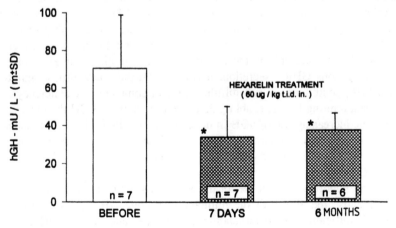

Figure 6 Peak response of serum hGH (mean ± SD) to an intranasal bolus of hexarelin (1 μg/kg) before and at the end of a long-term intranasal hexarelin treatment and after 3 months without treatment. *$p < 0.002$. (From Ref. 31.)

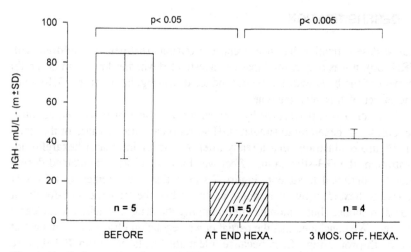

Figure 7 Peak response of serum hGH (mean ± SD) to an intranasal bolus of hexarelin (20 µg/kg) before and during long-term intranasal hexarelin treatment. (From Ref. 31.)

necessitate daily injections. Therefore, it was natural that the availability of intranasal or orally active GH-stimulating drugs be tested in short children. Thrice daily intranasal administration of hexarelin, one of the secretagogues, for almost a year, stimulated linear growth in those children with ISS who had a normal response to pharmacological stimuli. The reported findings prove that hexarelin and probably other similar GHSs have not only an immediate GH-secreting effect, but are able to increase endogenous hGH and IGF-I production during long-term administration, leading to anabolic and growth-promoting effects. As it has been demonstrated that the GHS hexarelin, as other similar peptide GHRPs (36) act only in the presence of a functioning pituitary and a minimal endogenous GHRH secretion, it is clear that these new drugs cannot act when these conditions are not fulfilled (i.e., in complete GH deficiency). This may be a limiting factor, enabling their use only in partial GH deficiency. However, the increasing use of hGH in nonconventional indications (34), such as short stature, Turner's syndrome, chronic renal failure, and catabolic states, opens great possibilities of the use of this new class of drugs in the pediatric age group (37).

Another important clinical use for the new GHSs should be the "one-bolus test" to diagnose the ability of the pituitary to secrete GH, to differentiate between GH deficiency of hypothalamic origin (adequate or good hGH response), from pituitary origin (no hGH response to GHS), and to assess completeness of medical or surgical hypophysectomy. In IGHD, a negative test should raise the a possibility of GH gene deletion; a repeated negative bolus test is proof that long-term administration of a secretagogue would be noneffective. The intranasal or orally active drugs may thus replace GHRH employed so far, for the foregoing purposes (38–40).

ACKNOWLEDGMENTS

The author wishes to thank Dr. R. Deghenghi (Europeptides Argenteuil, France) for the close cooperation, generous supply of hexarelin, and support for the hexarelin studies. We are indebted to Dr. C. Y. Bowers (New Orleans) for the supply of GHRP-1 and many interesting talks.

ZL is incumbent of the Irene and Nicholas Marsh Chair of Endocrinology and Juvenile Diabetes, Sackler School of Medicine, Tel Aviv University, Israel.

REFERENCES

1. Bowers CY, Momany F, Reynolds GA, Hong A, Newlander K. Conformational energy studies and in vitro activity of a new synthetic hexapeptide that acts on the pituitary to specifically release growth hormone. Endocrinology 1984; 14:1537–1545.
2. Ilson BE, Jorkasky DK, Curnow RT, Stote RM. Effect of a new synthetic hexapeptide to selectively stimulate growth hormone release in healthy human subjects. J Clin Endocrinol Metab 1989; 69:212–214.
3. Bowers CY. GH releasing peptides—structure and kinetics. J Pediatr Endocrinol 1993; 6:21–31.
4. Deghenghi R, Cananzi MM, Torsello A, Battisti C, Muller EE, Locatelli V. GH-releasing activity of hexarelin. A new growth hormone releasing peptide, in infants and adult rats. Life Sci 1994; 54:1321–1328.
5. Smith RG, Cheng K, Schoen WR, Pong SS, Hickey G, Jacks T. A nonpeptidyl growth hormone secretagogue. Science 1993; 260:1640–1643.
6. Smith RG, Pong SS, Hickey G, Jacks T, Cheng K, Leonard R. Modulation of pulsatile GH release through a novel receptor in hypothalamus and pituitary gland. Rec Prog Horm Res 1996; 51:261–285.
7. Howard AD, Fieghner SD, Cully DF, Arena JP, Liberator PA, Rosenblum CI. A receptor in pituitary and hypothalamus that functions in growth hormone release. Science 1996; 273:974–977.
8. Bowers CY, Reynolds GA, Durham D, Barrera CM, Pezzoli SS. Growth hormone (GH) releasing peptide stimulates GH release in normal men and acts synergistically with GH releasing hormone. J Clin Endocrinol Metab 1990; 70:975–982.
9. Dickson SL, Leng G, Robinson ICAF. Systemic administration of growth hormone releasing peptide activates hypothalamic arcuate neurons. Neurosci Lett 1993; 53:303–306.

10. Dickson SL, Leng G, Dyball REJ, Smith RG. Central actions of peptide and non-peptide growth hormone secretagogues in the rat. Neuroendocrinology 1995; 61:36–43.

11. Bowers CY, Alster DK, Frentz JM. The growth hormone-releasing activity of a synthetic hexapeptide in normal men and short statured children after oral administration. J Clin Endocrinol Metab 1992; 74:292–298.

12. Bellone J, Ghigo E, Almaretti E, Bartolotta F, Benson L. GH releasing activity of hexarelin, a new synthetic hexapeptide after intravenous administration in normal children. Horm Res 1993; 39:81–91.

13. Penalva A, Pombo M, Carballo A, Barreiro J, Casanueva FF, Dieguez C. Influence of sex, age and adrenergetic pathways on the growth hormone response to GHRP-6. Clin Endocrinol 1993; 38:87–91.

14. Hayashi S, Hidesuke K, Shinichiro O, Hiromi A, Kazuo CH. Effect of intravenous administration of growth hormone-releasing peptide on plasma growth hormone in patients with short stature. Clin Pediatr Endocrinol 1993; 2(suppl 2):69–74.

15. Tuilpakov N, Bulatov AA, Peterkova VA, Elizarova GP, Volvedoz NN, Bowers CY. Growth hormone (GH)-releasing effects of synthetic peptide GH-releasing peptide-2 and GH-releasing hormone $(1-29NH_2)$ in children with GH insufficiency and idiopathic short stature. Metabolism 1995; 44:1199–1204.

16. Laron Z, Bowers CY, Hirsch D, Selman-Almonte A, Pelz M, Keret R, Gil-Ad I. Growth hormone-releasing activity of growth hormone-releasing peptide-1 (a synthetic heptapeptide) in children and adolescents. Acta Endocrinol 1993; 129:424–426.

17. Laron Z, Frenkel J, Silbergeld A. Growth hormone-releasing peptide—hexarelin—in children: biochemical and growth promoting effects. In: Bercu BB, Walker RF, eds. Growth Hormone Secretagogues. New York: Springer Verlag, 1996:379–387.

18. Frenkel J, Silbergeld A, Deghengi R, Laron Z. Short term effect of intranasal administration of hexarelin—a synthetic growth hormone-releasing peptide. Preliminary communication. J Pediatr Endocrinol Metab 1995; 8:43–45.

19. Merimee T, Laron Z. Growth hormone, IGF-I and growth: new views of old concepts. In: Modern Endocrinology and Diabetes, vol. 4. London-Tel Aviv: Freund Publishing House, 1996:266.

20. Laron Z, Topper E, Gil-Ad I. Oral clonidine—a simple, safe and effective test for growth hormone secretion. In: Laron Z, Butenandt O, eds. Evaluation of Growth Hormone Secretion. Pediatr Adolesc Endocrinol 1983; 12:103–115.

21. Josefsberg Z, Kauli R, Keret R, Brown M, Bilaik O, Greenberg D, Laron Z. Tests for hGH secretion in childhood. In: Laron Z, Butenandt O, eds. Evaluation of Growth Hormone Secretion. Pediatr Adolesc Endocrinol 1983; 12:66–74.

22. Laron Z, Klinger B, Jensen LT, Erster B. Biochemical and hormone changes induced by one week of administration of rIGF-I to patients with Laron type dwarfism. Clin Endocrinol 1991; 35:145–150.

23. Greulich WW, Pyle SI. Radiographic Atlas of Skeletal Development of Hand and Wrist. Stanford: Stanford University Press, 1959.

24. Tanner JM, Whitehouse RH, Takaishi M. Standards from birth to maturity for

height, weight, height velocity and weight velocity in British children. Arch Dis Child 1966; 41:613–635.

25. Laron Z, Frenkel J, Gil-Ad I, Linger B, Lubin E, Wuthrich P, Boutignon F, Lenaerts V, Dehgenghi R. Growth hormone releasing activity by internasal administration of synthetic hexapeptide (hexarelin). Clin Endocrinol 1994; 41:539–541.

26. Laron Z, Klinger B. IGF-I treatment of adult patients with Laron syndrome: preliminary results. Clin Endocrinol 1994; 41:631–638.

27. Karp M, Laron Z, Doron M. Insulin secretion in children with constitutional familial short stature. J Pediatr 1973; 83:241–246.

28. Klinger B, Jensen LT, Silbergeld A, Laron Z. Insulin-like growth factor-I raises serum procollagen levels in children and adults with Laron syndrome. Clin Endocrinol 1996; 45:423–429.

29. Pihoker C, Badger TM, Reynolds GA, Bowers CY. Intranasal growth hormone (GH) releasing peptide-2 (GHRP-2) in children with GH deficiency. Growth effects and GH secretion studies during treatment (abstr). Proceeding Endocrine Society, vol 2. San Francisco, June 14–15, 1996: P3 54.

30. Kirk JMW, Trainer PJ, Majrowski WH, Murphy J, Savage MO, Besser GH. Treatment with $GHRH_{(1-29)}NH_2$ in children with idiopathic short statute induces a sustained increase in growth velocity. Clin Endocrinol 1994; 41:487–493.

31. Klinger B, Silbergeld A, Deghenghi R, Frenkel J, Laron Z. Desensitization from long-term intranasal treatment with hexarelin does not interfere with the biological effects of this growth hormone-releasing peptide in short children. Eur J Endocrinol 1996; 134:716–719.

32. Stabler B, Underwood LE. Growth, Stature and Adaptation. Behavioral, Social and Cognitive Aspects of Growth Delay. Chapel Hill, NC: University of North Carolina at Chapel Hill Press, 1994:258.

33. Aran O, Laron Z, Galatzer A, Nagelberg N, Rubicsek Y. Is short stature a handicap—legal and rehabilitation aspects. In: Laron Z, Mastragostino S, Romano C, Cohen A, Boero S, eds. Limb Lengthening—for Whom, When and How? London–Tel Aviv: Freund Publishing House, 1995:131–135.

34. Laron Z, Butenandt O. Optimum use of growth hormone in children. Drugs 1991; 42:1–8.

35. Laron Z. Somatomedin I in clinical use: facts and potential. Drugs 1993; 45:1–8.

36. Aloi JA, Gertz BJ, Hartman ML, Huhn WC, Pezzoli SS, Wittreich JM, Krupa DA, Thorner MO. Neuroendocrine responses to a novel growth hormone secretagogue, L-692,429, in healthy older subjects. J Clin Endocrinol Metab 1994; 79:943–949.

37. Laron Z. Growth hormone secretagogues: clinical experience and therapeutic potential. Drugs 1995; 50:595–601.

38. Laron Z, Keret R, Bauman B, Pertzelan A, Ben-Zeev Z, Olsen DB, Comaru-Schally AM, Schally AV. Differential diagnosis between hypothalamic and pituitary hGH deficiency with the aid of synthetic $GH\text{-}RH_{1-44}$. Clin Endocrinol 1984; 21:9–12.

39. Laron Z, Bauman B. Growth hormone releasing hormone (GH-RH, GRF) and important new clinical tool. Eur J Pediatr 1986; 45:6–9.

40. Laron Z. Usefulness of the growth hormone-releasing hormone test regardless of which fragment is used ($GHRH_{1-44}$, $_{1-40}$ or $_{1-29}$). Isr J Med Sci 1991; 27:343–345.

23

Endocrine Response to Growth Hormone-Releasing Peptides Across Human Life Span

Ezio Ghigo, Emanuela Arvat, Laura Gianotti, Lidia Di Vito, Fabio Broglio, Giampiero Muccioli, and Franco Camanni
University of Turin, Turin, Italy

I. INTRODUCTION

The first growth hormone-releasing peptides (GHRPs) were invented, rather than isolated, in 1977 (1) and GHRP-6, a hexapeptide, was the first one active in vivo (2–4). The dose-dependent GH-releasing activity of GHRP-6 has been demonstrated in several species and in humans after intravenous, subcutaneous, intranasal, and even oral administration (2–4). Now the GHRP family includes more potent analoges of GHRP-6, such as GHRP-1, a heptapeptide, and GHRP-2 and hexarelin*, two hexapeptides the activity of which has been extensively studied in animals and in humans (2–6). Other GHRP mimetics, having structures more amenable to chemical modifications and optimization of oral bio-availability, have been synthetized. Among them, nonpeptidyl GHRPs (L-692,429, L-692,585, L-700,653, L-163,191 or MK-0677) have been studied in animals and even in humans (7–12); more recently, some cyclic peptides as well as penta-, tetra-, and pseudotripeptides have also been synthesized and studied in animals (13–15).

Although a clear-cut GH-releasing effect is common to all these molecules, the activity of peptidyl and nonpeptidyl GHRPs is not fully specific. In fact, it has been clearly demonstrated that they also possess a slight, but significant, stimulatory effect on prolactin (PRL) as well as on corticotropin (adreno-

---

*Hexarelin is a trademark for examorelin (I.N.N.).

corticotropic hormone; ACTH) and cortisol secretion in both animals (7,16–21) and humans (9–11,22–25).

The aim of this chapter is to focus on the endocrine effects of the now known GHRPs, with particular attention to the age-related variations and the mechanisms underlying these effects in humans. The results of the studies in different pathophysiological conditions and the possible clinical implications will also be taken into account. To better understand the possible mechanisms underlying the endocrine effects of GHRPs, the fundamentals of GHRP receptor distribution and results of the most important studies in animals will be considered before evaluating data in humans.

II. GHRP RECEPTORS

The GHRPs and nonpeptidyl GHRP mimetics act on specific receptors, the intracellular mechanisms of which are distinct from those of GHRH. In fact, the human GHRP receptor has recently been cloned. It is encoded by a rare mRNA, with a predicted open-reading frame of 366 amino acids, and a transmembrane topography typified by the G–protein-coupled receptor family. Notably, the receptor sequence does not show significant homology with other receptors yet known, although receptor transcripts are expressed in both the pituitary and the hypothalamus (26).

A specific high–affinity-binding site that mediates the activity of GHRPs and nonpeptidyl GHRP mimetics has been identified in the anterior pituitary and hypothalamic membranes of rat and pig (27–29). We have recently extended the study on the presence of GHRP receptors in humans (30,31), in whom these compounds have their highest GH-releasing effect (2–4). Among the various tissues tested, the hypothalamus and the pituitary gland of adult subjects showed the highest specific GHRP binding. Well detectable specific GHRP binding was also found in the choroid plexus, cerebral cortex, hippocampus, and pons–medulla, but not in thalamus, striatum, substantia nigra, cerebellum, and callous body (Fig. 1). No significant sex-related difference in GHRP binding to various tissues was found.

Altogether, the existence of specific GHRP-binding sites at the pituitary level and within the central nervous system (CNS) strengthens the hypothesis that GHRPs mimic an unidentified endogenous ligand that could be involved in neuroendocrine and even extraneuroendocrine activities, such as the control of sleep and food intake (32,33).

III. ENDOCRINE EFFECTS OF GHRPs IN ANIMALS

At the pituitary level, GHRPs stimulate GH release from somatotroph cells in vitro (34–39), although they have even been reported to be able to induce GH

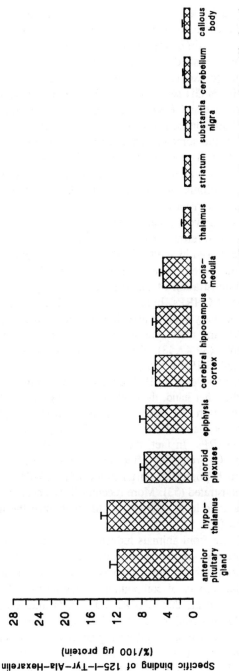

Figure 1 Specific binding of ^{125}I-Tyr-Ala-hexarelin in human pituitary gland and different brain regions of adult subjects (% of total radioactivity added). Values represent mean ± SEM.

synthesis (40). The GH-releasing activity of GHRPs in vitro is lower than that of GHRH (36), whereas their interaction is additive or truly synergistic (34–37,41). Generally, the stimulatory effect of GHRPs on GH secretion from somatotroph cells is abolished by specific GHRP, but not by GHRH antagonists (36–38). Although exogenous somatostatin is able to inhibit the stimulatory effect of GHRPs on GH secretion from the pituitary in vitro (34,36), there is evidence suggesting that GHRPs could act by antagonizing the inhibitory activity of somatostatin on somatotroph release by counteracting its hyperpolarizing effect on the somatotroph cell membrane (38).

Noteworthy, the GH-releasing activity of GHRPs on a hypothalamic–pituitary preparation was higher than that on the pituitary alone (36). It then became clear that the stimulatory effect of GHRPs on GH secretion is clearly higher in vivo than it is in vitro.

In vivo, GHRPs have a clear synergistic effect with GHRH (36); they prevent the normal cyclic refractoriness to GHRH (42), and their activity is age-dependent, being decreased in aged animals (43–45). The evidence that the GH response to direct intracerebroventricular (icv) GHRP injection is higher than that to systemic administration of the same dose (46), points to an important hypothalamic action of GHRPs. In agreement with this assumption, in animals after lesions of the pituitary stalk or after hypothalamic ablation, the GH-releasing effect of GHRPs and their synergism with GHRH are strongly reduced, although not abolished (39,47,48). Thus, normal hypothalamic–pituitary function is needed for full GHRP activity.

At the hypothalamic level, GHRPs do not negatively influence somatostatin (36,40,49,50). On the other hand, the evidence showing that GHRPs and GHRH have synergistic effects on GH secretion in vitro and in vivo (36,37) does not rule out the possibility that GHRH activity at least partly mediates the GH-releasing effect of GHRPs. In fact, the integrity of GHRH activity is needed for the GHRPs' full GH-releasing effect (36,42,49,51,52), and an increased release of GHRH in hypophysial portal blood after GHRP administration in sheep has been demonstrated (53). More recently, it has been hypothesized that, even at the hypothalamic level, GHRPs could antagonize the inhibitory influence of somatostatin (46).

There is evidence from animals indicating that the activity of GHRPs is also under the influences of neurotransmitters, neuropeptides, gonadal and adrenal steroids, and metabolic fuels, such as lipids (3,4). Studies addressing a desensitization to the activity of compounds belonging to GHRP family, both in vitro and in vivo, indicate that GHRPs and GHRH induce homologous, but not heterologous desensitization (34,36,37,42,43). Clear desensitization to GHRP activity has been demonstrated during continuous, but not during intermittent, GHRP infusion (16,42,43,54). Prolonged GHRP administration in animals increases IGF-I levels (16,20,43,44), indicating that GHRP-induced GH secretion is biologically active and that GHRP treatment is able to enhance the activity of the GH–IGF-I axis.

The activity of both peptidyl and nonpeptidyl GHRPs is not fully specific for GH. In fact, they also possess slight PRL, ACTH, and cortisol secretory activity (9–11,22–25). The mechanisms underlying these effects have not yet been studied extensively. The stimulatory effect on PRL seems to include a direct effect on somatomammotroph cells (55). On the other hand, there is evidence showing that the stimulatory effect on cortisol secretion is due to the ACTH-releasing activity of GHRPs which, in turn, seems dependent on central mechanisms (39,48). In fact, the stimulatory effect of GHRPs on the hypothalamic–pituitary–adrenal (HPA) axis is lost in rats with hypothalamic–pituitary disconnection (48) and GHRPs do not stimulate ACTH secretion from the rat pituitary as well as they do from human ACTH-secreting adenoma cells in culture (Ong H, personal communication). Interestingly, in rats, GHRP-6 and CRH have no interaction, whereas GHRP-6 and vasopressin (AVP) show a synergistic effect on ACTH secretion (21). Given this evidence, a CRH-mediated action for GHRPs has been suggested. Whatever the mechanism underlying the ACTH-releasing activity of GHRPs may be, it has been reported that prolonged treatment with GHRPs in young, obese, diabetic rats worsens their diabetic and metabolic state, probably by increasing the activity of the HPA axis (19).

IV. ENDOCRINE EFFECTS OF GHRPs IN HUMANS

A. Physiological Conditions

1.   Growth Hormone-Releasing Activity of GHRPs

The GH-releasing effect of GHRPs is dose-related (2–4). The GH response to 1 µg/kg iv of these peptides is generally higher than that elicited by 1 µg/kg iv GHRH (2–4). However, 1 µg/kg is not the maximal effective dose of GHRPs. In fact, 2 µg/kg iv of hexarelin or GHRP-2 have the same GH-releasing effect and elicit a further GH rise relative to the 1 µg/kg dose (24,56). Also nonpeptidyl GHRPs have a dose-related effect, but their potency is lower than that of peptides (9–11). A dose-related effect of GHRPs has also been shown after subcutaneous, intranasal, and even oral routes of administration (2–4). Notably, the GH response to GHRPs shows good intraindividual reproducibility (24), different from what is observed with GHRH (57).

The GHRP activity does not depend on sex (24,58,59), but undergoes marked age-related variations (59–64). The GH-releasing activity of hexarelin is present at birth and persists in a manner similar to that in prepubertal children; then, it clearly increases at puberty, persists in young adults, and then decreases in aging (4,60–63), when its effect is still higher than that of GHRH (60–62). Similar results have been observed when studying the effect of GHRP-6, GHRP-1, GHRP-2 and L-692,429 in aging (10–12,56,59,65). The age-related effect of GHRPs is different from that of GHRH (Fig. 2), the effect of which seems maximal in birth, and then progressively decreases to very low activity in aging (4,66).

The mechanisms underlying the age-related effects of GHRPs are poorly understood. Some data suggest that the enhanced activity of GHRPs at puberty could depend on gonadal steroids. In fact, the GH response to hexarelin is more marked in pubertal girls than in boys, being positively related to estradiol levels (63); moreover, the hexarelin-induced GH rise in prepubertal children is enhanced by testosterone, as well as by ethinyl estradiol, but not by oxandrolone pretreatment (67,68). Although estradiol could play a role in the increased GH-releasing activity of GHRPs at puberty, the fall of estrogen levels in menopause does not play a role in the reduction of the somatotroph response to GHRPs. In fact, 3-month treatment with transdermal estradiol does not modify the GH response to hexarelin in postmenopausal women (69). On the other hand, concomitant reduction of GHRH activity and somatostatinergic hyperactivity could play a role in the reduced effect of GHRPs in aging. In elderly subjects, the GH response to GHRPs is enhanced by both GHRH and arginine (60,62), which acts by somatostatin inhibition, but only the coadministration of these three substances restores the GH response to levels recorded in young adults (4). Beside age-related variations in the neural control of somatotroph function, the possibility has also to be considered that the reduced GH-releasing activity of GHRPs in aging could partly depend on an impairment of receptor or post-receptor mechanisms. In fact, we found that the GH response to hexarelin is

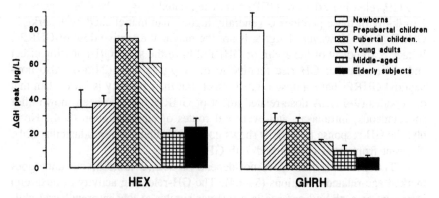

Figure 2 GH-releasing effect of hexarelin (2.0 µg/kg iv) and GHRH (1.0 µg/kg iv) across the human life span.

improved, but not restored, by supramaximal doses of the hexapeptide (70). To clarify the hypothalamic–pituitary mechanisms underlying the GH-releasing activity of GHRPs in man, several studies have been performed in young adults.

The GHRPs and GHRH have a synergistic effect (22,71,72), and even a very low GHRP dose is able to potentiate the GHRH-induced GH rise in humans (22). These data agree with the evidence in animals pointing to different mechanisms of action for these peptides (see Sec. III). The GH-releasing effect of GHRPs is increased by insulin-induced hypoglycemia, but is not modified by pyridostigmine, an indirect cholinergic agonist; arginine, or atenolol, a β_1-adrenergic antagonist; and clonidine, an α_2-adrenergic agonist, which are known to potentiate the GHRH-induced GH rise (71–73), as well as by naloxone (25). Notably, pyridostigmine and arginine are also unable to modify the synergistic effect of GHRPs and GHRH (74,75). On the other hand, pirenzepine, a muscarinic M_1-receptor antagonist, and albuterol (salbutamol), a β_2-adrenergic agonist, which are known to nearly abolish the somatotroph response to GHRH by stimulation of hypothalamic somatostatin, only blunt the hexarelin-induced GH response (72,73), as well as atropine does for the somatotroph response to GHRP-6 plus GHRH (71).

The GH response to GHRPs is also partially resistant to inhibition by other substances known to abolish the GHRH-induced GH rise. In fact, the GH response to GHRPs is only blunted by glucose load, glucocorticoids, and rhGH, which likely stimulate hypothalamic somatostatin release (75–78). Similarly, the increase of circulating free fatty acids, which could act directly at the pituitary level antagonizing depolarization of the somatotroph cell membrane, only blunts the GH response to GHRPs (76) which, on the other hand, is enhanced by acipimox, a lypolysis inhibitor (79). Notably, even the infusion of exogenous somatostatin, administered at a dose that can abolish the GHRH-induced GH rise, inhibits, but does not abolish the somatotroph responsiveness to hexarelin (72).

All together, these findings in humans agree with others in animals favoring the hypothesis that the mechanisms underlying the GH-releasing activity of GHRPs include antagonization of somatostatinergic activity both at the pituitary and the hypothalamic level (38,46). This could also explain the good intraindividual reproducibility of the GH response to GHRPs (24).

Interestingly, to emphasize that the activity of GHRPs depends mainly on the functional hypothalamic integrity, the GH-releasing effect of GHRP-6 and hexarelin, either alone or in combination with GHRH, is strongly reduced in patients with pituitary stalk disconnection (80,81). Moreover, the reduced effect of hexarelin in aging is fully restored to young levels only when it is coadministered with GHRH and arginine (4). These data confirm that the releasable GH pool in the pituitary does not vary with age (82) and reinforce the hypothesis that the activity of GHRPs is at least partly dependent on both GHRH and somatostatin activity. This agrees with the existence of an unknown endogenous U-factor that could mediate the effects of GHRPs (36) and implies that U-factor activity could be impaired in aging (Fig. 3).

Similar to animals (2,34,36,37,42,43), in humans GHRPs and GHRH induce homologous, but not heterologous desensitization (83–85). The homologous desensitization is marked during continuous GHRP infusion (83–85), whereas it is attenuated during intermittent administration (61,86,87). A key point is that the prolonged administration of GHRPs and, even more so, that of nonpeptidyl GHRP mimetics—namely, MK-0677—by parenteral, intranasal, and oral administration enhances spontaneous GH pulsatility over 24 h and increases IGF-I levels in normal young adults, as well as in short children and even in elderly subjects (11,12,65,83,84,86–88). Thus, it is clear that long-term treatment with GHRPs is able to augment the activity of the GH–IGF-I axis.

2. Prolactin-Releasing Activity of GHRPs

The stimulatory effect of GHRPs on PRL secretion in humans is slight and dose-dependent. In fact, the PRL rise after GHRP administration is within the normal range of basal PRL levels (22–24). Moreover, the PRL response to hexarelin

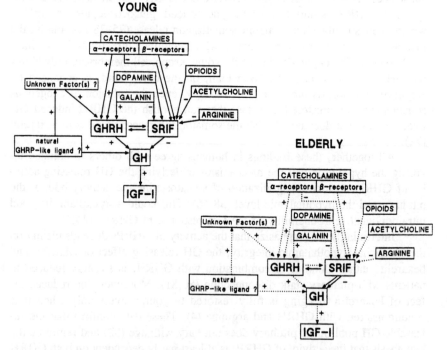

Figure 3 Neuroregulatory control of GH secretion across the human life span.

is markedly lower than that recorded after TRH and even after arginine administration (56; Ghigo et al., personal communication), agents which, in turn, are endowed with a PRL-releasing activity clearly lower than dopaminergic antagonists (Fig. 4). Differently from the somatotroph responsiveness, the lactotroph responsiveness to GHRPs does not vary with age. In fact, the PRL response to hexarelin administration in prepubertal and pubertal children, young adults, and elderly subjects is similar (89); also gender does not seem to influence the GHRP-induced PRL rise (23,24).

The mechanisms underlying the PRL-releasing activity of GHRPs are still unclear. At least in humans, there is now evidence showing that it is not mediated by opioidergic, serotoninergic, or histaminergic pathways (25,90). On the other hand, GHRPs and galanin do not show any interaction in their stimulatory effect on lactotoph secretion (Ghigo et al., personal communication). The evidence that the PRL-releasing effect of GHRPs is maintained in acromegalic patients bearing somatomammotroph pituitary adenoma, but not in patients bearing pure prolactinoma agrees with the hypothesis that the increase in PRL secretion after GHRP administration comes from direct stimulation of somatomammotroph cells (91).

Corticotropin- and Cortisol-Releasing Activity

The stimulatory effect of GHRPs on the activity of HPA axis in human, is more clear than that of lactotroph secretion (56). In fact, the ACTH and cortisol

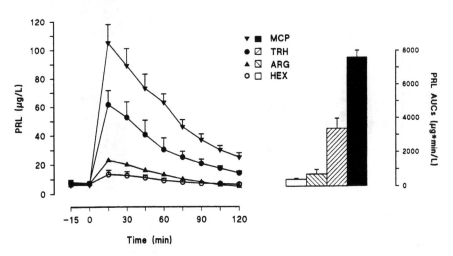

Figure 4   Mean (± SEM) PRL responses to hexarelin (2.0 µg/kg iv), arginine (ARG; 0.5 g/kg iv), TRH (400 µg iv), or metoclopramide (MCP; 10 mg iv) in normal young adults.

response to GHRPs overlaps with that after AVP or naloxone administration, and it is similar to that after CRH administration in young adults (25,92; Fig 5). However, the stimulatory effect of GHRPs seems only a brief neuroendocrine effect, being lost during prolonged treatment (11).

The stimulatory effect of GHRPs on HPA axis seems independent of sex, whereas it shows peculiar age-related variations. In fact, the ACTH, but not the cortisol, response to hexarelin increases at puberty; then shows a reduction in adulthood and, again, a trend toward increase in aging (89). This age-related pattern of ACTH response to GHRPs is different from that to CRH, the effect of which does not show increase at puberty, although it does show a trend toward increase in aging (93). On the other hand, the age-related pattern of adrenocorticotropic responsiveness to GHRPs is dissociated from that of somatotropic cells in aging; together with the age-independent effect on lactotroph secretion, these findings indicate that GHRPs act at different levels or on different receptors to induce different endocrine responses (Fig. 6). Also, the mechanisms underlying the stimulatory effect of GHRPs on HPA axis in humans are largely

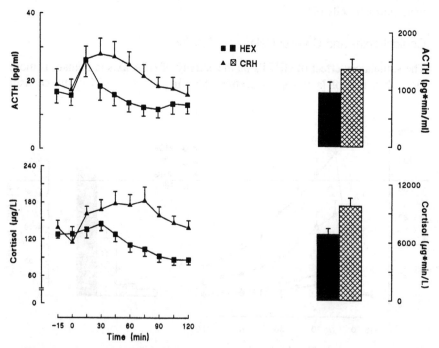

Figure 5 Mean (± SEM) ACTH and cortisol response curves (*left panel*) and AUCs (*right panel*) after hexarelin (2.0 µg/kg iv) and hCRH (2.0 µg/kg iv) in normal young subjects.

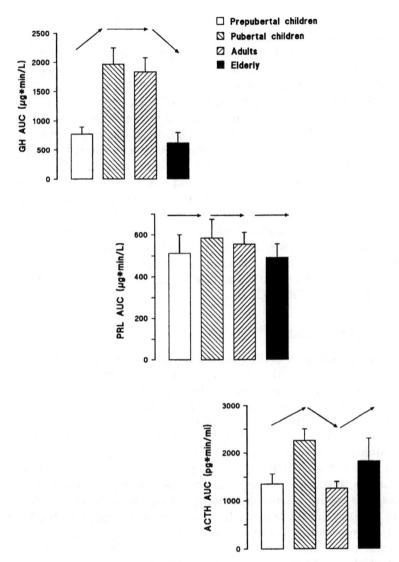

Figure 6 Age-related variations in the GH-, PRL-, and ACTH-releasing activity of hexarelin across the human life span.

unclear. Similar to animals, the stimulatory effect of GHRPs on cortisol secretion in humans is due to their ACTH-releasing activity which, in turn, seems dependent on CNS-mediated mechanisms. In fact, the ACTH and cortisol response to GHRPs are lacking in patients with hypothalamic–pituitary disconnection, whereas GHRPs do not stimulate ACTH release from pituitary cells (5,80; Ong H, personal communication). Interestingly, preliminary data suggest that the mechanisms mediating the ACTH-releasing activity of GHRPs in humans are different from those in animals. In fact, in young adults the co-administration of hexarelin and CRH or naloxone, a CRH-mediated stimulus, has an effect less than additive on ACTH and cortisol secretion, and hexarelin and AVP have no interaction (92); this in spite of the well-known synergistic effect of CRH and AVP on ACTH secretion (93). Other data indicate that the ACTH-releasing activity of hexarelin is not mediated by serotoninergic and histaminergic mechanisms, which play a major role in the neural control of the HPA axis (90). On the other hand, the ACTH and cortisol response to hexarelin is abolished by dexamethasone pretreatment as well as by alprazolam, a benzodiazepine that likely acts by inhibition of hypothalamic CRH release (92). In all, these data suggest that the ACTH-releasing activity of GHRPs could be, at least partly, dependent on CRH- and AVP-mediated mechanisms. In agreement with this hypothesis, in patients with Cushing's syndrome caused by pituitary ACTH-secreting adenoma, the ACTH and cortisol response to hexarelin is five-fold higher than that to CRH (92).

B. Pathological Conditions

The GH-releasing activity of GHRPs has also been tested in some human pathophysiological conditions, with particular attention to GH hyper- and hypo-secretory states. The stimulatory effect of GHRPs on GH secretion is preserved in the presence of somatotropic adenoma both in vitro (55,94) and in vivo (91,95,96). In vitro, GHRP-6 releases more GH than GHRH from somatotropinoma cells and is active even on cells refractory to GHRH stimulation (55,94). Moreover, the mean somatotropic responsiveness to GHRP-6, both alone and combined with GHRH, in acromegalic patients is similar to that in normal subjects, although a dramatic interindividual variability is present (96). In acromegalics bearing a somatomammotropic adenoma, both GH and PRL responses to hexarelin have been similar to those in normal subjects (91). Interestingly, this is not true in patients with idiopathic hyperprolactinemia nor in those with micro- and macroprolactinoma, in whom GHRPs are unable to modify the high PRL levels, whereas they elicit a GH response clearly lower than that in controls (91). Thus, it appears that hyperprolactinemic patients are partially refractory to the activity of GHRPs.

A preserved GH-releasing effect of GHRPs has also been reported in patients with functional GH hypersecretory states, such as anorexia nervosa (97), and in hyperthyroid patients (98) as well as in critically ill patients (99). On the other hand, GHRPs alone and in combination with GHRH induce marked GH response in children with idiopathic short stature (63,67,100,101) and even in patients with GH deficiency either in childhood or in adulthood (102,103). The possible usefulness of GHRPs to test the maximal secretory capacity of somatotropic cells in the diagnosis of GH deficiency has been hypothesized by some (67,68,100) but not by other authors (101,102). Notably, in children, as well as in adults with pituitary stalk lesion (80,81), the GH response to GHRPs alone and combined with GHRH has been markedly impaired.

The possible therapeutic usefulness of GHRPs in children with short stature has also been tested (6,104). Recent data demonstrate that, in open-label studies, prolonged treatments with intranasal hexarelin (or with sc GHRP-2) are able to increase IGF-I levels and growth velocity in children with idiopathic short stature or with GH insufficiency (87,88,105). During prolonged treatment with MK-677, a clear-cut IGF-I increase has been found, even in adult patients with GH deficiency, in spite of small increase of spontaneous GH secretion (106). GHRPs alone, and combined with GHRH, also elicit marked GH release in obese patients in whom there is a well known reduction of somatotroph function (74,107,108). The GH response to GHRPs in obesity is marked, but remains lower than that in controls (74,107,108); notably, the inhibition of lypolysis by acipimox is able to further enhance the GH response to the combined administration of GHRP-6 and GHRH in obese patients (107), further pointing to the role of metabolic alterations in the pathogenesis of GH insufficiency in obesity.

The GH-releasing effect of GHRP-6 both alone and combined with GHRH is strongly reduced in hypothyroidism (109) and nearly absent in Cushing's syndrome (110), pointing to the severity of GH insufficiency in this latter condition. On the other hand, although the ACTH–cortisol-releasing effect of hexarelin is lost in patients with Cushing's syndrome caused by a cortisol-secreting adrenal adenoma or by ectopic ACTH-secreting tumor, that in patients with Cushing's disease caused by pituitary ACTH-secreting adenoma is strikingly exaggerated and is fivefold higher than that elicited by CRH (92,111). These findings point to the possible clinical usefulness of GHRPs for the differential diagnosis between pituitary and ectopic ACTH-dependent hypercortisolism.

V. SUMMARY

Growth hormone-releasing peptides and nonpeptidyl GHRP mimetics are synthetic molecules that possess strong, dose-related and reproducible GH-releasing activity after intravenous, subcutaneous, intranasal, and oral administration in humans. GHRPs stimulate GH, acting at both the pituitary and the hypothalamic level, where specific receptors are present. In young adults GHRPs release more GH than GHRH, and their coadministration has a synergistic effect, pointing to mechanisms of actions that are at least partly different. However, normal activity of GHRH-secreting neurons is needed for the full GH-releasing effect of GHRPs. Unlike that of GHRH, the GH-releasing activity of GHRPs is not further increased by substances acting by inhibition of hypothalamic somatostatin and is only blunted by substances that stimulate hypothalamic somatostatin, and even by free fatty acids and exogenous somatostatin, which act directly on somatotroph cells. Thus, the GH-releasing activity of GHRPs is partially refractory to inhibitory influences. Actually, GHRPs act, at least partly, by antagonizing somatostatin activity, both at the pituitary and the hypothalamic level. The GH-releasing effect of GHRPs is not dependent on sex, but undergoes age-related variations. In fact, it is low at birth, strikingly increases at puberty, persists similar in adulthood, and decreases thereafter; in middle-aged persons, this is similar to that in elderly subjects. Variations in gonadal steroid levels seem to influence the activity of GHRPs only in childhood, but not in aging. In fact, the reduced GH response to GHRPs in the elderly is probably due to concomitant GHRH hypoactivity and somatostatinergic hyperactivity, although an impairment of GHRP receptors could also play a role. The GHRPs also possess a slight stimulatory effect on PRL and a more clear ACTH–cortisol-releasing activity that is similar to that of hCRH. Interestingly, although the PRL response to GHRPs does not vary with age, the ACTH and cortisol responses to GHRPs show an increase at puberty, a decrease in adulthood and, again, a trend toward increase in aging. The different age-related patterns of GH, PRL, and ACTH–cortisol response to GHRPs in humans suggest that actions at different levels, or on different receptor subtypes, mediate their effects. The PRL-releasing activity of GHRPs could be due to a direct effect at the pituitary level. On the other hand, the ACTH-releasing activity of GHRPs takes place by a CNS-mediated mechanism that could be partly dependent on both hypothalamic CRH and AVP. The ACTH-releasing activity of GHRPs must be taken into particular account; in fact, in patients with pituitary ACTH-dependent Cushing's syndrome, it is strikingly higher than that induced by hCRH.

AKNOWLEDGMENTS

Personal studies included in this chapter were supported by grants from CNR, MURST, FSMEM, and Europeptides. The authors wish to thank Dr. Romano Deghenghi and Dr. Muni Franklin Boghen, as well as Josefina Ramunni, Barbara Maccagno, Jaele Bellone, Gianluca Aimaretti, Mauro Maccario, Massimo Procopio, Silvia Grottoli, Corrado Ghe', Maria Cristina Ghigo, Lodovico Benso, Edoardo Bartolotta, Giovanni Farello, Sergio Bernasconi, Lucia Ghizzoni, Cecilia Volta, Giampaolo Ceda, Guido Rizzi, and Gianmario Boffano for participating to our studies.

REFERENCES

1. Bowers CY, Chang J, Momany F, Folkers K. Effects of the enkephalins and enkephalin analogs on release of pituitary hormones in vitro. In: Macintyne I, ed. Molecular Endocrinology. Amsterdam: Elsevier/North Holland Biochemical Press, 1977:287–292.
2. Bowers CY, Veeraragavan K, Sethumadhavan K. Atypical growth hormone releasing peptides. In: Bercu BB, Walker RF, eds. Growth Hormone II, Basic and Clinical Aspects. New York: Springer-Verlag, 1993:203–222.
3. Korbonits M, Grossman AB. Growth hormone-releasing peptide and its analogues. Novel stimuli to growth hormone release. Trends Endocrinol Metab 1996; 6:43–49.
4. Ghigo E, Arvat E, Gianotti L, Grottoli S, Rizzi G, Ceda GP, Deghenghi R, Camanni F. Aging and GH-releasing peptides. In: Bercu B, Walker R, eds. Growth Hormone Secretagogues. New York: Springer-Verlag, 1996:415–431.
5. Deghenghi R. Growth hormone releasing peptides. In: Bercu B, Walker R, eds. Growth Hormone Secretagogues. New York: Springer-Verlag, 1996:85–102.
6. Laron Z. Growth hormone secretagogues. Drugs 1995; 50:595–601.
7. Smith RG, Cheng K, Pong SS, Hickey G, Jacks T, Butler B, Chan WWS, Chaung LYP, Judith F, Taylor J, Schoen WR, Wyvratt MJ, Fisher MH. A non-peptidyl GH secretagogue. Science 1993; 260:1640–1643.
8. DeVita RJ, Wyratt MJ. Benzolactam growth hormone secretagogues. Drugs Future 1996; 21:273–281.
9. Gertz BJ, Barrett JS, Eisenhandler R, Krupa DA, Wittreich JM, Seibold JR, Schneider SH. Growth hormone response in man to L-692,429, a novel non peptide mimic of growth hormone-releasing peptide-6. J Clin Endocrinol Metab 1993; 77:1393–1397.
10. Aloi JA, Gertz BJ, Hartman ML, Huhn WC, Pezzoli SS, Wittreich M, Krupa DA, Thorner MO. Neuroendocrine responses to a novel growth hormone secretagogue, L-692,429, in healthy older subjects. J Clin Endocrinol Metab 1994; 79:943–949.
11. Copinschi G, van Onderbergen A, L'hermite-Baleriaux M, Mendel CM, Caufriez A, Leproult R, Bolognese JA, De Smet M, Thorner MO, van Cauter E. Effects of 7-day treatment with a novel, orally active, growth hormone (GH) secretagogue, MK-677, on 24-hour GH profiles, insulin-like growth factor I, and adrenocortical function in normal young men. J Clin Endocrinol Metab 1996; 81:2776–2782.
12. Chapman IM, Hartman ML, Pezzoli SS, Thorner MO. Enhancement of pulsatile growth hormone secretory by continuous infusion of a growth hormone-releasing peptide mimetic, L-692,429, in older adults—a clinical research center study. J Clin Endocrinol Metab 1996; 81:2874–2880.
13. Deghenghi R, Boutignon F, Luoni M, Grilli R, Guidi M, Locatelli V. Small peptides as potent releasers of growth hormone. J Pediatr Endocrinol Metab 1995; 8:311–313.
14. McDowell RS, Elias KA, Stanley MS, Burdick DJ, Burnier JP, Chan KS, Fairbrother WJ, Hammonds RG, Ingle GS, Jacobsen NE, Mortensen DL, Rawson

TE, Won WB, Clark RG, Somers TC. Growth hormone secretagogues: characterization, efficacy, and minimal bioactive conformation. Biochemistry 1995; 92:11165–11169.

15. Elias KA, Ingle GS, Burnier JP, Hammonds RG, McDowell RS, Rawson TE, Somers TC, Stanley MS, Cronin MJ. In vitro characterization of four novel classes of growth hormone-releasing peptide. Endocrinology 1995; 136:5694–5699.

16. Jacks T, Hickey G, Judith F, Taylor J, Chen H, Krupa D, Feeney W, Schoen W, Ok D, Fisher M, Wyratt M, Smith R. Effects of acute and repeated intravenous administration of L-692,585, a novel non-peptidyl growth hormone secretagogue, on plasma growth hormone, IGF-I, ACTH, cortisol, prolactin, insulin, and thyroxine levels in beagles. J Endocrinol 1994; 143:399–406.

17. Hickey GJ, Jacks T, Judith F, Taylor J, Schoen WR, Krupa D, Cunningham P Clark J, Smith RG. Efficacy and specifity of L-692,429, a novel nonpeptidyl growth hormone secretagogue, in beagles. Endocrinology 1994; 134:695–701.

18. Chang CH, Rickes EL, Marsilio F, McGuire L, Cosgrove S, Taylor J, Chen H, Feighner S, Clark JN, DeVita R, Schoen W, Wyvratt M, Fischer M, Smith RG, Hickey GJ. Activity of a novel non-peptidyl GH secretagogue, L-700,653, in swine. Endocrinology 1995; 136:1065–1071.

19. Clark RG, Mortenesen DL, Won, WB, Ma YH, Stanley MS, Somers TC, Fairhall KM, Thomas GB, Robinson ICAF. Novel GHRP analogs acutely activate the HPA axis and are diabetogenic in ZDF rats. 10th International Congress of Endocrinology. San Francisco, CA, June 12–15, 1996.

20. Hickey GJ, Jacks T, Schleim K, Frazier E, Chen H, Krupa D, Feeney W, Narpunet R, Patchett A, Smith RG. Repeated administration of the GH secretagogue MK-0677 increases and maintains elevated IGF-1 levels. 10th International Congress of Endocrinology. San Francisco, CA, June 12–15, 1996.

21. Thomas GB, Fairhall KM, Robinson ICAF. Activation of the hypothapamo-pituitary-adrenal axis by the growth hormone (GH) secetagogue, GH-releasing peptide-6, in rats. Endocrinology 1997; 138:1585–1591.

22. Bowers CY, Reynolds GA, Durham D, Barrera CM, Pezzoli SS, Thorner MO. Growth hormone (GH)-releasing peptide stimulates GH release in normal men and acts synergistically with GH-releasing hormone. J Clin Endocrinol Metab 1990; 70:975–982.

23. Imbimbo BP, Mant T, Edward M, Amin D, Froud A, Lenaerts U, Boutignon F, Deghenghi R. Growth hormone releasing activity of hexarelin in humans: a dose-response study. Eur J Clin Pharmacol 1994; 46:421–425.

24. Ghigo E, Arvat E, Gianotti L, Imbimbo BP, Lenaerts V, Deghenghi R, Camanni F. Growth hormone-releasing activity of hexarelin, a new synthetic hexapeptide, after intravenous subcutaneous, intranasal and oral administration in man. J Clin Endocrinol Metab 1994; 78:693–698.

25. Korbonits M, Trainer PJ, Besser GM. The effect of an opiate antagonist on the hormonal changes induced by hexarelin. Clin Endocrinol 1995; 43:365–371.

26. Howard AD, Feighner SD, Cully DF, Arena JP, Liberator PA, Rosenblum CI, Uamelin M, Ureniuk Dl, Palyhe OC, Anderson J, Paress PS, Diaz C, Chou N, Liu KK, McKee KK, Pong SS, Cheung LY, Elbrecht A, Dashkevicz M, Keavehs R, Rishy M, Siringjhsinghji DJS, Dean DC, Melillo DG, Patchett AA, Nergund

R, Srilllin PR. A receptor in pituitary and hypothalamus that functions in growth hormone release. Science 1996; 273:974–977.

27. Codd EE, Shu AYL, Walker RF. Binding of a growth hormone releasing hexapaptide to a specific hypothalamic and pituitary binding sites. Neuropharmacology 1989; 28:1139–1344.

28. Sethumadhavan K, Veeraragavan K, Bowers CY. Demonstration and characterization of the specific binding of growth hormone-releasing peptide to rat anterior pituitary and hypothalamic membranes. Biochem Biophys Res Commun 1991; 178:31–37.

29. Pong SS, Chaung LYP, Dean DC, Nargunt RP, Patchett AA, Smith RG. Identification of a new G–protein-linked receptor for growth hormone secretagogues. Mol Endocrinol 1996; 10:57–61.

30. Muccioli G, Ghe' C, Ghigo MC, Arvat E, Papotti M, Boghen MF, Deghenghi R, Ong H, Ghigo E. Presence of specific receptors for hexarelin, a GH releasing hexapeptide, in human brain and pituitary gland. International Symposium on Growth Factors in Endocrinology and Metabolism. Berlin, 1995.

31. Muccioli G, Ghe' C, Ghigo MC, Arvat E, Papotti M, Boghen MF, Deghenghi R, Ong H, Ghigo E. Growth hormone-releasing peptide receptors in human brain and pituitary gland. Fundam Clin Pharmacol 1996; 10(suppl 2):205.

32. Locke W, Kirgis HD, Bowers CY, Abdo AA. Intracerebroventricular growth-hormone-releasing peptide-6 stimulates eating without affecting plasma growth hormone responses in rats. Life Sci 1995; 56:1347–1352.

33. Frieboes RM, Murck H, Maier P, Schier T, Holsboer F, Steiger A. Growth hormone-releasing peptide-6 stimulates sleep, growth hormone, ACTH and cortisol release in normal man. Neuroendocrinology 1995; 61:584–589.

34. Badger TM, Millard WY, McCormik GF, Bowers CY, Martin JB. The effects of growth hormone (GH)-releasing peptides on GH secretion in perfused pituitary cells of adult male rats. Endocrinology 1984; 115:1432–1438.

35. Sartor O, Bowers CY, Chang D. Parallel studies of His-D-Trp-Ala-Trp-D-Phe-Lys-NH$_2$ and human pancreatic growth hormone-releasing factor-44-NH$_2$ in rat primary pituitary cell monolayer culture. Endocrinology 1985; 116:952–957.

36. Bowers CY, Sartor AO, Reynolds GA, Badger TM. On the action of the growth hormone-releasing hexapeptide, GHRP. Endocrinology 1991; 128:2027–2035.

37. Cheng K, Chan WWS, Barreto A, Convey EM, Smith RG. The synergistic effects of His-D-Trp-Ala-Trp-D-Phe-Lys-NH$_2$ on growth hormone (GH)-releasing factor-stimulated GH release and intracellular adenosine 3′,5′-monophosphate accumulation in rat primary cell culture. Endocrinology 1989; 124:2791–2798.

38. Goth MI, Lyons CE, Canny BY, Thorner MO. Pituitary adenylate cyclase activating polypeptide, growth hormone (GH)-releasing peptide and GH-releasing hormone stimulate GH release through distinct pituitary receptors. Endocrinology 1992; 130:939–944.

39. Mallo F, Alvarez CV, Benitez L, Burguera B, Coya R, Casanueva FF. Regulation of His-DTRP-Ala-Trp-DPhe-Lys-NH$_2$ (GHRP6)-induced GH secretion in the rat. Neuroendocrinology 1992; 57:247–251.

40. Locatelli V, Grilli R, Torsello A, Cella SG, Wehrenberg WB, Muller EE. Growth hormone-releasing hexapeptide is a potent stimulator of growth hormone gene

expression and release in the growth hormone-releasing hormone-deprived infant rat. Pediatr Res 1994; 36:169–174.

41. Chen K, Chan WWS, Butler B, Wei L, Schoen WR, Wyratt MJ, Fisher MH, Smith RG. Stimulation of growth hormone release from rat pituitary cells by L-692,429, a novel non-peptidyl GH secretagogue. Endocrinology 1993; 132:2729–2731.

42. Clark RG, Carlsson MS, Trojnar J, Robinson ICAF. The effects of a growth hormone-releasing peptide and growth hormone-releasing factor in conscious and anaesthetized rats. J Neuroendocrinol 1989; 1:252–255.

43. Sartor O, Bowers CY, Reynolds GA, Momany FA. Variables determining the growth hormone response of His-D-Trp-Ala-Trp-D-Phe-Lys-NH₂ in the rat. Endocrinology 1985; 117:1441–1447.

44. Walker RF, Yang AW, Masuda R, Hu CS, Bercu BB. Effects of growth hormone releasing peptides on stimulated growth hormone secretion in old rats. In: Bercu BB, Walker RF, eds. Growth Hormone II, Basic and Clinical Aspects. New York: Springer-Verlag, 1993:167–192.

45. Cella S, Locatelli V, Poratelli M, De Gennaro Colonna V, Imbimbo BP, Deghenghi R, Muller EE. Hexarelin, a potent GHRP analogue: interaction with GHRH and clonidine in young and aged dogs. Peptides 1995; 16:81–86.

46. Fairhall KM, Mynett A, Robinson ICAF. Central effects of growth hormone-releasing hexapeptide (GHRP-6) on growth hormone release are inhibited by central somatostatin action. J Endocrinol 1995; 144:555–560.

47. Fletcher TP, Thomas GB, Willoughby JO, Clarke IJ. Constitutive growth hormone secretion in sheep after hypothalamopituitary disconnection and the direct in vivo pituitary effect of growth hormone releasing peptide 6. Neuroendocrinology 1994; 60:76–86.

48. Hickey GJ, Drisko J, Faidley T, Chang C, Anderson L, Nicolich S, McGuire L, Rickes E, Krupa D, Feeney W, Friscino B, Cunningham P, Frazier E, Chen H, Laroque P, Smith RG. Mediation by the central nervous system is critical to the in vivo activity of the GH secretagogue L-692,585. J Endocrinol 1996; 148:371–380.

49. Conley LK, Teik JA, Deghenghi R, Imbimbo BP, Giustina A, Locatelli V, Wehremberg WB. Mechanism of action of hexarelin and GHRP-6: analysis of the involvement of GHRH and somatostatin in the rat. Neuroendocrinology 1995; 61:44–50.

50. Muruais J, Penalva A, Dieguez C, Casanueva FF. Influence of endogenous cholinergic tone and alpha-adrenergic pathways on growth hormone responses to His-D-Trp-Ala-Trp-D-Phe-Lys-NH₂ in the dog. J Endocrinol 1993; 138:211–218.

51. Dickson SL, Leng G, Robinson ICAF. Systemic administration of growth hormone-releasing peptide activates hypothalamic arcuate neurons. Neurosci Lett 1993; 53:303–307.

52. Bercu BB, Yang SW, Masuda R, Walker RF. Role of selected endogenous peptides in growth hormone-releasing hexapeptide activity: analysis of growth hormone-releasing hormone, thyroid hormone-releasing hormone, and gonadotropin-releasing hormone. Endocrinology 1992; 130:2579–2586.

53. Guillame V, Magnan E, Cataldi M, Dutour A, Sauze N, Renard M, Razafindraibe

H, Conte-Devolx B, Deghenghi R, Lenaerts V, Oliver C. Growth hormone (GH)-releasing hormone secretion is stimulated by a new GH-releasing hexapeptide in sheep. Endocrinology 1994; 135:1073–1076.

54. Fairhall KM, Mynett A, Smith RG, Robinson ICAF. Consistent GH responses to repeated injections of GH-releasing hexapeptide (GHRP-6) and the non-peptide GH secretagogue, L-692,585. J Endocrinol 1995; 145:417–426.

55. Adams EF, Petersen B, Lei T. Hunchfelder M., Fanlbusch R. The growth hormone secretagogue, L-692,429, induces phosphatidylinositol hydrolysis and hormone secretion by human pituitary tumors. Biochem Biophys Res Commun 1995; 214:454–460.

56. Arvat E, Di Vito L, Maccagno B, Broglio F, Boghen MF, Deghenghi R, Camanni F, Ghigo E. Effects of GHRP-2 and hexarelin, two synthetic GH-releasing peptides, on GH, prolactin, ACTH and cortisol level in man. Comparison with the effects of GHRH, TRH and hCRH. Peptides 1997; 18:885–891.

57. Mazza E, Ghigo E, Goffi S, Procopio M, Imperiale E, Arvat E, Bellone J, Boghen MF, Muller EE, Camanni F. Effect of the potentiation of cholinergic activity on the variability in individual GH response to GHRH. J Endocrinol Invest 1989; 12:795–799.

58. Penalva A, Pombo M, Carballo A, Barreiro J, Casanueva FF, Dieguez C. Influence of sex, age and adrenergic pathways on the growth hormone response to GHRP-6. Clin Endocrinol (Oxf) 1993; 38:87–91.

59. Bowers CY. GH releasing peptides. Structure and kinetics. J Pediatr Endocrinol 1993; 6:21–31.

60. Ghigo E, Arvat E, Rizi G, Bellone J, Nicolosi M, Boffano GM, Mucci M, Boghen MF, Camanni F. Arginine enhances the growth hormone (GH)-releasing activity of a synthetic hexapeptide (GHRP-6) in elderly but not in young subjects after oral administration. J Endocrinol Invest 1994; 17:157–162.

61. Ghigo E, Arvat E, Rizzi G, Goffi S, Mucci M, Boghen MF, Camanni F. Growth hormone-releasing activity of growth hormone-releasing peptide-6 is maintained after short-term oral pretreatment with the hexapapetide in normal aging. Eur J Endocrinol 1994; 131:1–4.

62. Arvat E, Gianotti L, Grottoli S, Imbimbo BP, Lenaerts V, Deghenghi R, Camanni F, Ghigo E. Arginine and growth hormone-releasing hormone restore the blunted growth hormone-releasing activity of hexarelin in elderly subjects. J Clin Endocrinol Metab 1994; 79:1440–1443.

63. Bellone J, Aimaretti G, Bartolotta E, Benso L, Imbimbo BP, Lenaerts V, Deghenghi R, Camanni F, Ghigo E. Growth hormone-releasing activity of hexarelin, a new synthetic hexapeptide, before and during puberty. J Clin Endocrinol Metab 1995; 80:1090–1094.

64. Micic D, Popovic V, Kendereski A, Macul D, Casanueva FF, Dieguez C, Laron Z. Growth hormone secretion after the administration of GHRP-6 or GHRH combined with GHRP-6 does not decline in late adulthood. Clin Endocrinol 1995; 42:191–194.

65. Chapman IM, Bach MA, Van Cauter E, Farmer M, Krupa D, Taylor AM, Schilling LM, Cole KY, Skiles EH, Pezzoli SS, Hartman ML, Veldhuis JD, Gormley GJ, Thorner MO. Stimulation of the growth hormone (GH)-insulin-like

growth factor axis by daily oral administration of a GH secretagogue (MK-677) in healthy elderly subjects. J Clin Endocrinol Metab 1996; 81:4249–4257.

66. Chatelain P, Clairs O, Lapillonne A, Dimayo M, Salle B. Fetal and neonatal somatotroph function. Front Paediatr Neuroendorinol 1994; 20:127–130.

67. Loche S, Cambiaso P, Carta D, Setzu S, Imbimbo BP, Borrelli P, Pintor C, Cappa M. The growth hormone-releasing activity of hexarelin, a new synthetic hexapeptide, in short normal and obese children and in hypopituitary subjects. J Clin Endocrinol Metab 1995; 80:674–678.

68. Loche S, Colao A, Cappa M, Bellone J, Aimaretti G, Farello G, Faedda A, Lombardi G, Deghenghi R, Ghigo E. The growth hormone response to hexarelin in children: reproducibility and effect of sex steroids. J Clin Endocrinol Metab 1997; 82:861–864.

69. Arvat E, Gianotti L, Broglio F, Maccagno B, Bertagna A, Deghenghi R, Camanni F, Ghigo E. Oestrogen replacement does not restore the reduced GH-releasing activity of hexarelin, a synthetic hexapeptide, in postmenopausal women. Eur J Endocrinol 1997; 136:483–487.

70. Arvat E, Ceda GP, Ramunni J, DiVito L, Gianotti L, Boghen MF, Deghenghi R, Ghigo E. Supramaximal doses of GHRH or hexarelin, a synthetic hexapeptide, enhance the low somatotrope responsiveness in aging. J Endocrinol Invest 1996; 19(suppl 5):3.

71. Penalva A, Carballo A, Pombo M, Casanueva FF, Dieguez C. Effect of growth hormone (GH)-releasing hormone (GHRH), atropine, pyridostigmine, or hypoglycemia on GHRP-6-induced GH secretion in man. J Clin Endocrinol Metab 1993; 76:168–171.

72. Arvat E, Gianotti L, Di Vito L, Imbimbo BP, Lenaerts V, Deghenghi R, Camanni F, Ghigo E. Modulation of growth hormone-releasing activity of hexarelin in man. Neuroendocrinology 1995; 61:51–56.

73. Arvat E, Gianotti L, Ramunni J, DiVito L, Deghenghi R, Camanni F, Ghigo E. Influence of beta-adrenergic agonists and antagonists on the GH-releasing effect of hexarelin in man. J Endocrinol Invest 1996; 19:25–29.

74. Cordido F, Penalva A, Peino R, Casanueva FF, Dieguez C. Effect of combined administration of growth hormone (GH)-releasing hormone, GH-releasing peptide-6 and pyridostigmine in normal and obese subjects. Metabolism 1995; 44:745–748.

75. Arvat E, Di Vito L, Gianotti L, Ramunni J, Boghen MF, Deghenghi R, Camanni F, Ghigo E. Mechanisms underlying the negative growth hormone (GH) autofeedback on the GH-releasing effect of Hexarelin in man. Metabolism 1997; 46:83–88.

76. Maccario M, Arvat E, Procopio M, Gianotti L, Grottoli S, Imbimbo BP, Lenaerts V, Deghenghi R, Camanni F, Ghigo E. Metabolic modulation of the growth-hormone-releasing activity of hexarelin in man. Metabolism 1995; 44:134–138.

77. Gertz BJ, Sciberras DG, Yogendran L, Christie K, Bador K, Krupa D, Wittreich JM, James I. L-602,429, a nonpeptide growth hormone (GH) secretagogue reverses glucocorticoid suppression of GH secretion. J Clin Endocrinol Metab 1994; 79:745–749.

78. Massoud AF, Hindmarsh PC, Brook CG. Hexarelin induced growth hormone release is influenced by exogenous growth hormone. Clin Endocrinol 1995; 43:617–621.

79. Peino R, Cordido F, Penalva A, Alvarez V, Dieguez C, Casanueva FF. Acipimox-mediated plasma free fatty acid depression per se stimulates growth hormone (GH) secretion in normal subjects and potentiates the response to other GH-releasing stimuli. J Clin Endocrinol Metab 1996; 81:909–913.

80. Popovic V, Damjanovic S, Micic D, Djurovic M, Dieguez C, Casanueva FF. Blocked growth hormone-releasing peptide (GHRP-6)-induced GH secretion and absence of the synergic action of GHRP-6 plus GH-releasing hormone in patients with hypothalamopituitary disconnection: evidence that GHRP-6 main action is exerted at the hypothalamic level. J Clin Endocrinol Metab 1995; 80:942–947.

81. Pombo M, Harreiro J, Penalva A, Peino R, Dieguez C, Casanueva FF. Absence of growth hormone (GH) secretion after the administration of either GH releasing hormone (GHRH), GH-releasing peptide (GHRP-6), or GHRH plus GHRP-6 in children with neonatal pituitary stalk transection. J Clin Endocrinol Metab 1995; 80:3180–3184.

82. Ghigo E, Arvat E, Gianotti L, Ramunni J, Di Vito L, Maccagno B, Grottoli S, Camanni F. Human aging and the GH–IGF-I axis. J Pediatr Endocrinol Metab 1996; 9:271–278.

83. Jaffe CA, Ho PJ, Demott-Friberg R, Bowers CY, Barkan AL. Effects of a prolonged growth hormone (GH)-releasing peptide infusion on pulsatile GH secretion in normal men. J Clin Endocrinol Metab 1993; 77:1641–1647.

84. Huhn WC, Hartman ML, Pezzoli SS, Thorner MO. Twenty-four-hour growth hormone (GH)-releasing peptide (GHRP) infusion enhances pulsatile GH secretion and specifically attenuates the response to a subsequent GHRP bolus. J Clin Endocrinol Metab 1993; 76:1202–1208.

85. DeBell WK, Pezzoli, SS, Thorner MO. Growth hormone (GH) secretion during continous infusion of GH-releasing peptide: partial response attenuation. J Clin Endocrinol Metab 1991; 72:1312–1316.

86. Ghigo E, Arvat E, Gianotti L, Grottoli S, Rizzi G, Ceda GP, Boghen MF, Deghenghi R, Camanni F. Short term administration of intranasal or oral hexarelin, a synthetic hexapeptide, does not desensitize the GH responsiveness in human aging. Eur J Endocrinol 1996; 135:407–412.

87. Klinger B, Silberld A, Deghenghi R, Frenkel J, Laron Z. Desensitization from long-term intranasal treatment with hexarelin does not interfere with the biological effects of this growth hormone-releasing peptide in short children. Eur J Endocrinol 1996; 134:716–719.

88. Laron Z, Frenkel J, Deghenghi R, Anin S, Klinger B, Silbergeld A. Intranasal administration of the GHRP hexarelin accelerates growth in short children. Clin Endocrinol 1995; 43:631–635.

89. Arvat E, Di Vito L, Ramunni J, Gianotti L, Maccagno B, Broglio F, Deghenghi R, Camanni F. Age-related variations in the GH-, PRL- and ACTH-releasing activities of hexarelin in man. Endocrinol Metabol 1997; (suppl A):34.

90. Arvat E, Maccagno B, Ramunni J, Gianotti L, Di Vito L, Deghenghi R, Camanni

F, Ghigo E. Effects of histaminergic antagonists on the GH-releasing activity of GHRH or hexarelin, a synthetic hexapetide, in man. J Endocrinol Invest 1997; 20:122–127.

91. Ciccarelli E, Grottoli S, Razzore P, Gianotti L, Arvat E, Deghenghi R, Camanni F, Ghigo E. Hexarelin, a synthetic growth hormone releasing peptide, stimulates prolactin secretion in acromegalic but not in hyperprolactinemic patients. Clin Endocrinol 1996; 44:67–71.

92. Ghigo E. Effects of GHRPs during lifespan in humans. 10th International Congress of Endocrinology. San Francisco, CA, June 12–15, 1996.

93. Orth DN. Corticotropin-releasing hormone in humans. Endocr Rev 1992; 13:164–191.

94. Renner U, Brockmeier S, Strasburger CJ, Lange M, Schopohl J, Muller OA, von Werder K, Stalla GK. Growth hormone (GH)-releasing peptide stimulation of GH release from human somatotroph adenoma cells: interaction with GH-releasing hormone, thyrotropin-releasing hormone, and octreotide. J Clin Endocrinol Metab 1994; 78:1090–1096.

95. Popovic V, Damjanovic S, Micic D, Petakov M, Dieguez C, Casanueva FF. Growth hormone (GH) secretion in active acromegaly after the combined administration of GH-releasing hormone and GH-releasing peptide-6. J Clin Endocrinol Metab 1994; 79:456–460.

96. Alster DK, Bowers CY, Jaffe CA, Ho PJ, Barkan AL. The growth hormone (GH) response to GH-releasing peptide (His-D-Trp-Ala-Trp-D-Phe-Lys-NH$_2$), GH-releasing hormone, and thyrotropin-releasing hormone in acromegaly. J Clin Endocrinol Metab 1993; 77:842–845.

97. Popovic V, Micic D, Damjanovic S, Djurovic M, Simic M, Gligorovic M, Dieguez C. Evaluation of pituitary GH reserve with GHRP-6. J Pediatr Endocrinol Metab 1996; 9:289–298.

98. Ramos-Dias JC, Pimentel Filho F, Reis AF, Lengyel AMJ. Different growth hormone (GH) responses to GH-releasing peptide and GH-releasing hormone in hyperthyroidism. J Clin Endocrinol Metab 1996; 81:1343–1346.

99. Van der Berghe G, de Zegher F, Bowers CY, Wouters P, Muller P, Soetens F, Vlasselaers D, Schetz M, Verwaest C, Lauwers P, Bouillon R. Pituitary responsiveness to GH-releasing hormone, GH-releasing peptide-2 and thyrotropin-releasing hormone in critical illness. Clin Endorinol 1996; 45:341–351.

100. Laron Z, Frenkel J, Gil-Ad, Klinger B, Imbim E, Wuthrich P, Boutignon F, Lengerts V, Deghenghi R. Growth hormone releasing activity by intranasal administration of a synthetic hexapeptide (hexarelin). Clin Endocrinol 1994; 41:539–541.

101. Bellone J, Ghizzoni L, Aimaretti G, Volta C, Boghen MF, Bernasconi S, Ghigo E. Growth hormone-releasing effect of oral growth hormone-releasing peptide 6 (GHRP-6) administration in children with short stature. Eur J Endocrinol 1995; 133:425–429.

102. Pombo M, Barreiro J, Penalva A, Mallo F, Casanueva FF, Dieguez C. Plasma growth hormone response to growth hormone releasing hexapeptide (GHRP-6) in children with short stature. Acta Paediatr 1995; 84:904–908.

103. Leal-Cerro A, Garcia E, Astorga R, Casanueva FF, Dieguez C. Growth hormone (GH) responses to the combined administration of GH-releasing peptide 6 in adults with GH deficiency. Eur J Endocrinol 1995; 132:712–715.

104. Thorner MO, Hartman ML, Gaylinn BD. Current status of therapy with growth hormone-releasing neuropeptides. In: Savage MO, Bourgulgon J, Grossman AB, eds. Frontiers in Paediatric Neuroendocrinology. Oxford: Blakwell Scientific, 1994:161–167.

105. Mericq F, Cassorla T, Salazar A, Avila A, Iniquez G, Bowers CY, Merriam GR. Increased growth velocity during prolonged GHRP-2 administration to growth hormone deficient children. 77th Annu Meet Endocr Soc. Washington, DC, June 14–17, 1995.

106. Chapman IM, Pescovitz OH, Treep T, Murphy G, Cerchio K, Krupa D, Gertz B, Polvino W, Skiles EH, Pezzoli SS, Thorner MO. Oral administration of the GHRP-mimetic MK-0677 stimulates the GH/IGF-I axis in GH deficient adults. 10th International Congress of Endocrinology. San Francisco, CA, June 12–15, 1996.

107. Cordido F, Peino R, Penalva A, Alvarez CV, Casanueva FF, Dieguez C. Impaired growth hormone secretion in obese subjects is partially reversed by acipimox-mediated plasma free fatty acid depression. J Clin Endocrinol Metab 1996; 81:914–918.

108. Grottoli S, Maccario M, Procopio M, Oleandri S, Razzore P, Taliano M, Camanni F, Ghigo E. Somatotrope responsiveness to hexarelin, a synthetic hexapeptide, is refractory to the inhibitory effect of glucose in obesity. Eur J Endocrinol 1996; 135:678–682.

109. Edwards CA, Dieguez C. Scanlon MF. Effects of hypothyroidism, tri-iodo-thyronine and glucocorticoids on growth hormone responses to growth hormone-releasing hormone and His-D-Trp-Ala-Trp-D-Phe-Lys-NH$_2$. J Endocrinol 1989; 121:31–36.

110. Leal Cerro A, Pumar A, Garcia-Garcia E, Dieguez C, Casanueva FF. Inhibition of growth hormone release after the combined administration of GHRH and GHRP-6 in patients with Cushing's syndrome. Clin Endocrinol 1994; 41:649–654.

111. Ghigo E, Aruat E, Ramunni J, Walao A, Gianotti L, Deghenghi R, Lombardi G, Cambanni F. Adrenocorticotropin- and cortisol-releasing effect of hexarelin, a synthetic growth hormone–releasing peptide, in normal subjects and patients with Cushing's syndrome. J Clin Endocrinol Metab 1997; 82:2439–2444.

24

GHRH and GHRP-6 Exert Different Effects on Sleep Electroencephalogram and Nocturnal Hormone Secretion in Normal Men

Axel Steiger, Ralf-Michael Frieboes, and Harald Murck
Max Planck Institute of Psychiatry, Munich, Germany

I. INTRODUCTION

In humans, sleep is characterized by the cyclic occurrence of non-rapid eye movement (–REM) and REM sleep, and the patterns of hormone secretion (1,2). During the first half of the night, slow-wave sleep (SWS) and the growth hormone (GH) peak predominate, whereas cortisol secretion reaches its nadir. In contrast, during the early morning hours, cortisol increases and the amounts of SWS and GH are low. During the acute episode of depression and during aging similar changes of sleep–endocrine activity occur. A decrease of SWS and an increase of shallow sleep are found, GH release is blunted, and more cortisol is secreted (3–5). These observations suggest that during the first half of the night, there are at least two different factors that regulate both the sleep electroencephalogram (EEG) and nocturnal hormone secretion, and that there is a reciprocal interaction between these factors.

Studies in laboratory animals showed that the neuropeptides, growth hormone-releasing hormone (GHRH) and corticotropin-releasing hormone (CRH), are these factors. After intracerebroventricular (icv) administration of GHRH to rats and rabbits (6–8) and intravenous (iv) administration of GHRH to rats (9), SWS increased. In contrast, after icv CRH, SWS decreased in rats (6). Furthermore, inhibition of endogenous GHRH by a GHRH antagonist and by GHRH antibodies induced more shallow sleep (10). These data demonstrated that, in the rat, GHRH and CRH, besides their stimulating effect on the release of GH or corticotropin (adrenocorticotropic hormone; ACTH) and cortisol, respectively, exert opposite effects on sleep. GHRH promotes sleep, whereas CRH disturbs sleep.

In humans the sleep-EG effects of CRH correspond to those in the rat. This was shown by a previous study from our laboratory. We mimicked the physiological pulsatile release of neuropeptides and administered CRH in a repetitive fashion to young normal men (11). After CRH administration, REM sleep and, during the second half of the night, SWS decreased, the nocturnal GH surge was blunted, and cortisol increased. To clarify whether the sleep–EEG effects of GHRH in humans are similar to those in laboratory animals, we chose a protocol similar to that of our previous study on the effects of CRH (11). Furthermore, in this study we examined the influence of somatostatin, which is the endogenous inhibitor of GH release. In previous studies, somatostatin had no effect on sleep in young human subjects (12,13). In rats, REM sleep was either diminished (14) or enhanced (15) by somatostatin. In another study, non-REM sleep was decreased by somatostatin in rats (16).

Furthermore, we were interested in the effects of growth hormone-releasing peptide (GHRP-6) on sleep and nocturnal hormone release. This synthetic peptide stimulates GH release in humans and animals (17,18). In most of the previous studies GHRP was given during daytime to healthy volunteers (19–21) or to short-statured children (22). The GH stimulation after GHRP is the effect of binding on non-GHRH pituitary receptors (19,23). GHRP mimicked the effect of GHRH to stimulate pituitary GH release (24), this without binding to the GHRH receptor. The question arises, whether GHRP also shares other regulatory effects of GHRH, such as blunting of cortisol and enhancing of SWS. To clarify this issue, in a second protocol, we examined the effects of GHRP on sleep EEG and hormone secretion in young normal male controls.

II. METHODS

Two studies were performed in paid normal male controls. In study 1 the effects of GHRH and somatostatin on sleep EEG and nocturnal hormone secretion were compared with placebo. In study 2 the influence of GHRP was examined. Each of the study groups consisted of seven paid healthy male control subjects of normal weight and height, with a mean age of 25.1 years (SD ± 1.8) in study 1 and 25.3 years (SD ± 1.3) in study 2. Before entering the study the subjects underwent extensive examinations to exclude factors that could render study results ambiguous. The study was approved by the Ethics Committee for Human Experiments of the Max Planck Institute of Psychiatry.

III. PROTOCOL

In study 1, the subjects had three sessions in the sleep laboratory at 1-week intervals. In study 2, they had two sessions. Each session consisted of two nights. The first served for adaptation to the laboratory setting. On the second day an indwelling forearm catheter was inserted at 1930 h and connected to a plastic tubing extension, placed through a soundproof lock into the adjacent room, and kept patent with a 0.9% saline drip. Electrodes for sleep-EEG recordings were fixed until 2000 h. Sleep EEG was recorded from 2300 h ("lights off") until 0700 h. Then the subjects were wakened if necessary. Outside of this period sleeping was not allowed. In the neighboring laboratory, blood was sampled for analysis of GH and cortisol (studies 1 and 2) and ACTH (study 2) every 30 min from 2000 to 2200 h and every 20 min from 2200 to 0700 h.

In study 1 during a given session, placebo or 50 µg human GHRH (hGHRH) (Bissendorf Peptide, Wedemark, Germany) or 50 µg somatostatin (Ferring, Kiel, Germany) was administered iv four times at 2200, 2300, 2400, and 0100 h. In study 2 the same schedule was used for iv administration four times of 50 µg GHRP (Clinalfa, Laeufelfingen, Switzerland) or placebo.

The plasma concentrations of GH, cortisol, and ACTH were determined by commercial radioimmunoassays (RIAs). Sleep EEG was scored visually by a rater who was unaware of the treatment, according to standard guidelines (25). In study 2, in addition, EEG spectral analysis was performed as described previously (26).

The statistics of all group values are expressed as means ± standard deviation (SD). Statistical evaluation of study 1 was based on repeated measure analysis of variance. By this analysis effects of the treatment on the various variables were examined for statistical significance, with averaged multivariate tests combined with univariate F tests of significance. Differences between the treatment levels (placebo, GHRH, and somatostatin) for the mean of each variable were tested for significance with simple contrasts. In study 2, we used Wilcoxon's paired rank test. Methods are described in more detail elsewhere (protocol 1, see Ref. 27; protocol 2, see Ref. 28).

IV. RESULTS

A. Study 1: Effects of GHRH and Somatostatin

Patterns of sleep EEG, cortisol, and GH secretion of one representative subject for placebo, GHRH, and somatostatin is given in Figure 1.

1. Sleep EEG

Table 1 shows sleep architecture variables. After GHRH administration, SWS increased significantly and, more particularly, stage 4 sleep was enhanced in

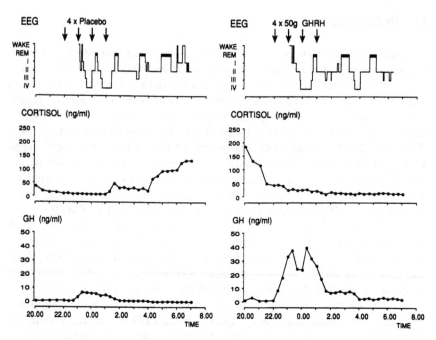

Figure 1 Sleep structure, cortisol, and GH secretion in a representative control subject under placebo and under pulsatile application of GHRH.

comparison with placebo and somatostatin. Also during the last third of the night, when SWS seldom occurs in normal subjects, SWS increased significantly after GHRH administration. No significant effects of somatostatin on sleep EEG were found.

2. Hormone Secretion

Table 2 gives the endocrine variables under placebo, GHRH, and somatostatin treatment. The nocturnal GH surge was enhanced significantly by GHRH. After GHRH administration, cortisol levels decreased significantly during the total night and particularly during the second half of the night. Somatostatin had no significant effect on endocrine variables.

B. Study 2: Effects of GHRP

1. Sleep EEG

Sleep architecture parameters after GHRP and after placebo are given in Table 3. Time spent in non-REM sleep stage 2 increased significantly with GHRP.

Table 1 Sleep-EEG Architecture Variables Under Placebo, GHRH, and Somatostatin (Mean ± SD)

	Placebo (1)	GHRH (2)	Somatostatin (3)	Test with contrasts 1:2	Test with contrasts 1:3	Test with contrasts 2:3	Univariate F test with df = 2, 12 for the effects F statistic	Univariate F test with df = 2, 12 for the effects Significance of F
Sleep architecture, min spent in each stage during total night								
Awake	17.50 ± 24.58	5.71 ± 12.54	7.07 ± 7.7	n.s.	n.s.	n.s.	1.04	0.379
Stage 1	17.07 ± 7.80	16.29 ± 9.89	15.14 ± 6.9	n.s.	n.s.	n.s.	0.10	0.903
Stage 2	247.36 ± 37.48	214.36 ± 32.60	232.43 ± 39.6	n.s.	n.s.	n.s.	2.77	0.102
Stage 3	36.14 ± 14.28	35.86 ± 7.49	27.57 ± 9.8	n.s.	n.s.	n.s.	1.32	0.303
Stage 4	26.14 ± 25.09	46.86 ± 22.09	35.43 ± 26.9	*	n.s.	n.s.	6.72	0.011
SWS	62.29 ± 24.58	82.71 ± 19.80	63.00 ± 31.4	*	n.s.	n.s.	6.37	0.013
REM	93.64 ± 22.74	84.50 ± 24.48	91.79 ± 16.7	n.s.	n.s.	n.s.	0.52	0.607

*, Significant differences; n.s., not significant; SWS, slow-wave sleep; REM, rapid-eye-movement sleep.

Table 2 Endocrine Variables Under Placebo, GHRH, and Somatostatin (Mean ± SD)

	Placebo (1)	GHRH (2)	Somatostatin (3)	Test with contrasts			Univariate F test with df = 2, 12 for the effects	
				1:2	1:3	2:3	F statistic	Significance of F
GH secretion								
Mean GH conc. (ng/mL)								
22.00–7.00	3.2 ± 2.0	10.8 ± 4.2	3.0 ± 1.7	*	n.s.	*	24.81	0.0005
22.00–3.00	5.4 ± 3.8	19.0 ± 8.2	4.8 ± 3.2	*	n.s.	*	22.18	0.0005
3.00–7.00	0.8 ± 0.3	2.0 ± 0.9	1.2 ± 1.0	*	n.s.	n.s.	4.07	0.0450
AUC (ng/mL × min)								
Peak	1.120 ± 863	3.845 ± 2.419	1.098 ± 904	*	n.s.	n.s.	10.51	0.0020
Cortisol secretion								
Mean cortisol conc. (ng/mL)								
22.00–7.00	61.4 ± 12.9	46.6 ± 19.7	70.8 ± 12.6	*	n.s.	*	9.08	0.0009
22.00–3.00	21.4 ± 7.1	22.2 ± 9.8	32.0 ± 18.6	n.s.	n.s.	n.s.	2.24	0.1300
3.00–7.00	98.3 ± 19.2	70.4 ± 34.7	109.0 ± 15.3	*	n.s.	*	9.75	0.0030

*, Significant differences; n.s., not significant; AUC, area under curve.

Table 3 Sleep Architecture Variables Under Placebo and GHRP (Mean ± SD)

	Placebo (1)	GHRP (2)	WRT 1:2
Minutes spent in each sleep stage during SPT			
Awake	19.9 ± 16.0	12.2 ± 13.2	n.s.
Stage 1	31.4 ± 14.7	26.3 ± 11.0	n.s.
Stage 2	254.4 ± 25.8	270.1 ± 25.3	$p < 0.02$
Stage 3	43.3 ± 9.0	38.1 ± 10.5	n.s.
Stage 4	21.4 ± 14.0	22.6 ± 17.4	n.s.
SWS	64.6 ± 17.7	60.7 ± 24.5	n.s.
REM	87.0 ± 19.8	83.9 ± 23.1	n.s.

SPT, sleep-period time; SWS, slow-wave sleep; REM, rapid-eye–movement sleep; WRT, Wilcoxon's paired-rank test.

Intermittent wakefulness showed a nonsignificant trend to decrease after the substance. Other sleep-EEG variables including SWS were not affected by the treatment. An EEG spectral analysis showed no difference between GHRP and placebo.

2. Endocrine Variables

Figure 2 depicts the profiles of the plasma concentrations (mean ± SEM) of GH, cortisol, and ACTH with GHRP and with placebo. Hormone parameters are given in Table 4. The GH surge was enhanced significantly by GHRP. Furthermore, the hypothalamic–pituitary–adrenocortical (HPA) system hormones, cortisol, and ACTH, were stimulated by GHRP. ACTH concentration increased significantly between 2200 and 0200 h. Cortisol levels were elevated significantly after GHRP during the total night. Also the cortisol nadir was increased after GHRP administration.

IV. DISCUSSION

Our data demonstrate that GHRH and GHRP modulate sleep EEG and sleep-associated hormone secretion in men. Both substances stimulate nocturnal GH release, whereas they exert different effects on sleep EEG and even opposite effects on HPA hormone secretion.

Both GHRH and GHRP promote sleep, whereas they modulate different components of non-REM sleep. After GHRH SWS was increased, after GHRP stage 2 sleep was stimulated. The increase of SWS after GHRH was evident not only after sleep onset, at the time of GHRH administration when SWS predominates, but also through the total night. Obviously, the influence of

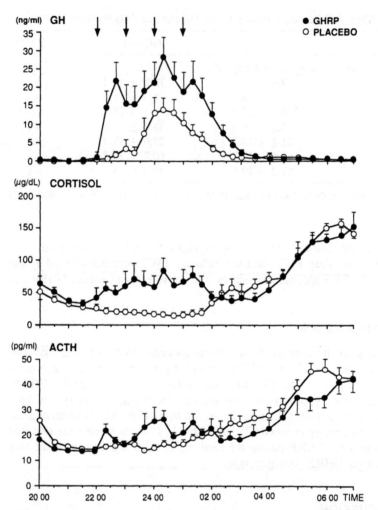

Figure 2 Time course of plasma concentrations of GH, cortisol, and ACTH (means ± SEM) in seven normal controls under repetitive iv injection (four times) of 50 μg GHRP-6 or placebo. Arrows indicate time of injection.

GHRH on SWS is temporarily dissociated from its stimulatory effect on GH, which is found only during the first half of the night. Our data show that GHRH promotes SWS in humans, similar to findings in other species after icv (6–8) and iv (9) administration. It contradicts previous studies in which GHRH was given as a single iv injection to normal controls before sleep onset (29,30). Two recent studies support our findings. Kerkhofs et al. (31) injected single boluses

Table 4 Endocrine Variables Under Placebo and GHRP (Mean ± SD)

	Placebo (1)	GHRP (2)	WRT 1:2
GH secretion, mean concentration (ng/mL)			
2200–0700 h	3.9 ± 2.8	10.4 ± 6.6	$p < 0.02$
2200–0300 h	5.5 ± 4.0	15.4 ± 9.6	$p < 0.02$
0300–0700 h	0.7 ± 0.6	0.7 ± 0.6	n.s.
Maximum	17.2 ± 11.0	30.7 ± 13.1	$p < 0.02$
Cortisol secretion, mean concentration (ng/mL)			
2200–0700 h	53.3 ± 10.6	71.2 ± 20.6	$p < 0.02$
2200–0300 h	25.2 ± 9.0	56.0 ± 31.0	$p < 0.02$
0300–0700 h	102.7 ± 15.8	94.8 ± 22.7	n.s.
Nadir	9.5 ± 5.2	21.4 ± 8.8	$p < 0.05$
ACTH secretion, mean concentration (pg/mL)			
2200–0700 h	24.6 ± 4.4	24.5 ± 4.6	n.s.
2200–0200 h	16.6 ± 3.1	21.0 ± 5.3	$p < 0.02$

WRT = Wilcoxon's paired-rank test; n.s. - not significant.

of GHRH in three different protocols: (1) after the onset of the first SWS period, (2) after 60 s of the third REM period, and (3) after sleep deprivation until 0400 h. The results were for 1, unchanged SWS, but increased REM sleep; for 2, a decrease in intermittent wakefulness and an increase in SWS; and for 3, a decrease in intermittent wakefulness. Marshall et al. (32) compared pulsatile administration, as described here, with continuous infusion of GHRH in normal controls. They confirmed an increase of SWS and, furthermore, they found an increase of REM sleep after pulsatile administration, whereas the continuous infusion was not effective. This study corroborates our view that pulsatile administration of neuropeptides is a crucial methodological issue in the investigation of the effects of peptides on sleep.

The effects of GHRH on sleep and hormone secretion in our study were opposite to those we found previously after CRH administration in normal control subjects (11). After CRH treatment we observed a decrease of SWS during the second half of the night, a blunting of GH release, and an elevation of the cortisol level. These data support the hypothesis that GHRH and CRH play a key role in sleep regulation, and that changes of the GHRH/CRH ratio result in alterations of sleep–endocrine activity (33). This theory is further supported by recent studies from our laboratory. We found that the effects of GHRH on sleep EEG and cortisol secretion were weaker, or even absent, in three protocols in which a change of the GHRH/CRH ratio is suggested: (1) in young normal men at the early morning hours (high physiological activity of CRH, 34); (2) patients with major depression (pathological overactivity of CRH) (35); and (3) healthy elderly men and women (reduced activity of GHRH) (36). These data suggest that, in young normal men during the first half of the night, GHRH is active, resulting in high amounts of SWS and GH and the nadir of cortisol, whereas during the second night, CRH suppresses GH and SWS and stimulates cortisol release. Also, during depression and during aging the GHRH/CRH ratio is changed in favor of CRH. This mechanism may contribute to the well-known changes of sleep–endocrine activity during both states, such as shallow sleep, blunted GH release, and elevated cortisol levels (3–5).

The sleep-EEG changes after GHRH and GHRP cannot be explained by the changes of peripheral hormone secretion. Selective enhancement of stage 2 sleep after GHRP administration appears to be a unique effect. To our knowledge, no other substance shares this property of GHRP. When GH is given to normal human controls or animals, SWS decreases (probably by feedback inhibition of GHRH) and REM sleep increases (37,38). The latter effect can be explained as a direct central effect of GH, as demonstrated by experiments with iv administration of GHRH to intact and hypophysectomized rats (9). Brief administration of cortisol to young (39,40) and elderly normal controls (41) increased SWS, probably by feedback inhibition of CRH and suppressed REM sleep. After ACTH administration, decreases of REM sleep and SWS were found in normal controls (42). Therefore, we conclude that the effects of GHRH and GHRP on sleep EEG represent direct central effects that are not prevented by the blood–brain barrier.

In line with previous studies (12,13), we failed to find effects of somatostastin on sleep EEG in young normal controls. In elderly subjects, however, we recently observed deterioration of sleep after somatostatin administration (43). This observation suggests that, besides CRH, somatostatin exerts sleep-disturbing effects. Probably, these effects are counteracted by GHRH in young subjects. The decline of the activity of GHRH in the elderly probably facilitates the sleep-disturbing influence of somatostatin.

As expected the GH surge was enhanced after GHRH and after GHRP as well. The only difference from previous studies (44,45) is that, during sleep, the return of the plasma GH concentration to baseline is protracted. This finding may be due to enhanced responsiveness of the somatotropic system during the first half of the night.

In the present studies, GHRH and GHRP exert opposite influences on HPA activity. GHRH, when given during the first few hours of the night, induced a decrease of cortisol levels throughout the night and particularly during the interval with the highest activity of the HPA system in the second half of the night. In contrast cortisol and ACTH were stimulated after GHRP. The different effects of GHRH and GHRP on sleep EEG and HPA hormone secretion underline that these peptides act through different receptors (19,46).

The mechanisms involved in the cortisol suppression after GHRH are unclear. Whereas it is well documented that HPA hormones are capable of modulating GH release (11,47), only few data exist about the effects of GHRH and GH on this system. In patients with Cushing's disease cortisol secretion was suppressed after GH administration, whereas this effect was absent in normal controls (48). Because ACTH secretion was not investigated after GHRH, it is not possible to determine which level of the HPA system mediated cortisol suppression after administration of this peptide. A reciprocal interaction of the HPA and the GHRH–GH systems appears possible at the pituitary or suprapituitary level. The increase of cortisol and ACTH after GHRP is similar to a previous finding by Hayashi et al. (21). These authors found a slight but significant increase of cortisol after iv administration in the morning, in normal controls. The elevation of GH levels after GHRP may lead to a feedback inhibition of endogenous GHRH, which causes the CRH-regulated HPA system to dominate. An alternative explanation would be a direct stimulation of the HPA system by the GHRP receptor.

V. PERSPECTIVES

Our data demonstrate that endogenous GHRH and the synthetic peptide GHRP not only act as GH secretagogues, but also exert influences on sleep EEG and HPA hormone release: GHRH and GHRP promoted sleep. Currently, in most of the world, the most frequently used hypnotics are benzodiazepines. These drugs are not capable of inducing natural sleep, for they suppress SWS, EEG-delta power, and REM sleep (49,50). Furthermore, at least after brief administration of benzodiazepines, blunted GH release and stimulation of prolactin were found (51,52). Moreover, there is a well-known risk of addiction in benzodiazepine treatment. It appears possible that, in the future, substances related to GHRH and GHRP may be used as hypnotics. A preliminary study from our laboratory also revealed an increase of stage 2 sleep after intranasal administration of GHRP (Frieboes RM, Murck H, Holsboer F, Steiger A, unpublished data). In a similar vein, 1 week of oral treatment with another GH secretagogue, MK-677 prompted an increase of SWS in young normal controls (53). The changes of cortisol secretion and of sleep EEG after GHRH suggest that this peptide has CRH-antagonistic properties. CRH overactivity plays a key role in the pathophysiology of depression (54). Therefore, analogs of GHRH may be useful for the treatment of affective disorders. In contrast with GHRH, iv GHRP stimulated HPA hormone secretion. Further preliminary results from our laboratory also indicate an increase of ACTH after oral administration of GHRP, whereas ACTH release was blunted after intranasal GHRP (Frieboes and colleagues, unpublished data). The GHRP-6 analog hexarelin*, which is a potent

---

*Hexarelin is a trademark for examorelin (I.N.N.).

GH secretagogue (55,56), stimulated cortisol release after bolus iv administration in normal controls (57). On the other hand, after short-term treatment with intranasal or oral hexarelin in healthy elderly controls, cortisol levels remained unchanged (58). It appears necessary to perform long-term studies on the effects of prolonged treatment with GH secretagogues, such as GHRP-6 and hexarelin, particularly after oral and intranasal administration. In these protocols the effects on sleep EEG, cortisol, and ACTH release should be monitored to delineate benefits and possible sequelae of such treatment.

ACKNOWLEDGMENTS

Our studies were supported by a grant from the Deutsche Forschungsgemeinschaft (Ste 486/1-2).

REFERENCES

1. Weitzman ED. Circadian rhythms and episodic hormone secretion in man. Annu Rev Med 1976; 27:225–243.
2. Steiger A, Herth T, Holsboer F. Sleep-electroencephalography and the secretion of cortisol and growth hormone in normal controls. Acta Endocrinol 1987; 116:36–42.
3. Reynolds CF 3rd, Kupfer DJ. Sleep research in affective illness: state of the art circa 1987. Sleep 1987; 10:199–215.
4. Steiger A, von Bardeleben U, Herth T, Holsboer F. Sleep EEG and nocturnal secretion of cortisol and growth hormone in male patients with endogenous depression before treatment and after recovery. J Affect Disord 1989; 16:189–195.
5. van Coevorden A, Mockel J, Laurent E, Kerkhofs M, L'Hermite-Baleriaux M, Decoster C, Neve P, Van Cauter E. Neuroendocrine rhythms and sleep in aging men. Am J Physiol 1991; 260:E651–E661.
6. Ehlers CL, Reed TK, Henriksen S, J. Effects of corticotropin-releasing factor and growth hormone-releasing factor on sleep and activity in rats. Neuroendocrinology 1986; 42:467–474.
7. Nistico G, De Sarro GB, Bagetta G, Muller EE. Behavioural and electrocortical spectrum power effects of growth hormone releasing factor in rats. Neuropharmacology 1987; 26:75–78.
8. Obàl F Jr, Alföldi P, Cady AB, Johannsen L, Sary G, Krueger JM. Growth hormone-releasing factor enhances sleep in rats and rabbits. Am J Physiol 1988; 255:310–316.
9. Obàl F Jr, Floyd R, Kapas L, Bodos B, Krueger JM. Effects of systemic GHRH on sleep in intact and in hypophysectomized rats. Am J Physiol 1996; 263:230–237.
10. Obàl F Jr, Payne L, Kapas L, Opp M, Krueger JM. Inhibition of growth hormone-

releasing factor suppresses both sleep and growth hormone secretion in the rat. Brain Res 1991; 557:149–153.

11. Holsboer F, von Bardeleben U, Steiger A. Effects of intravenous corticotropin-releasing hormone upon sleep-related growth hormone surge and sleep EEG in man. Neuroendocrinology 1988; 48:32–38.

12. Parker DC, Rossman LG, Siler TM, Rivier J, Yen SS, Guillemin R. Inhibition of the sleep-related peak in physiologic human growth hormone release by somatostatin. J Clin Endocrinol Metab 1974; 38:496–499.

13. Kupfer DJ, Jarrett DB, Ehlers CL. The effect of SRIF on the EEG sleep of normal men. Psychoneuroendocrinology 1992; 17:37–43.

14. Rezek M, Havlicek V, Hughes KR, Friesen H. Behavioural and motor excitation and inhibition induced by the administration of small and large doses of somatostatin into the amygdala. Neuropharmacology 1977; 16:157–162.

15. Danguir J. Intracerebroventricular infusion of somatostatin selectively increases paradoxical sleep in rats. Brain Res 1986; 367:26–30.

16. Beranek L, Obàl F, Bodosi B, Taishi P, Laczi F, Krueger JM. Inhibition of non-REM sleep in response to a long-acting somatostatin analogue, sandostatin, in the rat. J Sleep Res 1996; 5:14.

17. Bowers CY. GH releasing peptides—structure and kinetics. J Pediatr Endocrinol 1993; 6:21–31.

18. Huhn WC, Hartman ML, Pezzoli SS, Thorner MO. Twenty-four-hour growth hormone (GH)-releasing peptide (GHRP) infusion enhances pulsatile GH secretion and specifically attenuates the response to a subsequent GHRP bolus. J Clin Endocrinol Metab 1993; 76:1202–1208.

19. Bowers CY, Sartor AO, Reynolds GA, Badger TM. On the actions of the growth hormone-releasing hexapeptide, GHRP. Endocrinology 1991; 128:2027–2035.

20. Hartman ML, Farello G, Pezzoli SS, Thorner MO. Oral administration of growth hormone (GH)-releasing peptide stimulates GH secretion in normal men. J Clin Endocrinol Metab 1992; 74:1378–1384.

21. Hayashi S, Okimura Y, Yagi H, Uchiyama T, Takeshima Y, Shakutsui S, Oohashi S, Bowers CY, Chihara K. Intranasal administration of His-D-Trp-Ala-Trp-D-Phe-LysNH$_2$ (growth hormone releasing peptide) increased plasma growth hormone and insulin-like growth factor-I levels in normal men. Endocrinol Jpn 1991; 38:15–21.

22. Bowers CY, Alster DK, Frentz JM. The growth hormone-releasing activity of a synthetic hexapeptide in normal men and short statured children after oral administration. J Clin Endocrinol Metab 1992; 74:292–298.

23. Bowers CY, Momany FA, Reynolds GA, Hong A. On the in vitro and in vivo activity of a new synthetic hexapeptide that acts on the pituitary to specifically release growth hormone. Endocrinology 1984; 114:1537–1545.

24. Bercu BB, Yang SW, Masuda R, Walker RF. Role of selected endogenous peptides in growth hormone-releasing hexapeptide activity: analysis of growth hormone-releasing hormone, thyroid hormone-releasing hormone, and gonadotropin-releasing hormone. Endocrinology 1992; 130:2579–2586.

25. Rechtschaffen A, Kales A. A Manual of Standardized Terminology, Techniques and Scoring System for Sleep Stages of Human Subjects. Bethesda, MD: US Department of Health, Education and Welfare, 1968.

26. Steiger A, Trachsel L, Guldner J, Hemmeter U, Rothe B, Rupprecht R, Vedder H, Holsboer F. Neurosteroid pregnenolone induces sleep EEG changes in man compatible with inverse agonistic $GABA_A$-receptor modulation. Brain Res 1993; 615:267–274.

27. Steiger A, Guldner J, Hemmeter U, Rothe B, Wiedemann K, Holsboer F. Effects of growth hormone-releasing hormone and somatostatin on sleep EEG and nocturnal hormone secretion in male controls. Neuroendocrinology 1992; 56:566–573.

28. Frieboes RM, Murck H, Maier P, Schier T, Holsboer F, Steiger A. Growth hormone-releasing peptide-6 stimulates sleep, growth hormone, ACTH and cortisol release in normal man. Neuroendocrinology 1995; 61:584–589.

29. Garry P, Roussel B, Cohen R, Biot-Laporte S, Charfi AE, Jouvet M, Sassolas G. Diurnal administration of human growth hormone-releasing factor does not modify sleep and sleep-related growth hormone secretion in normal young men. Acta Endocrinol 1985; 110:158–163.

30. Kupfer DJ, Jarrett DB, Ehlers CL. The effect of GRF on the EEG sleep of normal males. Sleep 1991; 14:87–88.

31. Kerkhofs M, Van Cauter E, Van Onderbergen A, Caufriez A, Thorner MO, Copinschi G. Sleep-promoting effects of growth hormone-releasing hormone in normal men. Am J Physiol 1993; 264:E594–E398.

32. Marshall L, Moelle M, Boeschen G, Steiger A, Fehm HL, Born J. Greater efficacy of episodic than continuous growth hormone releasing hormone (GHRH) administration in promoting slow wave sleep (SWS). J Clin Endocrinol Metab 1996; 81:1009–1013.

33. Ehlers CL, Kupfer DJ. Hypothalamic peptide modulation of EEG sleep in depression: a further application of the S-process hypothesis. Biol Psychiatry 1987; 22:513–517.

34. Schier T, Guldner J, Colla M, Holsboer F, Steiger A. Changes in sleep–endocrine activity after growth hormone-releasing hormone depend on time of administration. J Neuroendocrinol 1997; 9:201–205.

35. Steiger A, Guldner J, Colla-Muller M, Friess E, Sonntag A, Schier T. Growth hormone-releasing hormone (GHRH)-induced effects on sleep EEG and nocturnal secretion of growth hormone, cortisol and ACTH in patients with major depression. J Psychiatr Res 1994; 28:225–238.

36. Guldner J, Friess E, Colla-Mueller M, Schier T, Holsboer F, Steiger A. Influence of growth hormone-releasing hormone (GHRH) on sleep EEG and on nocturnal secretion of cortisol, ACTH and growth hormone in elderly normal controls [abstr]. Exp Clin Endocrinol 1994; 102:69.

37. Mendelson WB. Studies of human growth hormone secretion in sleep and waking. Int Rev Neurobiol 1982; 23:367–389.

38. Stern WC, Jalowiec JE, Shabshelowitz H, Morgane PJ. Effects of growth hormone on sleep-waking patterns in cats. Horm Behav 1975; 6:189–196.

39. Born J, Späth-Schwalbe E, Schwakenhofer H, Kern W, Fehm HL. Influences of corticotropin-releasing hormone, adrenocorticotropin, and cortisol on sleep in normal man. J Clin Endocrinol Metab 1989; 68:904–911.

40. Friess E, von Bardeleben U, Wiedemann K, Lauer C, J, Holsboer F. Effects of

pulsatile cortisol infusion on sleep-EEG and nocturnal growth hormone release in healthy men. J Sleep Res 1994; 3:73–79.

41. Bohlhalter S, Murck H, Holsboer F, Steiger A. Cortisol enhances non-REM sleep and growth hormone secretion in elderly subjects. Neurobiol Aging 1997; 18:423–429.

42. Gillin JC, Jacobs LS, Snyder F, Henkin RI. Effects of ACTH on the sleep of normal subjects and patients with Addison's disease. Neuroendocrinology 1974; 15:21–31.

43. Frieboes RM, Murck H, Schier T, Holsboer F, Steiger A. Somatostatin impairs sleep in elderly human subjects. Neuropsychopharmacology 1997; 16:339–345.

44. Losa M, Bock L, Schophol J, Stalla GK, Muller OA, von Werder K. Growth hormone releasing factor infusion does not sustain elevated GH-levels in normal subjects. Acta Endocrinol 1984; 107:462–470.

45. Vance ML, Kaiser DL, Evans WS, Furlanetto R, Vale W, Rivier J, Thorner MO. Pulsatile growth hormone secretion in normal man during a continuous 24-hour infusion of human growth hormone releasing factor (1–40). Evidence for intermittent somatostatin secretion. J Clin Invest 1985; 75:1584–1590.

46. Ilson BE, Jorkasky DK, Curnow RT, Stote RM. Effect of a new synthetic hexapeptide to selectively stimulate growth hormone release in healthy human subjects. J Clin Endocrinol Metab 1989; 69:212–214.

47. Wiedemann K, von Bardeleben U, Holsboer F. Influence of human corticotropin-releasing hormone and adrenocorticotropin upon spontaneous growth hormone secretion. Neuroendocrinology 1991; 54:462–468.

48. Schteingart DE. Suppression of cortisol secretion by human growth hormone. J Clin Endocrinol Metab 1980; 50:721–725.

49. Borbély AA, Mattmann P, Loepfe M, Fellmann I, Gerne M, Strauch I, Lehmann D. A single dose of benzodiazepine hypnotics alters the sleep EEG in the subsequent drug-free night. Eur J Pharmacol 1983; 89:157–161.

50. Borbély AA, Achermann P. Ultradian dynamics of sleep after a single dose of benzodiazepine hypnotics. Eur J Pharmacol 1991; 195:11–18.

51. Copinschi G, Van Onderbergen A, L'Hermite-Baleriaux M, Szyper M, Caufriez A, Bosson D, L'Hermite M, Robyn C, Turek FW, Van Cauter E. Effects of the short-acting benzodiazepine triazolam, taken at bedtime, on circadian and sleep-related hormonal profiles in normal men. Sleep 1990; 13:232–244.

52. Steiger A, Guldner J, Lauer CJ, Meschenmoser C, Pollmächer T, Holsboer F. Flumazenil exerts intrinsic activity on sleep EEG and nocturnal hormone secretion in normal controls. Psychopharmacology 1994; 113:334–338.

53. Copinschi G, Van Onderbergen A, L'Hermite-Baleriaux M, Mendel CM, Caufriez A, Leproult R, Bolognese JA, De Smet M, Thorner MO, Van Cauter E. Effects of a 7-day treatment with a novel orally active nonpeptide growth hormone secretagogue, MK-677, on 24-hour growth hormone profiles, insulin-like growth factor-I and adrenocortical function in normal young men. J Clin Endocrinol Metab 1996; 81:2776–2782.

54. Holsboer F. Neuroendocrinology of affective disorders. In: Bloom F, Kupfer DJ, eds. Neuropsychopharmacology. Fourth Generation of Progress. New York: Raven Press, 1995:957–970.

55. Deghenghi R, Cananzi MM, Torsello A, Battisti C, Muller EE, Locatelli V. GH-releasing activity of hexarelin, a new growth hormone releasing peptide, in infant and adult rats. Life Sci 1994; 54:1321–1328.
56. Ghigo E, Arvat E, Gianotti L, Imbimbo BP, Lenaerts V, Deghenghi R, Camanni F. Growth hormone-releasing activity of hexarelin, a new synthetic hexapeptide, after intravenous, subcutaneous, intranasal, and oral administration in man. J Clin Endocrinol Metab 1994; 78:693–698.
57. Massoud AF, Hindmarsh PC, Brook CGD. Hexarelin-induced growth hormone, cortisol and prolactin release: a dose–response study. J Clin Endocrinol Metab 1996; 81:4338–4341.
58. Ghigo E, Arvat E, Gianotti L, Grottoli S, Rizzi G, Ceda GP, Boghen MF, Deghenghi R, Cammani F. Short-term administration of intranasal or oral hexarelin, a synthetic hexapeptide, does not desensitize the growth hormone responsiveness in human aging. Eur J Endocrinol 1996; 135:407–412.

25

Reproducibility of the Growth Hormone Response to Hexarelin* in Children and the Effect of Sex Steroids

Sandro Loche and Antonella Faedda
Opsedale Regionale per le Microcitemie, Cagliari, Italy

Annamaria Colao, Bartolomeo Merola, and Gaetano Lombardi
Università Federico II, Naples, Italy

Marco Cappa
Ospedale Pediatrico Bambino Gesù, IRCCS, Rome, Italy

Jaele Bellone, Gianluca Aimaretti, and Ezio Ghigo
University of Turin, Turin, Italy

Giovanni Farello
Università di L'Aquila, L'Aquila, Italy

Romano Deghenghi
Europeptides, Argenteuil, France

I. INTRODUCTION

Spontaneous growth hormone (GH) secretion (1) as well as the response of GH to several stimuli (2) are increased in puberty and after administration of testosterone (3) or estrogen (2). Sex steroid priming is widely used to test pituitary function before the onset of the growth spurt, when, without priming, pharmacological and physiological tests may yield subnormal responses (4). The GH increase in puberty is due to an increase in GH pulse amplitude, rather than its frequency (1), occurs earlier in girls than in boys, and is seen in children

---

*Hexarelin is a trademark for examorelin (I.N.N.).

with precocious puberty (1). A similar pulse amplitude-dependent increase in GH secretion is also seen after the pharmacological induction of puberty with estrogens, androgens, human chorionic gonadotropin (hCG) or gonadotropin-releasing hormone (GnRH) (1), as well as after androgen treatment in hypo-gonadal men (5,6). Conversely, suppression of ovarian function by GnRH analogs in girls with central precocious puberty results in decreased GH secretion (1).

Comparison of the effects of testosterone (T) and the nonaromatizable androgens, oxandrolone (Ox; 7,8) and dihydrotestosterone (DHT; 9,10), as well as studies on the effects of the antiestrogen tamoxifen (6,11), on the somato-tropic axis have provided convincing evidence that the androgen-dependent increase in GH secretion is due to the conversion of T to estradiol after aro-matization.

Hexarelin (Hex) is a synthetic hexapeptide analog of growth hormone-releasing peptide-6 (GHRP-6; 12), with potent GH-releasing activity in both adults (13,14) and children (15,16). The GH response to a maximal dose of Hex is consistently higher than that elicited by a maximal dose of GH-releasing hormone (GHRH; 13,15), shows a limited intrasubject variability in both adults (13) and children (16,17), and increases at puberty (16,17), as well as after sex steroid administration (15,17).

In this chapter, we report the results of our studies on the reproducibility of the GH response to Hex in children, as well as on the effects of sex steroids on the Hex-induced GH secretion. All children studied were referred to our institutions for evaluation of short stature, and were found to have familial short stature or constitutional delay of growth.

II. REPRODUCIBILITY OF THE GROWTH HORMONE RESPONSE TO HEXARELIN

Twenty-five children (17 boys and 8 girls, age 9.5–13.5 years, all prepuber-tal) were tested on two occasions with Hex (2 µg/kg iv; prepared and supplied by Europeptides, Argenteuil, France), with an interval of 3–7 days. Their in-dividual GH responses are shown in Table 1. The GH peaks and the areas under the curve (AUC) after the first and second test sessions were significantly cor-related (peak, $r = 0.86, p < 0.0001$; AUC, $r = 0.91$, p < 0.0001). The mean ($\pm$ SD) coefficient of variation of the peak and AUC GH response to Hex was 22.7 $\pm$ 21.0% and 24.0 $\pm$ 20.7%, respectively, indicating a limited intra-individual variability.

A. Effects of Androgens on the Growth Hormone Response to Hexarelin

The GH response to Hex was reevaluated in ten boys (ages 10.0–13.7 years) 1 week after administration of testosterone (testosterone enanthate, 100 mg im),

Table 1   Individual GH Peak (μg/L) and AUC (μg · min/L)
Responses to iv Bolus Injection of Hex (2 μg/kg) in 25 Short
Normal Children Tested on Two Separate Days

	Test 1		Test 2	
Case No.	Peak	AUC	Peak	AUC
1	28.1	1347	31.9	1481
2	30.0	1646	54.0	2587
3	27.1	1493	27.5	1388
4	26.1	1383	43.0	2406
5	66.4	2984	59.5	2708
6	130.3	6479	61.1	2518
7	32.2	1130	34.3	1390
8	149.3	9570	123.9	8665
9	11.7	471	19.0	798
10	26.5	1136	24.3	1075
11	66.9	5091	63.2	3542
12	25.0	1834	26.0	2166
13	51.0	3592	34.0	2047
14	45.0	3627	43.0	3097
15	81.0	3273	59.0	2538
16	23.0	1256	45.0	1783
17	26.0	1307	20.0	988
18	44.0	1920	38.0	1936
19	28.8	1298	38.0	1792
20	93.0	3122	81.0	2282
21	48.0	3627	45.0	3330
22	49.0	2452	52.0	2872
23	66.0	3031	71.0	3069
24	38.0	2045	51.3	2711
25	53.0	3131	51.0	3349
Mean	50.6	2730	47.8	2501
SD	33.4	1983	22.5	1498

and in eight boys (age 9.5-13 years) 1 week after administration of oxandrolone (2.5 mg/day po). Mean ($\pm$ SD) GH peak and mean AUC before testosterone administration were 41.8 ± 21.0 μg/L and 1967 ± 799 μg · min/L, respectively. After priming with testosterone the GH response to Hex was significantly increased (peak = 71.1 ± 28.3 μg/L, $p < 0.001$; AUC = 3753 ± 1344 μg · min/L, $p < 0.005$; Fig. 1). The mean GH peak (45.1 ± 14.1 μg/L) and mean AUC (2107 ± 637 μg · min/L) in the eight children after oxandrolone administration were not significantly different from those before treatment (peak = 49.0 ± 25.7 μg/L; AUC = 2424 ± 1079 μg · min/L; see Fig. 1).

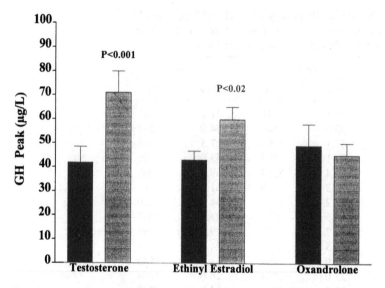

Figure 1 Mean (± SEM) GH peak responses to iv bolus injection of Hex before (closed bars) and after (shaded bars) pretreatment with testosterone (ten boys), ethinyl extradiol (five boys and ten girls), and oxandrolone (eight boys).

B. Effects of Estrogens on the Growth Hormone Response to Hexarelin

In 15 subjects (5 boys and 10 girls; ages 8.1–12.4 years) the GH response to Hex was reevaluated after 3 days of ethinyl estradiol administration (0.1 mg/day po). Mean (± SD) GH peak (60.0 ± 20.0 µg/L) and mean AUC (3010 ± 1428 µg · min/L) after ethinyl estradiol were significantly higher than those before treatment (peak = 43.0 ± 14.5 µg/L, $p < 0.005$; AUC = 2176 ± 951 µg · min/L, $p < 0.02$; see Fig. 1). The GH response to Hex before ethinyl estradiol was slightly higher in girls (peak = 46.8 ± 15.6 µg/L; AUC = 2512 ± 957 µg · min/L) than in boys (peak = 35.2 ± 8.8, $p < 0.1$; AUC = 1486 ± 459 µg · min/L, $p < 0.05$). After ethinyl estradiol the GH response to Hex was similar between girls (peak = 58.7 ± 22.7 µg/L; AUC = 3328 ± 1584 µg · min/L) and boys (peak = 62.3 ± 14.9 µg/L; AUC = 2960 ± 1401 µg · min/L).

III. DISCUSSION

We have shown that the GH response to Hex has limited intraindividual variability. In two previous reports in which the issue of the reproducibility of the

GH stimulation tests in children was addressed, the coefficient of variation (CV) for the arginine and L-dopa tests was high, averaging 86.5 and 77.6%, respectively (18), whereas the CV for the GHRH test averaged 73.4%, a value that was slightly improved to 47.5% when the children were pretreated with pyridostigmine (19). Thus, the intrasubject CV of the GH response to Hex is much lower (22.7%) than that previously reported for other pharmacological stimuli. The availability of a GH stimulation test that is potent and quite reproducible could be of value in the clinical setting. In children with short stature and poor growth, the diagnosis of GH deficiency is classically established when GH concentrations do not reach an arbitrary cutoff value after two pharmacological stimuli. One of the major disadvantage of provocative tests is the poor reproducibility and the great number of false-negative responses frequently observed, even in normal children (18–20). This variability has been attributed to the periodic secretion of somatostatin, which may influence the somatotroph response to the stimulus (21). Furthermore, the GH responses to stimulation may also be influenced by the pattern of GH secretion preceding the stimulus (i.e., whether the latter is administered during a spontaneous trough or peak of GH secretion; 21). However, recent observations indicate that the GH response to Hex is partially refractory to variations of hypothalamic somatostatin tone (22,23), or to the feedback action of GH (24). These findings might explain the limited variability of the GH response to Hex observed in both adults and children (13,17).

Pretreatment with both testosterone or ethinyl estradiol, but not with oxandrolone, increases the GH response to Hex. These findings are in agreement with other investigators, who showed that the effects of testosterone on the somatotropic axis are dependent on its aromatization to estradiol. In fact, oxandrolone (7,8) and dihydrotestosterone (9,10), both nonaromatizable androgens, fail to increase spontaneous GH secretion in boys. Furthermore, estrogen receptor blockade with tamoxifen reduces GH secretion in late pubertal boys (11) and in normal adult men (6), and inhibits GH secretion induced by testosterone treatment in hypogonadal men (6). Thus, our data indicate that the enhancing effect of testosterone on the GH response to Hex is mediated by estradiol. Indirect support to this conclusion comes from the observations that the GH response to Hex is greater in pubertal girls than in pubertal boys, and that it correlates with circulating estradiol in girls, but not with testosterone concentrations in boys (16). Furthermore, in animals the GH response to GHRP-6 is greater in the female than in the male rat (25).

Although the exact mechanism of action of GHRPs is still unknown, a number of findings indicate that their principal site of action is the hypothalamus. In fact, their effectiveness is far greater in vivo than in vitro or when the in vitro experiments are carried out on hypothalamic–pituitary incubates (26). In addition, the GH response to either GHRP-6 or Hex is absent or markedly blunted in animals (27) and in humans with hypothalamic–pituitary disconnection (28,29). Furthermore, GHRPs act synergistically with GHRH to release GH both in vitro (30) and in vivo (28,31). The synergistic effect of GHRH and GHRP-6 is also lost in patients with hypothalamic–pituitary disconnection (28). A specific receptor for GHRP-6, the parent compound of Hex (12), has been recently cloned (32), suggesting that Hex may represent a synthetic analog of an endogenous ligand.

The stimulatory effects of sex steroids on the somatotropic axis cannot be explained by an increased pituitary responsiveness to GHRH, because the GH response to GHRH does not change with puberty (16) or after sex steroid administration (10). Furthermore, the GH response to GHRH after reduction of the hypothalamic somatostatin tone by pyridostigmine also does not change with puberty (34) or after testosterone administration (10), suggesting that a decreased somatostatin tone does not mediate the testosterone-induced GH increase. It has been suggested that the mechanism involves an increase in hypothalamic GHRH release (10). Because Hex and GHRH are synergistic in vivo (13), it is possible that the augmentation of the GH-releasing effect of Hex induced by sex steroid priming is due to the ability of the latter to increase the release of endogenous GHRH. Alternatively, it could be speculated that sex steroids increase the number or activity of the hypothalamic or pituitary GHRP receptors.

In conclusion, Hex is a potent and reproducible stimulus for GH secretion in children. In addition, sex steroids markedly augment the GH-releasing effect of Hex. The results of our studies suggest that the sex steroid-induced increase of the GH response to Hex is mediated by estrogens.

ACKNOWLEDGMENTS

This work was supported in part by a grant from the Ministero della Università e della Ricerca Scientifica e Tecnologica.

REFERENCES

1. Kerrigan JR, Rogol AD. The impact of gonadal steroid hormone action on growth hormone secretion during childhood and adolescence. Endocr Rev 1992; 13:281–298.
2. Marin G, Domene HM, Barnes KM, et al. The effect of estrogen priming and puberty on the growth hormone response to standardized treadmill exercise and arginine-insulin in normal girls and boys. J Clin Endocrinol Metab 1994; 79:537–541.

3. Martin L, Clark J, Conner T. Growth hormone secretion enhanced by androgens. J Clin Endocrinol Metab 1968; 28:425–428.

4. Gourmelen M, Pham-Huu-Trung M, Girard F. Transient partial hGH deficiency in prepubertal children with delay of growth. Pediatr Res 1979; 13:221–224.

5. Liu L, Merriam GR, Sherins RJ. Chronic sex steroid exposure increases mean plasma growth hormone concentration and pulse amplitude in men with isolated hypogonadotropic hypogonadism. J Clin Endocrinol Metab 1987; 64:651–656.

6. Weissberger AJ, Ho KKY. Activation of the somatotropic axis by testosterone in adult males: evidence for the role of aromatization. J Clin Endocrinol Metab 1993; 76:1407–1412.

7. Malhotra A, Poon E, Tse WY, et al. The effects of oxandrolone on the growth hormone and gonadal axes in boys with constitutional delay of growth and puberty. Clin Endocrinol 1993; 38:393–398.

8. Link K, Blizzard RM, Evans WS, Kaiser DL, Parker MW, Rogol AD. The effect of androgens on the pulsatile release and the twenty-four hour mean concentration of growth hormone in peripubertal males. J Clin Endocrinol Metab 1986; 52:159–164.

9. Keenan BS, Richards GE, Ponder SW. Androgen-stimulated pubertal growth: the effects of testosterone and dihydrotestosterone on growth hormone and insulin-like growth factor-I in the treatment of short stature and delayed puberty. J Clin Endocrinol Metab 1993; 76:996–1001.

10. Eakman GD, Dallas JS, Ponder SW, Keenan BS. The effects of testosterone and dihydrotestosterone on hypothalamic regulation of growth hormone secretion. J Clin Endocrinol Metab 1996; 81:1217–1223.

11. Metzger DL, Kerrigan JR. Estrogen receptor blockade with tamoxifen diminishes growth hormone secretion in boys: evidence for a stimulatory role of endogenous estrogens during male adolescence. J Clin Endocrinol Metab 1994; 79:513–518.

12. Deghenghi R, Cananzi MM, Torsello A, Battisti C, Müller EE, Locatelli V. GH-releasing activity of hexarelin, a new growth hormone releasing peptide, in infant and adult rats. Life Sci 1994; 54:1321–1328.

13. Ghigo E, Arvat E, Gianotti L, et al. GH-releasing activity of hexarelin, a new synthetic hexapeptide, after intravenous, subcutaneous, intranasal and oral administration in man. J Clin Endocrinol Metab 1994; 78:693–698.

14. Imbimbo BP, Mont T, Edwards M, et al. Growth hormone releasing activity of hexarelin in humans: a dose response study. Eur J Clin Pharmacol 1994; 46:421–425.

15. Loche S, Cambiaso P, Carta D, et al. The growth hormone-releasing activity of hexarelin, a new synthetic hexapeptide, in short normal and obese children, and in hypopituitary subjects. J Clin Endocrinol Metab 1995; 80:674–678.

16. Bellone J, Aimaretti G, Bartolotta E, et al. Growth hormone-releasing activity of hexarelin, a new synthetic hexapeptide, before and during puberty. J Clin Endocrinol Metab 1995; 80:1090–1094.

17. Loche S, Colao A, Cappa M, et al. The growth hormone response to Hexarelin in children :reproducibility and effect of sex steroids. J Clin Endocrinol Metab 1997; 82:861–864.

18. Tassoni P, Cacciari M, Cau M, et al. Variability of growth hormone response to pharmacological and sleep tests performed twice in short children. J Clin Endocrinol Metab 1990; 71:230–234.

19. Mazza E, Ghigo E, Goffi S, et al. Effect of the potentiation of cholinergic activity on the variability in individual GH response to GH-releasing hormone. J Endocrinol Invest 1989; 12:795–798.

20. Ghigo E, Bellone J, Aimaretti J, et al. Reliability of provocative tests to assess GH secretory status. Study in 472 normally growing children. J Clin Endocrinol Metab 1996; 81:3323–3327.

21. Devesa J, Lima L, Lois N, Lechjga MJ, Arce V, Tresguerres JAF. Reasons for the variability in the growth hormone (GH) responses to GHRH challenge: the endogenous hypothalamic–somatotroph rhythm (HSR). Clin Endocrinol (Oxf) 1989; 30:367–377.

22. Arvat E, Gianotti L, Di Vito L, et al. Modulation of growth hormone-releasing activity of hexarelin in man. Neuroendocrinology 1995; 61:51–56.

23. Maccario M, Arvat E, Procopio M, et al. Metabolic modulation of the growth hormone-releasing activity of hexarelin in man. Metabolism 1995; 44:134–138.

24. Cappa M, Setzu S, Bernardini S, et al. Exogenous GH administration does not inhibit the GH response to hexarelin in normal men. J Endocrinol Invest 1995; 18:762–766.

25. Sartor O, Bowers CY, Reynolds GA, Momany FA. Variables determining the growth hormone response of His-D-Trp-Ala-Trp-D-Phe-Lys-NH$_2$ in the rat. Endocrinology 1985; 117:1441–1447.

26. Bowers CY, Sartor AO, Reynolds GA, Badger TM. On the actions of the growth hormone-releasing hexapeptide, GHRP. Endocrinology 1991; 128:2027–2035.

27. Fletcher TP, Thomas GB, Willoughby JO, Clarke IJ. Constitutive growth hormone secretion in sheep after hypothalamopituitary disconnection and the direct in vivo pituitary effect of growth hormone releasing peptide 6. Neuroendocrinology 1994; 60:76–86.

28. Popovic V, Damjanovic S, Micic D, Djurovic M, Dieguez, C, Casanueva FF. Blocked growth hormone-releasing peptide (GHRP-6)-induced GH secretion and absence of the synergic action of GHRP-6 plus GH-releasing hormone in patients with hypothalamopituitary disconnection: evidence that GHRP-6 main action is exerted at the hypothalamic level. J Clin Endocrinol Metab 1995; 80:942–947.

29. Loche S, Cambiaso P, Merola B, et al. The effect of hexarelin on growth hormone (GH) secretion in patients with GH deficiency. J Clin Endocrinol Metab 1995; 80:2692–2696.

30. Cheng K, Chan WWS, Barreto A, Convey EM, Smith RG. The synergistic effect of His-D-Trp-Ala-Trp-D-Phe-Lys-NH$_2$ on growth hormone (GH)-releasing factor-stimulated GH release and intracellular adenosine 3′,6′-monophosphate accumulation in rat pituitary cell culture. Endocrinology 1989; 124:2791–2798.

31. Bowers CY, Reynolds GA, Durham D, Barrera CM, Pezzoli SS, Thorner MO. Growth hormone (GH)-releasing peptide stimulates GH release in normal men and acts synergistically with GH-releasing hormone. J Clin Endocrinol Metab 1990; 70:975–982.

32. Howard AD, Feighner SD, Cully DF, et al. A receptor in pituitary and hypothalamus that functions in growth hormone release. Science 1996; 273:974–977.
33. Cappa M, Salvatori R, Loche S, et al. The growth hormone response to pyridostigmine plus growth hormone releasing hormone is not influenced by pubertal maturation. J Endocrinol Invest 1991; 14:41–45.

26

Lymphoblastoid Interferon-α Is Able to Reduce Growth Hormone Response to GHRH in Patients with Active Chronic Hepatitis: Preliminary Data

Domenico Valle, Laura De Marinis, Antonio Mancini, Lucia Puglisi, Gian Lodovico Rapaccini, and Giovanni Gasbarrini
Catholic University School of Medicine, Rome, Italy

I. INTRODUCTION

A. Interferons

Interferons (IFNs) are proteins produced by cells of the immune system, in response to definite stimuli, that are involved in defense against the viral infections and in the modulation of the immune response. At least three types of such proteins exist, respectively produced by fibroblasts (IFN-β; 1) leukocytes (IFN-α; 2), and T lymphocytes (IFN-γ; 3). The IFNs inhibit viral replication, promoting synthesis of both cellular ribonucleases, which inactivate the viral RNA, and protein kinases, which block protein synthesis (4). Other biological functions of IFNs are known: the antiproliferative and the hormone-like activity, which affect the cellular maturation as well as the cellular cycle.

Interferon-α or leukocytic, is a nonglycosolated protein with a molecular weight of about 20 kDa, produced by B lymphocytes and macrophages when afflicted by viruses, tumoral cells, or mitogens (5). The principal IFNs-α include that derived from buffy coats, that derived from a lymphoblastoid cell line (or aN1), and a recombinant form.

B.   The Interferon Receptor and Cytokine Receptors Superfamily

Interferon-α and IFN-β have a common receptor (6,7), the extracellular portion of which is designed to bond with the ligands. It comprises two similar portions of 200 amino acids (D200), each having two subdomains of 100 amino acids (SD100) (8,9). The portion of DNA coding for the receptor of IFN-α (IFNAR) and IFN-β (IFNBR) has been identified on chromosome 21 (10) and then cloned (11). The receptor for IFN-γ (IFNGR) comprises only a D200 domain (9), the sequence of which is very similar to that of IFNAR (12).

The family of IFN receptors shares extraordinary similarities with the so-called superfamily of the receptors for cytokines, growth hormone (GH), and prolactin (PRL). This family includes the receptors for PRL and GH, for interleukin (IL)-2, IL-3, IL-4, IL-6, IL-7, granulocyte colony-stimulating factor (G-CSF), granulocyte–macrophage colony-stimulating factor (GM-CSF), erytropoietin (13–15), tumor necrosis factor (TNF-α), leukemia inhibitory factor (LIF), and M-oncostatin (8,13).

C.   Interleukins and the Pituitary

Several studies suggest a neuroendocrine function for the immune system. Not only the receptor for IL-1 has been identified in the brain (16), but IL-1 also increases the production of hypothalamic corticotropin-releasing factor (CRF), then induces the secretion of pituitary corticotropin (adrenocorticotropic hormone; ACTH), and increases the plasma levels of glucocorticoids. In addition, glucocorticoids play a negative-feedback role in IL-1 secretion (17).

Interleukin-1 may induce GH secretion through GH-releasing hormone (GHRH; 18); in addition, in vitro, IL-1 also increases ACTH secretion (19); conversely, IL-1 at high doses blunts GH secretion, because CRF (18) itself is responsible for an increase in somatostatin (SRIF; 20).

In vitro IL-2 reduces the GH release for direct stimulation of SRIF secretion (21). Other studies have demonstrated that IL-2 induces CRF release (22). Conversely, in vitro IL-2 induces GH secretion from mononuclear cells of peripheral human blood (23). IL-6 can also induce CRF secretion (24,25). IFN-γ, at physiological concentrations, inhibits both the secretion of ACTH, PRL, and GH from pituitary cells and their response to hypothalamic factors (26).

II.   THE AIM OF THE STUDY

The data presented in the foregoing show a strict correlation between the immune and endocrine systems. These correlations have reciprocal implications;

in fact, the cytokine receptors superfamily not only shows structural analogies, but also similarity in the postreceptor events.

In the present study, the pattern of the GH–insulin-like growth factor I (IGF-I) axis was evaluated in patients with hepatitis C virus (HCV)-related active chronic hepatitis (ACH), before and during IFN-α treatment. The purpose of the study was to investigate whether GH dynamics could be affected by immune system products in vivo, and to elucidate whether IGF-I secretion could be affected by IFN-α treatment, considering the similarities between GH and IFN receptors.

III. PATIENTS AND METHODS

A. Study Protocol

After securing their informed consent, we recruited 12 patients affected by ACH (assessed by hematochemical and histological data), with a mean ($\pm$ SEM) age of 49.50 $\pm$ 3.35 years and a mean BMI of 25.44 $\pm$ 1.16 kg/m^2, before ($n = 12$) and during ($n = 6$) lymphoblastoid IFN-α treatment.

The patients underwent a baseline GHRH test before treatment with lymphoblastoid IFN-α (Wellferon, Glaxo Wellcome) at a dose of 3 MU three times a week for 6 months. The choice of this IFN-α was suggested by its extensive use in the treatment of patients with ACH.

Patients with known endocrinopathies and those taking drugs recognized as interfering with the function of the hypothalamic–pituitary axis (e.g., antidepressants, benzodiazepines) were excluded from the study.

The patients underwent a standard GHRH test (GHRH 50 μg [GEREF, Serono] as bolus iv at time zero) before ($n = 12$) and during (after 3 months; $n = 6$) the treatment. Blood samples were collected 15, 30, 45, 60, 90, and 120 min after the stimulus. At time zero, blood samples for plasma IGF-I and IGFBP-3 were collected. The tests were performed at 8:00 AM after an overnight fast. The samples were centrifuged within 150 min of their collection, and plasma aliquots were preserved at -20C° until assayed. As control subjects, a group of 12 healthy subjects (10 men and 2 women; mean age: 48.51 $\pm$ 3.23 years; mean BMI: 24.12 $\pm$ 0.90 kg/m^2) were also tested.

B. Assays

Growth hormone was measured in duplicate by immunoradiometric assay (IRMA) on a solid-phase (coated tube), based on monoclonal double-antibody technique, using commercial kits (Radim, Pomezia, Italy). Intra-assay and interassay coefficients of variation (CV) were 2.5 and 5.8%, respectively. The lowest amount of GH detected was 0.04 μg/L.

Plasma IGF-I was measured by radioimmunoassay (RIA) method, using kits from the Nichols Institute (San Juan Capistrano, CA). Soluble IGF-I was separated from interfering binding proteins by precipitation with ethanol–HCl. IGF-I normal values ranged from 170 to 330 mg/L (for age older than 26 years).

Plasma IGFBP-3 was measured by an RIA method using commercial kits by Mediagnost (Tubingen, Germany), previously described by Blum et al. (27). For molar comparison between IGF-I and IGFBP-3, we considered 30.5 kDa the molar weight of IGFBP-3, as suggested by Juul et al. (28). The intra- and interassay CV were below 8 and 15%, respectively. For each assay, all samples from the same patient were measured in the same assay.

C. Statistics

All results are expressed as mean $\pm$ standard error of the mean (SEM). GH plasma concentrations were also expressed as area under the curve (AUC) relative to zero, calculated by trapezoidal rule. The distribution of the data were tested by the Kolmogorov-Smirnov test to verify whether the samples come from a specified distribution: the data were not normally distributed. The significance of differences between the tests performed before and during IFN treatment has been assessed by the nonparametrical Mann-Whitney U-test. The level of statistical significance was set at $p < 0.05$. For statistical evaluation we used the software package Statistica (Statsoft Inc., release 4.1, 1993, for Windows 3.1).

IV. RESULTS

A. Before IFN Therapy

Before IFN therapy (pre-IFN), the GH response to GHRH in the group of patients with HCV-related ACH was similar to that shown in controls, both as peak (Fig. 1) and as AUC (Fig. 2).

B. After 3 Months of IFN Therapy

The GH response to GHRH during therapy (during IFN) was significantly reduced in both the pre-IFN and the control groups (see Figs. 1 and 2).

C. IGF-I, IGFBP-3, and the IGF-I/IGFBP-3 Ratio

The data are reported in Table 1. Both IFG-1 and IGFBP-3 are reduced during IFN therapy as well as compared with control subjects. The IGF-1/IGFBP-3 ratio remains constant during IFN treatment.

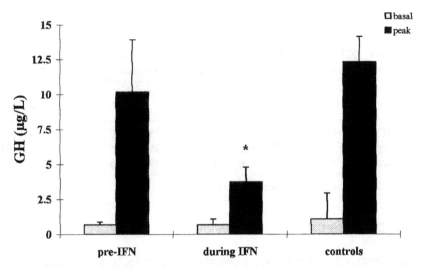

Figure 1 Mean (± SEM) GH (µg/L) basal and peak values in response to GHRH in 12 ACH patients before IFN treatment (pre-IFN) and during IFN (in 6 of 12 patients tested just previously). At right are the GH values of controls ($n = 12$). *p < 0.05 during-IFN versus pre-IFN and controls.

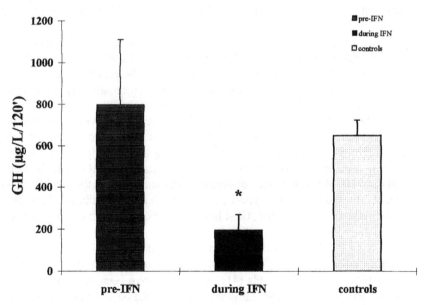

Figure 2 Mean (± SEM) GH-AUC (µg/L per 120 min) in response to GHRH in 12 ACH patients before IFN treatment (pre-IFN) and during IFN (in 6 of 12 patients tested just previously). On the right side are reported the GH values of controls ($n = 12$). *$p < 0.05$ during-IFN versus pre-IFN and controls.

Table 1 IGF-I and IGFBP-3 Plasma Levels Before and During Treatment with IFN in Patients Affected by ACH; a Control Group Is Reported in the Lowest Row

	IGF-I (µg/L)	IGFBP-3 (µg/mL)	IGF-I/IGFBP-3
Pre-IFN ($n = 12$)	124.17 ± 21.64	2.75 ± 0.60	0.17 ± 0.02
During IFN ($n = 6$)	70.50 ± 28.50*	1.65 ± 0.05*	0.18 ± 0.03
Controls ($n = 12$)	180.11 ± 25.12	4.29 ± 0.65	0.20 ± 0.02

*$p < 0.05$ during IFN versus pre-IFN and controls.

V. DISCUSSION

Interferon-α and IFN-β have a common receptor (6,7). Its extracellular portion, consigned to bond with the ligands, comprises two similar portions of 200 amino acids (D200) (8,9). IFNGR comprises only a D200 domain (9) very similar to that of IFNAR (12). The receptor for the GH (GHR) is a glycoprotein of 134 kDa from the cytokine receptor superfamily (8,29). In response to binding with the GH molecule, the receptor is quickly phosphorylated (30), activating JAK2 (31). IL-3, IL-6, and IFN-γ are also able to activate JAK2 (32–34).

Secretion of GH is, basically, the effect of a balance of a dual-control central system represented by GHRH (stimulatory) and by SRIF (inhibitory). Peripheral IGF-I also exerts an inhibitory effect. Secretion of the GH is also modified by estrogens (35), progesterone, testosterone, and thyroid hormones (36), and inhibited by high levels of glycocorticoids (37).

IGF-I circulates bound to IGFBP-3, a protein that probably prolongs the half-life (12–15 h in humans; 38). IGF-I acts at the level of the target organs promoting the anabolic pathways and activating the transcription of mRNA for several proteins. It has recently been demonstrated that IGF-I facilitates the proliferation of cellular lines of breast carcinoma and that this proliferation is blocked by the anti-IGF-I-type-1-receptor antibody; in addition, the presence of the receptor of type 1 for IGF-I has been demonstrated in cell lines of breast carcinoma (39).

IGFBP-3 is principally synthesized by the liver (40). Regulation of the hepatic production of IGFBP-3 is not well understood. Caloric restriction reduces plasma IGFBP-3 levels (41); low levels have been demonstrated in anorexia nervosa (42); IGF-I infusion normalizes IGFBP-3 levels, suggesting an important role of IGF-I in the control of IGFBP-3 production (43).

Moreover, several studies show a neuroendocrine function of the immune system. Not only has the receptor for IL-1 been identified in the brain (16), but IL-1 also increases the production of hypothalamic CRF and then induces the secretion of pituitary ACTH and increases plasma levels of glucocorticoids. Glucocorticoids also play a negative-feedback role for IL-1 secretion (17).

An original observation presented in this study is that patients with ACH show a normal GH secretory response to GHRH, including both peak values and AUC, and lower—but not significantly when compared with controls—plasma IGF-I and IGFBP-3 levels. Another original observation is the blunting effect of IFN-α treatment on the GH–IGF-I axis. The explanation for this is not easy, considering the strict correlations between the pituitary and the immune system. We hypothesize that IFN-α could increase the endogenous somatostatinergic tone, thereby inducing a direct release of SRIF, or through IL-1 or IL-6 and CRF. In fact, IL-1 may induce GH secretion through GHRH (18); in addition, in vitro IL-1 also increases ACTH secretion (19); conversely, high doses of IL-1 blunt GH secretion, as CRF (18) itself is responsible for an increase in SRIF (20).

In vitro, IL-2 reduces the secretion of GHRH because of an increase in SRIF release (21). In other studies, IL-2 has increased CRF release (22) and then SRIF. Conversely, IL-2 induces, in vitro, GH secretion from human blood mononuclear cells (23). In addition to IL-1 and 2, IL-6 could also induce hypothalamic CRF secretion (24,25). Moreover, IFN-γ, at physiological concentrations, inhibits both the secretion of ACTH, PRL, and GH from pituitary cells and their response to hypothalamic factors (26). A direct effect of IFN-α on the anterior pituitary cannot be excluded, but this hypothesis needs further study.

Another aspect of our study is that IGF-I is reduced during IFN-α treatment. This finding may suggest that a direct interaction between IFN-α and hepatic GH receptor probably can be excluded. Despite the structural similarities of the IFN-α and GH receptors, there does not appear to be a similarity in biological function in vivo. In fact, the reduction of IGF-I plasma levels observed during IFN-α treatment seems to rule out the possibility that IFN-α could effectively bind the GH receptor. Alternatively, it may be hypothesized that IFN-α, even by binding GH receptor, is unable to activate postreceptor events.

VI. CONCLUSIONS

In conclusion, these data show a slight, but significant, inhibitory effect of IFN-α treatment on GHRH-induced GH release. Different mechanisms could explain these effects: an increase in IFN-α–induced SRIF release or a direct SRIF-like effect of IFN-α and for an IL mediated SRIF effect.

Finally, the reduction of IGF-I observed during IFN-α treatment seems to exclude the idea that, in vivo, the similarity between GH and IFN receptor permits an overstimulation of GH receptor by IFN. This evidence suggests an important safety factor in the clinical use of this drug. In fact, activation of hepatic GH receptor could have involved an increase in IGF-I levels. The role of IGF-1 in HCV-related hepatocarcinogenesis needs further studies to be clarified.

REFERENCES

1. Knight E Jr. Interferon: purification and initial characterization from human diploid cells. Proc Natl Acad Sci USA 1976; 73:520-523.
2. Cantell K, Hirvonen S, Koistonen V. Partial purification of human leukocyte interferon on a large-scale. Methods Enzymol 1981; 78:499-505.
3. Trotta PP, Spiegel RJ. Interferon: current concepts of mechanisms of action. In: Muggia FM, ed. Concepts, Clinical Developments and Therapeutic Advances in Cancer Chemptherapy. The Hague: Martinus Nijhoff, 1987:141-159.
4. Samuel CE. The RNA-dependent P1/eIF-2α protein kinase. In: Baron S, Coppenhaver DH, Dianzani F, et al., eds. Interferon: Principles and Medical Applications. 1992:237-249.
5. Zoon KC, Bekisz J, Miller D. Human interferon alpha family: protein structure and function. In: Baron S, Coppenhaver DH, Dianzani F et al., eds. Interferon: Principles and Medical Applications. 1992:95-105.
6. Branca AA, Baglioni C. Evidence that types I and II interferons have different receptors. Nature 1981; 294:768-770.
7. Merlin G, Falcoff E, Aguet J. 125-I-labelled human interferons alpha, beta and gamma: comparative receptor binding data. J Gen Virol 1985; 66:1149-1152.
8. Bazan JF. Structural design and molecular evolution of a cytokine receptor superfamily. Proc Natl Acad Sci USA 1990: 87:6934-6938.
9. Thoreau E, Petridou B, Kelly PA, Mornon JP. Structural symmetry of the extracellular domain of the cytokine/growth hormone/prolactin receptor family and interferon receptors revealed by hydrophobic cluster analysis. FEBS Lett 1991; 282:26-31.
10. Lutfalla G, Roeckel N, Mogensen KE, Mattei MG, Uzè G. Assignment of the human interferon alpha receptor gene to chromosome 21q22.1 by in situ hybridization. J Interferon Res 1990; 10:515-517.
11. Uzè G, Lutfalla G, Gresser I. Genetic transfer of a functional human interferon alpha receptor into mouse cells: cloning and expression of its cDNA. Cell 1990; 60:225-234.
12. Gaboriaud C, Uzè G, Lutfalla G, Morgensen K. Hydrophobic cluster analysis reveals duplication in the external structure of human alpha interferon receptor and homology with gamma interferon receptor external domain. FEBS Lett 1990; 269:1-3.
13. Cosman D, Lyman SD, Idzerda RL, Beckmann MP, Park LS, Goodwin RG, March CJ. A new cytokine receptor superfamily. Trends Biochem Sci 1990; 15:265-270.
14. Fukunaga R, Ishizaka-Ikeda E, Seto Y, Nagata S. Expression cloning of a receptor for murine granulocyte colony-stimulating factor. Cell 1990; 341-350.
15. Bazan JF. A novel family of growth factor receptors: a common binding domain

in the growth hormone, prolactin, erythropoietin and interleukin-6 receptors, and the p75 interleukin-2 receptor. Biochem Biophys Res Commun 1989; 788-795.

16. Farrar WL, Kilian PL, Ruff MR, Hill JM, Pert CB. Visualization and characterization of interleukin 1 receptors in brain. J Immunol 1987; 139:459-463.

17. Buzzetti RL, McLoughli L, Scavo D, Rees LH. A critical assessment of the interactions between the immune system and the hypothalamo-pituitary-adrenal axis. J Endocrinol 1989; 120:183-187.

18. Payne LC, Obal F Jr, Opp MR, Krueger JM. Stimulation and inhibition of growth hormone secretion by interleukin-1 beta: the involvement of growth hormone releasing hormone. Neuroendocrinology 1992; 56:118-123

19. Bateman A, Singh A, Kral T, Solomon S. The immune-hypothalamic-pituitary-adrenal axis. Endocr Rev 1989; 10:92-112

20. Aguila MC, McCann SM. The influence of hGRF, CRF, TRH, and LHRH on SRIF release from median eminence fragments. Brain Res 1985; 348:180-182.

21. Karanth S, Aguila MC, McCann SM. The influence of interleukin-2 on the release of somatostatin and growth hormone releasing hormone by mediobasal hypothalamus. Neuroendocrinology 1993; 58:185-190.

22. Camronero JC, Rivas FJ, Borrell J, Guaza C. Interleukin-2 induces corticotropin releasing hormone release from superfused rat hypothalami: influences of glucocorticoids. Endocrinology 1992; 131:667-683.

23. Varma S, Sabharwal P, Shridan JF, Malarkey WB. Growth hormone secretion by human peripheral blood mononuclear cells detected by an enzyme-linked immunoplaque assay. J Clin Endocrinol Metab 1993; 76:49-53.

24. Sapolsky R, River C, Yamamoto G, Plotsky P, Vale W. Interleukin-1 stimulates the secretion of hypothalamic corticotropin releasing factor. Science 1987; 238:522-524.

25. Sharp BM, Matta SG, Peterson PK, Newton R, Chao C, McCallen K. Tumor necrosis factor-alpha is a potent ACTH secretagogue: comparison to interleukin-1. Endocrinology 1989; 124:3131-3133.

26. Vankelecom H, Carmeliet P, Heremans H, van Dammes J, Dijkmans R, Billiau A, Denef C. Interferon-γ inhibits stimulated adrenocorticotropin, prolactin, and growth hormone secretion in normal rat anterior pituitary cell cultures. Endocrinology 1990; 126:2919-2926.

27. Blum WF, Ranke MB, Kietzmann K, Gauggel E, Zeisel HJ, Bierich J. A specific radioimmunoassay for the growth hormone (GH)-dependent somatomedin binding protein: its use for diagnosis of GH deficiency. J Clin Endocrinol Metab 1990; 70:1292-1298.

28. Juul A, Main K, Blum WF, Lindholm J, Ranke MB, Skakkebaek NE. The ratio between serum insulin-like growth factor (IGF)-I and the IGF binding proteins (IGFBP-1,-2,-3) decreases with age and is increased in acromegalic patients. Clin Endocrinol (Oxf) 1994; 41:85-93.

29. Ihle J, Kerr IM. Jaks and Stats in signaling by the cytokine receptor superfamily. Trends Genet 1995; 11:69-74.

30. Stred SE, Stubbart JR, Argetsinger LS, Smith WC, Shafer JA, Talamantes F, Carter-Su C. Stimulation by growth hormone (GH) of GH-receptor-associated tyrosine kinase. Endocrinology 1992; 130:1626-1636.

31. Argetsinger LS, Campbell GS, Yang X, Witthun BA, Silvennoinen O, Ihle JN, Carter-Su C. Identification of JAK2 as a growth hormone receptor-associated tyrosine kinase. Cell 1993; 74:237–244.
32. Witthun BA, Quelle FW, Silvennoinen O, Yi T, Tang B, Miura O, Ihle JN. JAK2 associates with erythropoietin receptor is tyrosine phosphorilated and activated following stimulation with erythropoietin. Cell 1993; 74:227–236.
33. Silvennoinen O, Witthun BA, Quelle FW, Cleveland JL, Yi T, Ihle JN. Structure of the murine Jak2 protein-tyrosine kinase and its role in interleukin-3 signal transduction. Proc Natl Acad Sci USA 1993; 90:8429–8433.
34. Stahl N, Boulton TG, Farruggel T, Ip NY, Davis S, Witthun BA, Quelle FW, Silvennoinen O, Barbieri G, Pellegrini S, Ihle JN, Yancopoulos GD. Association and activation of Jak–Fyk kinases by CNTF-LIF-OSM-IL-6 beta receptor components. Science 1994; 263:92–95.
35. Dawson-Hughes B, Stern D, Goldman J. Regulation of growth hormone and somatomedin C secretion in postmenopausal women: effect of physiological estrogen replacement. J Clin Endocrinol Metab 1986; 63:424–432.
36. Martin JB. Regulation of growth hormone secretion. In: Raiti S, Tolman RA, eds. Human Growth Hormone. New York: Plenum Press, 1986:303–324.
37. Wehrenberg WB, Ling H, Bohlen P. Physiological roles of somatocrinin and somatostatin in the regulation of growth hormone secretion. Biochem Biophys Res Commun 1982; 109:562–567.
38. Guler HP, Schmid C, Zapf J, Foresch ER. Effects of recombinant insulin-like growth factor I on insulin secretion and renal function in normal human subjects. Proc Natl Acad Sci USA 1989, 86:2868–2872.
39. Yee D, Paik S, Lebovic GS, Marcus RR, Favoni RE, Cullen KJ, Lippman ME, Rosen N. Analysis of insulin-like growth factor I gene expression in malignancy: evidence for a paracrine role in human breast cancer. Mol Endocrinol 1989; 3:517–519.
40. Donovan SM, Atilano LC, Hintz RL, Wilson DM, Rosenfeld RG. Differential regulation of the insulin-like growth factor (IGF-I and II) and IGF-binding proteins during malnutrition in the neonatal rat. Endocrinology 1991; 129:149–157.
41. Thissen JP, Katelslegers JM, Underwood LE. Nutritional regulation of the insulin-like growth factors. Endocr Rev 1994; 15:80–101.
42. Counts DR, Gwritsman H, Carlsson LMS, Lesem M, Cutler GB. The effect of anorexia nervosa and refeeding on growth hormone-binding protein, the insulin-like growth factors (IGFs), and the IGF-binding proteins. J Clin Endocrinol Metab 1992; 75:762–767.
43. Thissen JP, Underwood LE, Maiter D, Maes M, Clemmons DR, Katelslegers JM. Evidence that pretranslational and translational defects decrease serum insulin-like growth factor I concentration during dietary protein restriction. Endocrinology 1991; 128:885–890.

27

Effects of Growth Hormone Secretagogues Inside and Outside the Hypothalamic–Hypophyseal System

**John-Olov Jansson, Johan Svensson, Bengt-Åke Bengtsson,
Eva Sjögren-Jansson, Anders Lindahl, Ola Nilsson,
B. Håkan Ahlman, and Bo Wängberg**
Sahlgrenska University Hospital, Göteborg, Sweden

Michael Nilsson
Institute of Neurobiology, Sahlgrenska University Hospital, Göteborg, Sweden

Lawrence A. Frohman and Rhonda D. Kineman
University of Illinois at Chicago, Chicago, Illinois

I. REGULATION OF PULSATILE GH SECRETION

In rodents, and maybe also in other species, episodic growth hormone (GH) secretion is important for several biological functions. These include sexual dimorphism of the liver (1,2) but, possibly, also functions of other organs (3). Studies in male rats have indicated that pulsatile GH secretion (4) is regulated at the pituitary level by an interplay between reciprocal and rhythmic release of GH-releasing hormone (GHRH) and somatostatin from the hypothalamus. Thus, each GH pulse is accompanied by increased release of GHRH and decreased release of somatostatin (5,6). At the hypothalamic level, there appears to be crosstalk between GHRH and somatostatin-containing neurons. Somatostatin appears to inhibit GHRH release (6), and GHRH may stimulate somatostatin release (7). It was shown by Clark and Robinson (8) that intravenous (iv) infusions of GHRH pulses at 3-h intervals to female rats, which normally have a more continuous GH secretory pattern than male rats (2), induce a male-type plasma GH pattern, with GH pulses following each GHRH infusion and, perhaps more importantly, low GH levels between the pulses (8). The latter effect could result from increased release of somatostatin owing to a direct stimulation at the hypothalamic level by GHRH. Alternatively, the GHRH-induced endogenous GH pulse could affect hypothalamic functions, including somatostatin release, as discussed in the following.

II.   THE POSSIBLE ROLE OF ACUTE GH FEEDBACK IN REGULATION OF PULSATILE GH SECRETION

The episodic secretion of GH at 3-h intervals in conscious male rats (4) has been used as a model for studies of the regulation of GH pulsatility. Two reports, using two different experimental paradigms, have provided evidence that the pulsatile GH secretory pattern in male rats is regulated by a reciprocal interaction between the hypothalamus and the somatotrophs (9,10). The rapid autofeedback of a GH pulse could contribute to the suppressed GH levels during the following trough. Because there are few indications of a direct effect of GH at the pituitary level, this influence is probably exerted at the hypothalamic level. There are also experimental data suggesting that increased release of somatostatin is of importance for this GH-induced suppression of GH during troughs (9). If this hypothesis is true, it implies that there is a lag period before the autofeedback effect sets in, preventing a GH pulse from immediately suppressing itself. Moreover, it is reasonable to assume that the feedback effect lasts for about 60–90 min (i.e., the ordinary time interval between two GH pulses), thereby allowing the next spontaneous GH pulse to occur (10). The brief feedback effect induced by an endogenous GH surge could also explain why several substances have caused an immediate GH-releasing action, followed by a blunting or postponement of the next spontaneous GH pulse (see Ref. 9 and references therein).

III.   A NEW GROUP OF ORALLY ACTIVE GH SECRETAGOGUES

Research initiated by Bowers and co-workers (11,12) indicated that a new group of orally active GH secretagogues, exemplified by the synthetic hexapetide GH-releasing peptide-6 (GHRP-6) exert actions at the levels of both the hypothalamus and the pituitary. Later, Smith and co-workers (13) reported nonpeptidyl GH secretagogues that act on the same receptors as GHRP, with higher potency and greater bioavailability after oral administration. The GH secretagogues may also stimulate the release of corticotropin (adrenocorticotropic hormone; ACTH) and prolactin (PRL), although these effects seem to be less accentuated and more transient than the GH-releasing effect (14,15). It has also been reported that intracerebroventricular (icv) injection of GHRPs can stimulate appetite (increased feeding behavior) in experimental animals and that this effect is independent of the previously reported orectic effect of GHRH after icv administration (16). The GHRH and GHRP-6 mimetics act through similar, but distinctive, membrane receptors present both at the pituitary level and within the central nervous system (CNS). Both belong to the superfamily of G–protein-activating receptors with seven transmembrane regions (17,18).

IV. POSSIBLE ROLE OF THE GH SECRETAGOGUES IN REGULATION OF PULSATILE GH RELEASE

The role of the GHRP-6–type GH secretatogues in regulation of GH release appears to be rather complex. First, it was assumed that the main effect was at the pituitary level, but later reports indicated the presence of GHRP-binding sites in the hypothalamus (19) and increased hypothalamic activity following iv administration of these compounds (20). At this level, the GHRP-6 mimetics may act by stimulating the release of GHRH (21,22). Moreover, the GH secretatogues seem to be dependent on the action of GHRH to exert their stimulatory effect on GH secretion, as shown by the fact that they cannot stimulate GH release in the dwarf mouse model (7), which is known to have a defective GHRH receptor (23,24). The GH-releasing effect of GHRP is also reduced in rats pretreated with GHRH-neutralizing antibodies (21). There is probably also a role for somatostatin in the GH-releasing effect of GHRP (12,21).

It was recently reported that GHRP-6 may activate neurons in the arcuate nucleus that produce neuropeptide Y (NPY) in addition to GHRH (25). The possible physiological significance of this is unknown, but an effect of GHRPs on NPY could be of importance for stimulation of feeding behavior, because NPY has potent orectic effects. Alternatively, the action on NPY-producing neurons could be of importance for the stimulatatory effect of GHRPs on ACTH release, because NPY neurons may project from the arcuate nucleus to cell bodies in the paraventricular nucleus, producing corticotropin-releasing factor (see Ref. 25 and references therein). Finally, GH secretagogue-induced changes in NPY secretion may mediate modulatory effects on GH secretion (26).

V. ACROMEGALY AND CUSHING'S SYNDROME CAUSED BY ECTOPIC HORMONE PRODUCTION

Occasionally, studies of a rare human disease can provide new insights into biological functions. We have studied a case of combined acromegaly and Cushing's syndrome caused by ectopic secretion of both GHRH and ACTH from a carcinoid tumor. This condition is extremely rare, with few suspected and even fewer definitive cases reported in the literature (27). The details of the case report have recently been described (28). Briefly, a 49-year-old man was referred to Sahlgrenska University Hospital because of chest symptoms and signs of Cushing's syndrome, with abdominal adiposity and thin extremities. There were no overt signs of acromegaly, but the patient had diabetes mellitus and hypokalemia. The patient also had an elevated urine concentration of 5-hydroxyindoleacetic acid (5-HIAA) and an elevated serum concentration of chromogranin A, as well as a slight elevation of serum prolactin. The levels of circulating NPY were clearly elevated (515 pg/L; reference interval < 130).

Chest computed tomography (CT) and octreotide scintigraphy showed a thymic carcinoid with somatostatin receptors in the mediastinum as well as metastases in the left suprclavicular fossa. As discussed in the following, the patient had increased levels of circulating human GH. Measurement with a specific radioimmunoassay (RIA; 29) also showed high peripheral levels of GHRH. The levels of immunoreactive GHRH in eight plasma samples from the patient were markedly elevated (range 600–1050 ng/L), whereas control plasma from a person with no carcinoid contained less than 20 ng/L (Fig. 1). The patient also had elevated levels of circulating ACTH and cortisol, whereas the levels of corticotropin-releasing hormone (CRH) were low and undetectable in several samples obtained on two difference days.

To relieve some of the clinical symptoms, the patient was treated with the somatostatin agonist octreotide and the glucocorticoid synthesis inhibitor ketoconazole. Repeated blood sampling during 24 h while off this treatment showed markedly elevated serum cortisol levels without diurnal variability (Fig. 2). However, after initiation of treatment, the serum cortisol concentra-

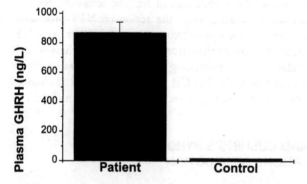

Figure 1 The levels of immunoreactive GHRH (P-GHRH) measured by RIA in eight plasma samples from a patient with a thymic carcinoid and in control plasma from a person with no carcinoid.

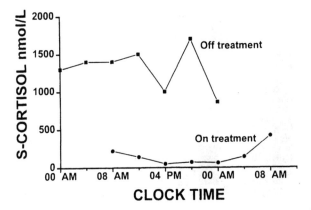

Figure 2 The 24-h serum cortisol profiles measured on and off treatment with octreotide and ketoconazole. The treatment with octreotide (Sandostatin; 100 μg tid, sc) and ketoconazole (400–1400 mg/day) had been given for 12 and 9 weeks, respectively.

tions were markedly suppressed, with a diurnal rhythmicity (i.e., enhanced levels at 0800 h compared with that at other time points) (see Fig. 2). There was also a slight decrease in plasma ACTH during this treatment, suggesting that an effect of octreotide on the production of ACTH by the tumor potentiated the direct suppressory effect of ketoconazole on cortisol production. The patient underwent surgical removal of most of the primary tumor, and the metastases of the supraclavicular fossa. There was clinical improvement, with decreased levels of plasma chromogranin A and serum cortisol, and it was possible to discontinue insulin treatment. The results of analysis with the PULSAR algorithm according to Merriam and Wachter (30), revealed a markedly elevated baseline GH secretion of 7.3 mU/L before treatment (Fig. 3; i.e., concentrations high even in relation to those found in patients with acromegaly; 31). During treatment with octreotide, the basal GH levels were suppressed by 40% and this effect was even more obvious (80%) after surgical removal of tumor tissue. The GH pulse height, however, remained largely unaffected during octreotide treatment and after surgery. Mean serum GH levels were suppressed by 35% after surgery (see Fig. 3).

Although the serum GHRH pattern was not measured in the present study, it seems reasonable to assume that it was rather constant, as detailed in earlier reports (32). It was concluded that hypothalamic influence, probably by rhythmic and reciprocal release of GHRH and somatostatin (5–7), can retain some GH pulsatility in the face of constantly high ectopic GHRH secretion (32).

The levels of serum insulin-like growth factor type I (IGF-I) were suppressed after surgery (from 748 to 254 μg/L). Thus, there were simultaneous octreotide-induced decreases in serum IGF-I levels and baseline plasma GH,

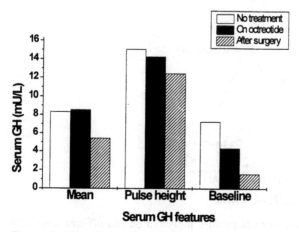

Figure 3   The effects of octreotide treatment and surgery on serum GH profiles in a patient with ectopic GHRH syndrome. Twenty-four-hour serum GH profiles were obtained when the patient received no treatment, after treatment with octreotide (Sandostatin; 100 μg tid, sc for 12 weeks, discontinued on the day of the GH measurement), and after surgery with removal of primary tumor tissue as well as metastases of the thymic carcinoid. The mean GH level, pulse height, and baseline level of the serum GH profiles were calculated with the PULSAR computer program according to Merriam and Wachter (30).

without effects on GH pulse height and with a small decrease in mean plasma GH. Because these data are from only one patient, the results should be interpreted cautiously. However, the basal nonpulsatile component of GH secretion may correlate better with serum IGF-I levels than does the mean GH secretion (31,33). The possibility that baseline GH secretion is a determinant of IGF-I production may gain support from the fact that continuous administration of GH enhances the levels of circulating IGF-I more efficiently than pulsatile treatment in GH-deficient rats (34) and humans (35,36). This, in turn, may be because the effect of continuous GH treatment on IGF-I production in the liver, which contributes most of the circulating IGF-I, may be more pronounced than in other organs (3). It is also possible, however, that the octreotide treatment given in this study directly inhibited IGF-I secretion (37).

VI.   THE RESPONSIVENESS OF THE TUMOR CELLS TO GH SECRETAGOGUES AND GHRH

Intracellular calcium influx appears to be a good marker of action by GHRP-6 mimetics and GHRH in somatotrophs (7,38). In the present study, we investigated the effects of GHRH and GHRP-6 on calcium influx in primary culture of tumor cells from the patient. Two examples of calcium influx into individual cells in primary cell culture from this carcinoid-producing GH, ACTH, and NPY are shown in Figure 4. The addition of GHRH (10 μM, cell 1) or GHRP-6 (10 μM, cell 2) increased the calcium influx, as measured by a 340/380 nm fluorescence ratio, using the calcium-sensitive dye fura-2 (28). All studied cells responsive to GHRH were nonresponsive to GHRP-6, and vice versa. Other cells from the same tumor were nonresponsive to both GHRH and GHRP-6. Cells from another thymic carcinoid that did not produce GHRH, ACTH, or NPY were completely unresponsive to GHRH or GHRP-6 (data not shown).

Histopathological examination of the tumor tissue revealed a neuroendocrine tumor with characteristics of a foregut carcinoid. The results of immunocytochemical analysis of consecutive sections of the tumor tissue indicate that most cells produce GHRH, ACTH, and NPY, and that these peptides were coexpressed. This is in contrast with our finding that the tumor cells were responsive to GHRH alone, to GHRP-6 alone, or to neither of these peptides (see foregoing). It seems unlikely, therefore, that there is a correlation between

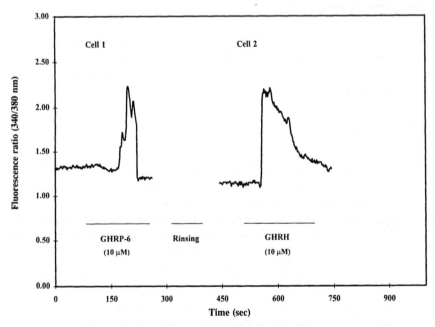

Figure 4 Cytosolic Ca^{2+} responses, as measured by 340/380-nm–fluorescence ratios, after stimulation of two cells with 10 μM GHRP-6 (left panel) and 10 μM GHRH (right panel). Cell 1 represents five cells responsive to GHRP-6 and cell 2 represents five cells responsive to GHRH.

the responsiveness in vitro to GHRP-type GH secretagogues and GHRH on one hand and the production of GHRH, ACTH, and NPY in vivo on the other.

At the hypothalamic level, GHRP-6 activates cells in the arcuate nucleus that express GHRH mRNA as well as cells that express NPY mRNA (25). Because there is a comparatively small degree of colocalization of GHRH in the NPY-containing cells in this region (39), it seems likely that most NPY-containing cells responsive to GHRP-6 do not express GHRH. This should be interpreted in the context that many GHRH cells express NPY, but few NPY cells express GHRH. This is different from the apparent colocalization of GHRH in the NPY-containing (and probably also GHRP-6-responsive) tumor cells of our own study (28).

In summary, the present results demonstrate responsiveness in this patient's cell tissue (presumably reflecting the presence through receptors) to the GHRP-type GH secretagogues outside the hypothalamus and the pituitary. It remains to be determined whether GHRP analog receptors are present in nontumor extrahypothalamic–hypophyseal tissues.

ACKNOWLEDGMENTS

We thank Dr. Sten Rosberg for help with the PULSAR program. The study was supported by the Swedish Medical Research Council (9894 and 5220), Kungliga Vetenskaps- och Vitterhetssamhället, Swedish Cancer Society (2998), and USPHS grant DK 30667.

REFERENCES

1. Gustafsson J-Å, Mode A, Norstedt G, Skett P. Sex steroid induced changes in hepatic enzymes. Annu Rev Physiol 1983; 45:51–60.
2. Jansson J-O, Edén S, Isaksson O. Sexual dimorphism in the control of growth hormone secretion. Endocr Rev 1985; 6:128–150.
3. Isgaard J, Carlsson L, Isaksson OGP, Jansson J-O. Pulsatile intravenous growth hormone (GH) infusion to hypophysectomized rats increases insulin-like growth factor I messenger ribonucleic acid in skeletal tissues more effectively than continuous GH infusion. Endocrinology 1988; 123:2605–2610.
4. Tannenbaum GS, Martin JB. Evidence for an endogenous ultradian rhythm governing growth hormone secretion in the rat. Endocrinology 1976; 98:562–570.
5. Tannenbaum G, Ling N. The interrelationship of growth hormone (GH)-releasing factor and somatostatin in generation of the ultradian rhythm of GH secretion. Endocrinology 1984; 115:1952–1957.
6. Plotsky P, Vale W. Patterns of growth hormone-releasing factor and somatostatin secretion into the hypophyseal–portal circulation of the rat. Science 1985; 230:461–463.

7. Frohman LA, Jansson J-O. Growth hormone releasing hormone. Endocr Rev 1986; 7:223–253.

8. Clark RG, Robinson ICAF. Growth induced by pulsatile infusion of an amidated fragment of human growth hormone releasing factor in normal and GHRF-deficient rats. Nature 1985; 314:281–283.

9. Sato M, Chihara K, Kita T, Kashio Y, Okimura Y, Kitajama N, Fujita T. Physiological role of somatostatin-mediated autofeedback regulation for growth hormone: importance of growth hormone in triggering somatostatin release during a trough period of pulsatile growth hormone release in conscious male rats. Neuroendocrinology 1989; 50:139–151.

10. Carlsson L, Jansson J-O. Endogenous growth hormone (GH) secretion in male rats is synchronized to pulsatile GH infusions given at 3-hour intervals. Endocrinology 1990; 126:6–10.

11. Bowers CY, Momany FA, Reynolds GH, Hong A. On the in vitro and in vivo activity of a new synthetic hexapeptide that acts on the pituitary to specifically release growth hormone. Endocrinology 1984; 114:1537–1545.

12. Bowers CY, Sartor AO, Reynolds GA, Badger TM. On the actions of the growth hormone-releasing hexapeptide, GHRP. Endocrinology 1991; 128:2027–2035.

13. Smith RG, Cheng K, Schoen WR, Pong S-S, Hickey G, Jacks T, Butler B, Chang W W-S, Chaung L-YP, Judith F, Taylor J, Wyvratt MJ, Fisher MH. A nonpeptidyl growth hormone secretagogue. Science 1993; 260:1640–1643.

14. Aloi JA, Gertz BJ, Hartman ML, Huhn WC, Pezzoli SS, Wittreich MJ, Krupa DA, Thorner MO. Neuroendocrine responses to a novel growth hormone secretagogue, L-692,429, in healthy older subjects. J Clin Endocrinol Metab 1994; 79:943–949.

15. Svensson J, Lönn L, Jansson J-O, Murphy G, Wyss D, Krupa D, Cerchio K, Gertz B, Bosaeus I, Sjöström L, Bengtsson B-Å. Two-month treatment of obese subjects with the oral growth hormone (GH) secretagogue MK-677 increases GH secretion, fat-free mass and energy expenditure. J Clin Endocrinol Metab. In press.

16. Okada K, Ishii S, Minami S, Sugihara H, Shibaski T, Wakabayashi I. Intracerebroventricular administration of the growth hormone-releasing peptide KP-102 increases food intake in free-feeding rats. Endocrinology 1996; 137:5155–5157.

17. Mayo K. Molecular cloning of an expression of a pituitary-specific receptor for growth hormone-releasing hormone. Mol Endocrinol 1992; 6:1734–1741.

18. Howard AD, Feighner SD, Cully DF, Arena JP, Liberator PA, Rosenblum CI, Hamelin M, Hreniuk DL, Palyha OC, Anderson J, Paress PS, Diaz C, Chou M, Liu KK, McKee KK, Pong S-S, Chaung L-Y, Elbrecht A, Dashkevicz M, Heavens R, Rigby M, Sirinathsinghji DJS, Dean DC, Melillo DG, Patchett AA, Nargund R, Griffin RP, DeMartino JA, Gupta SK, Schaeffer JM, Smith RG, Van der Ploeg LHT. A receptor in pituitary and hypothalamus that functions in growth hormone release. Science 1996; 273:974–977.

19. Codd EE, Shu AYL, Walker RF. Binding of a growth hormone releasing hexapeptide to specific hypothalamic and pituitary binding sites. Neuropharmacology 1989; 28:1139–1144.

20. Dickson SL, Leng G, Robinson ICAF. Systemic administration of growth hormone-releasing peptide (GHRP-6) activates hypothalamic arcuate neurones. Neuroscience 1993; 53:303–306.

21. Clark RG, Carlsson LMS, Trojnar J, Robinson ICAF. The effects of a growth hormone-releasing peptide and growth hormone-releasing factor in conscious and anaesthetized rats. J Neuroendocrinol 1989; 1:249–255.

22. Guillaume V, Magnan E, Cataldi M, Dutour A, Sauze N, Renard M, Razafindraibe H, Conte-Devolx B, Deghenghi R, Lenaerts V, Oliver C. Growth hormone (GH)-releasing hormone secretion is stimulated by a new GH-releasing hexapeptide in sheep. Endocrinology 1994; 135:1073–1076.

23. Clark RG, Robinson ICAF. Effects of a fragment of human growth hormone-releasing factor in normal and "little" mice. J Endocrinol 1985; 106:1–5.

24. Jansson J-O, Downs TR, Beamer WG, Frohman LA. Receptor-associated resistance to growth hormone-releasing factor in dwarf "little" mice. Science 1986; 232:511–512.

25. Dickson SL, Luckman SM. Induction of c-*fos* messenger ribonucleic acid in neuropeptide Y and growth hormone (GH)-releasing factor neurons in the rat arcuate nucleus following systemic injection of the GH secretagogue, GH-releasing peptide-6. Endocrinology 1997; 138:771–777.

26. Pierroz DD, Catzeflis C, Aebi AC, Rivier JE, Aubert ML. Chronic administration of neuropeptide Y into the lateral ventricle inhibits both the pituitary–testicular axis and growth hormone and insulin-like growth factor I secretion in intact adult male rats. Endocrinology 1996; 137:3–12.

27. Leveston S, McKeel JRD, Buckley PJ, et al. Acromegaly and Cushing's syndrome associated with a foregut carcinoid tumor. J Clin Endocrinol Metab 1981; 53:682–689.

28. Jansson J-O, Svensson J, Bengtsson B-Å, Frohman L, Nilsson O, Ahlman H, Wängberg B, Nilsson M. Combined Cushing's syndrome and acromegaly due to ectopic secretion of adrenocorticotrophic hormone (ACTH) and growth hormone releasing hormone (GHRH). Clin Endocrinol. In press.

29. Frohman L, Downs T. Measurement of growth hormone-releasing factor. Methods Enzymol 1986; 124:371–389.

30. Merriam G, Wachter K. Algorithms for the study of episodic hormone secretion. Am J Physiol 1982; 243:E310–E318.

31. Yoshida T, Shimatsu A, Sakane N, Hizuka N, Horikawa R, Tanaka T. Growth hormone (GH) secretory dynamics in a case of acromegalic gigantism associated with hyperprolactinemia: nonpulsatile secretion of GH may induce elevated insulin-like growth factor-I (IGF-I) and IGF-binding protein-3 levels. J Clin Endocrinol Metab 1996; 81:310–313.

32. Vance ML, Kaiser DL, Evans WS, Furlanetto R, Vale W, Rivier J, Thorner MO. Pulsatile growth hormone secretion in normal man during a continuous 24-hour infusion of human growth hormone releasing factor (1–40). J Clin Invest 1985; 75:1584–1590.

33. Hartman M, Veldhuis J, Vance M, Faria A, Furlanetto R, Thorner M. Somatotropin pulse frequency and basal concentrations are increased in acromegaly and are reduced by successful therapy. J Clin Endocrinol Metab 1990; 70:1375–1384.

34. Bick T, Hochberg Z, Amit T, Isaksson O, Jansson J-O. Roles of pulsatility and continuity of growth hormone (GH) administration in the regulation of hepatic GH-

receptors, and circulating GH-binding protein and insulin-like growth factor-I. Endocrinology 1992; 131:423–429.

35. Jörgensen J, Möller N, Lauritzen T, Christiansen J. Pulsatile versus continuous intravenous administration of growth hormone (GH) in GH-deficient patients: effects on circulating insulin-like growth factor-I and metabolic indices. J Clin Endocrinol Metab 1990; 70:1616–1623.

36. Johansson J-O, Oscarsson J, Bjarnason R, Bengtsson B-Å. Two weeks of daily injections and continuous infusion of recombinant human growth hormone (GH) in GH-deficient adults: I. Effects on insulin-like growth factor I (IGF-I), GH and IGF binding proteins, and glucose homeostasis. Metabolism 1996; 45:362–369.

37. Flyvberg A, Jorgensen K, Marshall S, Örskov H. Inhibitory effect of octreotide on growth hormone-induced IGF-I generation and organ growth in hypophysectomized rats. Am J Physiol 1991; 260:E568–E574.

38. Herrington J, Hille B. Growth hormone-releasing hexapeptide elevates intracellular calcium in rat somatotropes by two mechanisms. Endocrinology 1994; 135:1100–1108.

39. Coifu P, Croix D, Tramu G. Coexistence of hGHRF and NPY immunoreactivities in neurons of the arcuate nucleus of the rat. Neuroendocrinology 1987; 45:425–428.

28

Growth Hormone-Releasing Effect of Hexarelin*, a Synthetic GHRP, and GHRH in Children and Adults with Down Syndrome

Letizia Ragusa, Antonio Alberti, Corrado Romano, and Caterina Proto
OASI Institute for Research in Mental Retardation and Brain Aging, Troina, Italy

Fabio Colabucci
Catholic University, Rome, Italy

Maura Rosa Valetto, Laura Gianotti, Jaele Bellone, Gianluca Aimaretti, Emanuela Arvat, and Ezio Ghigo
University of Turin, Turin, Italy

Romano Deghenghi
Europeptides, Argenteuil, France

I. INTRODUCTION

There is clear evidence showing that in both animals and humans the activity of growth hormone–insulin-like growth factor type 1 (GH–IGF-I) axis undergoes age-related variations. In fact, both GH secretion and IGF-I levels increase at puberty, falling thereafter to very low levels in aging (1–4).

In spite of marked age-related variations in GH secretion, it has been demonstrated that the maximal secretory capacity of somatotroph cells does not

---

*Hexarelin is a trademark for examorelin (I.N.N.).

vary with age (5,6). Thus, the age-related reduction of somatotroph secretion depends on age-related alterations in the neural control of GH secretion (2,3,5,7); among them, the cholinergic impairment that connotes aging could play a major role (8–10).

On the other hand, there is evidence showing that GH secretion is impaired at an early age in Down syndrome (DS) compared with that in normal adults. In fact, the GH response to GHRH both alone and combined with cholinergic agonists (11,12), but not that with arginine (unpublished observations, submitted for publication), is already reduced in DS adults to the same extent observed in normal elderly subjects. Thus, the premature impairment of GH secretion in DS seems to be due to anticipated (cholinergic?) alterations in the neural control of GH secretion.

The GH-releasing peptides (GHRPs) and nonpeptidyl GHRP mimetics are synthetic, nonnatural molecules that possess a potent stimulatory effect on somatotroph secretion both in animal and in humans (13,14). Although synthetic and nonnatural, GHRPs act through specific receptors at the pituitary level as well as within the central nervous system (CNS), particularly at the hypothalamic level (15–18). It is now widely accepted that the GH-releasing effect of GHRPs takes place through actions at the pituitary and, mainly, at the hypothalamic level (14,15,18), the integrity of the hypothalamic–pituitary unit being needed for normal activity of GHRPs (19,20). Mechanisms underlying the stimulatory effect of GHRPs on somatotroph secretion include antagonism of somatostatinergic activity and stimulation of GHRH-secreting neurons (14,15,21). However, the likely existence of a putative endogenous GHRP-like ligand has to be taken into account (15). It has been already shown that the GH-releasing activity of GHRPs in humans undergoes important age-related variations (14,22–24). In fact, the GH response to GHRPs increases at puberty (25), persists in young adults, and is clearly reduced in elderly subjects (22,24). These variations likely reflect age-related changes in the neural control of GH secretion (5).

Given the foregoing, the aim of the present study was to verify the age-related variations in the GH response to hexarelin, a synthetic hexapeptide belonging to GHRP family (26), comparing them with those recorded after GHRH administration.

II. SUBJECTS AND METHODS

Seven prepubertal DS (PRE-DS, 5 boys and 2 girls, 11.5 ± 0.4 years of age), 8 pubertal DS children (PUB-DS; 5 boys and 3 girls, 14.9 ± 0.6 years of age), and 6 DS adults (A-DS, 4 men and 2 women, 24.8 ± 1.1 years of age) were studied. The diagnosis of DS was based on chromosomal analysis and typical clinical phenotype. Eight normal prepubertal (PRE-NC, 6 boys and 2 girls, 11.6 $\pm$ 0.8 years old), 6 pubertal children (PUB-NC, 4 boys and 2 girls, 14.5 $\pm$

0.6 years old), 15 normal adults (NA, 10 men and 5 women, 27.3 $\pm$ 0.9 years old) and 6 normal elderly subjects (NE, 3 men and 3 women, 78.7 $\pm$ 1.5 years old) were studied as controls. The protocol had been approved by the Ethical Committee of our department, and informed consent had been obtained from all subjects and parents of children.

All subjects underwent the following tests: (1) GHRH (GHRH-29; GEREF, Serono, Milan, Italy); 2 μg/kg as iv bolus at 0 min; (2) hexarelin (HEX; His-D-2-methyl-Trp-Ala-Trp-D-Phe-Lys-NH$_2$, Europeptides, Argenteuil, France); 2 μg/kg as iv bolus at 0 min. All tests were begun at 0830 after an overnight fasting and 30 min after an indwelling catheter had been inserted in an antecubital cubital vein kept patent by slow infusion of isotonic saline. Blood samples were taken basally at –60 and 0 min, and then every 15 min from 0 to +120 min. All samples from an individual subject were analyzed together.

Serum GH levels were measured in duplicate by immunoradiometric assay (hGH-CTK; Sorin, Italy). The sensitivity of the assay was 0.1 μg/L. The inter- and intra-assay coefficients of variation (CV) were 4.9–6.5% and 1.5–2.9%, respectively.

Serum IGF-I levels were measured in duplicate by radioimmunoassay (Nichols Institute Diagnostics, San Juan Capistrano, CA). To avoid interference by binding proteins, all samples were treated with acid ethanol. The sensitivity of the assay was 0.1 μg/L. The inter- and intra-assay coefficients of variation were 10.1–15.7% and 7.6–15.5%, respectively.

Results are expressed as mean $\pm$ SEM of absolute values (μg/L) or of areas under curves (AUC; μg/L·h). Statistical analysis of the data was carried out using a nonparametric ANOVA (Kruskal-Wallis or Wilcoxon's test).

III. RESULTS

The IGF-I levels in PRE-NC (173.4 $\pm$ 13.0 μg/L) were lower ($p < 0.01$) than those in PUB-NC (358.7 $\pm$ 73.0 μg/L), similar to those in NA (175.5 $\pm$ 11.2 μg/L) and higher ($p < 0.0001$) than those in NE (66.9 $\pm$ 14.5 μg/L). IGF-I levels in PRE-DS (193.7 $\pm$ 60.6 μg/L), PUB-DS (418.0 $\pm$ 26.4 μg/L), and A-DS (215.3 $\pm$ 29.1 μg/L) were similar to those in age-matched controls (Fig. 1). The GH responses to GHRH in PRE-NC (AUC $\pm$ SEM: 920.8 $\pm$ 218.0 μg/L·h), PUB-NC (994.8 $\pm$ 260.3 μg/L·h) and NA (800.8 $\pm$ 124.5 μg/L·h) were similar and higher ($p < 0.01$) than that in NE (278.1 $\pm$ 128.5 μg/L·h).

The GH response to GHRH in PRE-DS (1050.8 $\pm$ 248.6 μg/L·h) was similar to that in PUB-DS (1194.4 $\pm$ 417.0 μg/L·h), and both were higher ($p < 0.05$) than that in A-DS (520.0 $\pm$ 117.1 μg/L·h). The GH responses to GHRH in PRE-DS and PUB-DS were similar to those in normal children, whereas that in A-DS was lower ($p < 0.05$) than that in NA and similar to that in NE (Figs. 2 and 3).

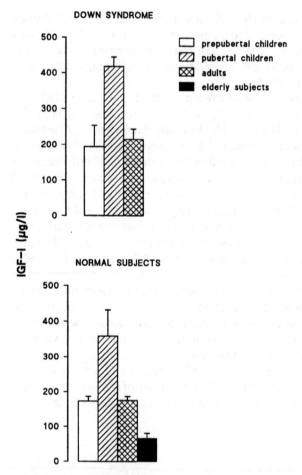

Figure 1 The IGF-I (µg/L) levels in Down syndrome patients (*upper panel*) and in age-matched normal subjects.

The HEX-induced GH response in PRE-NC (1382.4 ± 267.3 µg/L·h) was lower than those in PUB-NC (2933.7 ± 203.0 µg/L·h; $p < 0.001$) and NA (2209.7 ± 295.6 µg/L·h; $p < 0.05$); these two latter GH responses were similar and higher ($p < 0.002$) than that in NE (655.3 ± 130.8 µg/L·h).

The GH response to HEX in PRE-DS (1274.9 ± 328.3 µg/L·h) was lower ($p < 0.002$) than that in PUB-DS (3068.6 ± 375.0 µg/L·h), whereas that in A-DS (1403.9 ± 477.7 µg/L·h) was lower ($p < 0.01$) than that in PUB-DS and similar to that in PRE-DS. The GH responses to HEX in PRE-DS and PUB-DS were similar to those in normal children, whereas that in A-DS was lower ($p < 0.05$) than that in NA and similar to that in NE (see Figs. 2 and 3). HEX

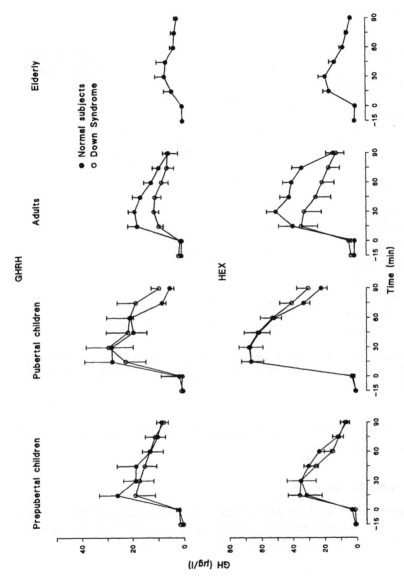

Figure 2 The GH response curves (mean ± SEM, µg/L) to GHRH (*upper panel*) and to hexarelin (*lower panel*) in Down syndrome patients and in age-matched normal subjects.

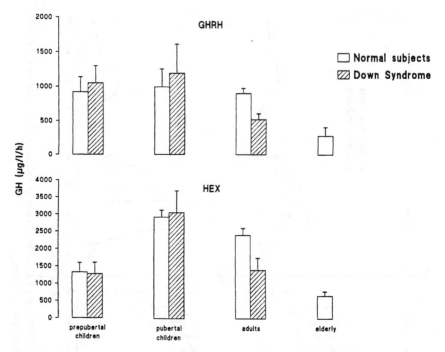

Figure 3 The AUCs (mean ± SEM, µg/L·h) of the GH response to GHRH (*upper panel*) and to hexarelin (*lower panel*) in Down syndrome patients and in age-matched normal subjects.

released more GH than GHRH in pubertal children, adults, and elderly, but not in prepubertal children.

A. Side Effects

Hexarelin as well as GHRH administration induced transient facial flushing. Mild sleeplessness was also recorded in some subjects after HEX administration. No medication was required, and no test had to be stopped.

IV. DISCUSSION

The present data confirm the age-related variations in the somatotroph responsiveness to GHRPs and GHRH in normal subjects and demonstrate that the GH-releasing activity of GHRPs as well as that of GHRH undergoes premature reduction in DS. In fact, in DS adults the GH response to hexarelin as well as that to GHRH is already reduced relative to age-matched controls and overlaps that observed in normal elderly subjects.

That the activity of GH–IGF-I axis undergoes age-related variations is widely accepted (1–4). In fact, both GH-secretion and IGF-I levels increase at puberty (1) decreasing thereafter, up to very low levels in aging (1,3,5). In spite of marked age-related variations in GH secretion, the maximal secretory capacity of somatotrophs does not vary with age (5,6).

Our present findings confirm that the GH-releasing activity of hexarelin undergoes clear age-related variations in normal subjects, which is in agreement with other data showing that the stimulatory effect of GHRPs on GH secretion varies across animal and human life spans (15,22–25). In fact, according to our present data, the GH response to GHRPs increased at puberty (25), persisting in a similar manner in early adulthood, and decreased thereafter, being reduced to the same extent by the sixth decade of life (10,14,15,22,23,25).

The age-related variations in the GH-releasing activity of GHRPs are dissociated, at least partially, from those in GHRH-induced GH secretion (14,24). In fact, across the human life span, the most marked GHRH-induced GH secretions are present at birth (5,27). Moreover, the GH response to GHRPs is clearly increased at puberty, whereas the GHRH-induced GH rise in prepubertal and pubertal children is similar (25; and present data). Interestingly, GHRPs release more GH than GHRH by puberty, but not before (14,24). On the other hand, by adulthood, the somatotroph responsiveness to GHRPs as well as that to GHRH undergoes associated age-related variations, being clearly reduced in aging (22,23). This evidence indicates that the GH-releasing activity of GHRPs takes place by a mechanism that is at least partly different from those of GHRH, in agreement with other data in animals and in humans (14,15).

The mechanisms underlying the age-related variations in the GH-releasing effect of GHRPs across human life span likely depend on the age-related variations in the neurohormonal and neurotransmitter control of somatotroph function (5) as well as in gonadal steroid levels (25,28), although the role of the putative endogenous GHRP-like ligand has to be taken into account (15). Particularly, the enhanced somatotroph responsiveness at puberty could depend on the increase in estradiol levels, as indicated by evidence that estradiol and testosterone, but not oxandrolone, augment the GH response to hexarelin in peripubertal children (28). On the other hand, in aging, the reduced responsiveness to GHRPs is independent of the fall of estrogens levels (29), whereas it is likely due to the concomitant increase of somatostatinergic activity and reduction GHRH activity (5,14).

Because of short stature and neuroendocrine alterations, namely cholinergic alterations, similar to those in normal and demented patients with Alzheimer's disease (30,31), the activity of GH–IGF-I axis in DS is of particular interest. So far, it has been demonstrated that GH secretion is prematurely impaired in DS. In fact, the GH response to GHRH both alone and combined with cholinergic agonists (11,12), is already reduced in DS adults to the same extent observed in normal elderly subjects. Interestingly, the maximal secretory capacity of somatotropic cells in DS adults is fully preserved, as shown by the fact that the GH response to GHRH combined with arginine does not decrease and is similar to that in normal adults (unpublished observations). Thus, the early impairment of GH secretion in DS seems to be due to the anticipated (cholinergic?) alterations in the neural control of GH secretion.

Our present findings show that, as in normal subjects, in DS patients the GH-releasing activity of GHRPs, namely hexarelin, undergoes age-related variations. The stimulatory effect of hexarelin, similar to that of GHRH alone (11,12; and present data) and combined with pyridostigmine (11), but not when combined with arginine (unpublished observations), is already reduced in DS adults to an extent similar to that usually recorded in normal aging. Given the evidence showing that the maximal secretory capacity of somatotrophs is preserved in DS adults (unpublished observations), our findings indicate that age-related alterations in the neural control of GH secretion also underly the premature reduction of the GH response to GHRPs in DS.

In conclusion, our present data further confirm that somatotropic function in Down syndrome undergoes anticipated aging, probably because of early age-related variations in the neural control of the GH–IGF-I axis.

ACKNOWLEDGMENTS

We wish to thank Dr. J. Ramunni for her participation in the study, and Mrs. A. Barberis and Mrs. M. Talliano for their skillfull technical assistance.

REFERENCES

1. Zadik Z, Chalew SA, McCarter RJ, Meistas M, Kowarski AA. The influence of age on the 24-hour integrated concentration of growth hormone in normal individuals. J Clin Endocrinol Metab 1985; 60:513–516.
2. Ghigo E. Neurotransmitter control of growth hormone secretion. In: De la Cruz LF, ed. Growth Hormone and Somatic Growth. Amsterdam: Excerpta Medica, 1992; 103–136.
3. Dieguez C, Page MD, Scanlon MF. Growth hormone neuroregulation and its alterations in disease states. Clin Endocrinol 1988; 28:109–121.
4. Corpas E, Harman SM, Blackman S. Human growth hormone and human aging. Endocr Rev 1993; 14:20–39.
5. Ghigo E, Arvat E, Gianotti L, Ramunni J, DiVito L, Maccagno B, Grottoli S, Camanni F. Human aging and the GH–IGF-I axis. J Pediatr Endocrinol Metab 1996; 9:271–278.

6. Ghigo E, Goffi S, Nicolosi M, Arvat E, Valente F, Mazza E, Ghigo MC, Camanni F. GH responsiveness to combined administration of arginine and GHRH does not vary with age in man. J Clin Endocrinol Metab 1990; 71:1481–1485.

7. Ghigo E, Arvat E, Goffi S, Bellone J, Nicolosi M, Procopio M, Camanni F. Neural control of growth hormone secretion in aged humans. In: Muller E, Cocchi D, Locatelli V, eds. Growth Hormone and Somatomedins During Lifespan. Berlin: Springer Verlag, 1993:275–287.

8. Gibson GE, Peterson C, Jender DJ. Brain acetylcholine synthesis declines with senescence. Science 1981; 213:675–678.

9. Barthus RT, Dean RL, Beer B, Lippa AS. The cholinergic hypothesis of geriatric memory dysfunction. Science 1982; 217:408–417.

10. Ghigo E, Goffi S, Arvat E, Imperiale E, Boffano GM, Valetto MR, Mazza E, Santi I, Magliona A, Boghen MF, Boccuzzi G, Camanni F. A neuroendocrinological approach to evidence an impairment of central cholinergic function in aging. J Endocrinol Invest 1992; 15:665–670.

11. Arvat E, Gianotti L, Ragusa L, Valetto MR, Cappa M, Aimaretti G, Ramunni J, Grottoli S, Camanni F, Ghigo E. The enhancing effect of pyridostigmine on the GH response to GHRH undergoes an accelerated age-related reduction in Down syndrome. Dementia 1996; 7:288–292.

12. Ragusa L, Alberti A, Romano C, Valetto MR, Aimaretti G, Proto C, Imbimbo BP, Deghenghi R, Colabucci F, Ghigo E. Effect of hexarelin on growth hormone secretion in Down syndrome. Dev Brain Dysfunct 1996; 9:204–210.

13. Bowers CY, Veeraragavan K, Sethumadhavan K. Atypical growth hormone releasing peptides. In: Bercu BB, Walker RF, eds. Growth Hormone II, Basic and Clinical Aspects. New York: Springer-Verlag, 1993:203–222.

14. Ghigo E, Arvat E, Muccioli G, Camanni F. Growth hormone (GH)-releasing peptides: a review. Eur J Endocrinol. 1997; 136:445–460.

15. Bowers CY, Momamy FA, Reynolds GA, Hong A. On the in vivo and in vitro activity of a new synthetic hexapeptide that acts on the pituitary to specifically release growth hormone. Endocrinology 1994; 114:1537–1545.

16. Codd EE, Shu AYL, Walker RF. Binding of a growth hormone releasing hexapeptide to specific hypothalamic and pituitary binding sites. Neuropharmacology 1989; 28:1139–1144.

17. Pong SS, Chaung LYP, Dean DC, Nargunt RP, Patchett AA, Smith RG. Identification of a new G-protein-linked receptor for growth hormone secretagogues. Mol Endocrinol 1996; 10:57–61.

18. Smith RG, Cheng K, Pong SS, Hickey G, Jacks T, Butler B, Chang WWS, Chaung LYP, Judith F, Taylor J, Schoen WR, Wyvratt MJ, Fisher MH. A non-peptidyl GH secretagogue. Science 1993; 260:1640–1643.

19. Mallo F, Alvarez CV, Benitez L, Burguera B, Coya R, Casanueva FF. Regulation of His-D-Trp-Ala-Trp-D-Phe-Lys-NH$_2$ (GHRP6)-induced GH secretion in the rat. Neuroendocrinology 1992; 57:247–251.

20. Popovic V, Damjanovic S, Micic D, Djurovic M, Dieguez C, Casanueva FF. Blocked growth hormone-releasing peptide (GHRP-6)-induced GH secretion and absence of the synergic action of GHRP-6 plus GH-releasing hormone in patients with hypothalamopituitary disconnection: evidence that GHRP-6 main action is exerted at the hypothalamic level. J Clin Endocrinol Metab 1995; 80:942–947.

21. Goth MI, Lyons CE, Canny BJ, Thorner MO. Pituitary adenylate cyclase activating polypeptide, growth hormone (GH)-releasing peptide and GH-releasing hormone stimulate GH release through distinct pituitary receptors. Endocrinology 1992; 130:939–944.

22. Arvat E, Gianotti L, Grottoli S, Imbimbo BP, Lenhaerts V, Deghenghi R, Camanni F, Ghigo E. Arginine and growth hormone-releasing hormone restore the blunted growth hormone-releasing activity of hexarelin in elderly subjects. J Clin Endocrinol Metab 1994; 79:1440–1443.

23. Aloi JA, Gertz BJ, Hartman ML, Huhn WC, Pezzoli SS, Wittreich M, Krupa DA, Thorner MO. Neuroendocrine responses to a novel growth hormone secretagogue, L-692,429, in healthy older subjects. J Clin Endocrinol Metab 1994; 79:943–949.

24. Ghigo E, Arvat E, Gianotti L, Grottoli S, Rizzi G, Ceda G, Deghenghi R, Camanni F. Aging and growth hormone releasing peptides. In: Bercu BB, Walker RF, eds. Growth Hormone Secretagogues. New York: Springer-Verlag, 1996:415–431.

25. Bellone J, Aimaretti G, Bartolotta E, Benso L, Imbimbo BP, Lenhaerts V, Deghenghi R, Camanni F, Ghigo E. Growth hormone-releasing activity of hexarelin, a new synthetic hexapeptide, before and during puberty. J Clin Endocrinol Metab 1995; 80:1090–1094.

26. Deghenghi R. Growth hormone releasing peptides. In: Bercu BB, Walker RF, eds. Growth Hormone Secretagogues. New York: Springer-Verlag, 1993:85–102.

27. Deiber M, Chatelain P, Naville D, Putet G, Salle B. Functional hypersomatotropism in small for gestational age (SGA) newborn infant. J Clin Endocrinol Metab 1989; 68:232–235.

28. Loche S, Colao A, Cappa M, Bellone J, Aimaretti G, Farello G, Faedda A, Lombardi G, Deghenghi R, Ghigo E. The growth hormone response to hexarelin in children: reproducibility and effect of sex steroids. J Clin Endocrinol Metab 1997; 82:861–864.

29. Arvat E, Gianotti L, Broglio F, Maccagno B, Bertagna A, Deghenghi R, Camanni F, Ghigo E. Oestrogen replacement does not restore the reduced GH-releasing activity of hexarelin, a synthetic hexapeptide, in postmenopausal women. Eur J Endocrinol 1997; 136:483–487.

30. Wisniewski KE, Howe J, Williams DG, Wisniewski HM. Precocious aging and dementia in patients with Down's syndrome. Biol Psychiatry 1978; 13:619–627.

31. Williams RS, Matthysse S, Swaab DF, Fliers E, Mirmiran M, Van Gool WA, van Haaren F. Age-related changes in Down syndrome brain and the cellular pathology of Alzheimer disease. Prog Brain Res 1986; 70(4):49–67.

29

The Critical Importance of Growth Hormone-Releasing Hormone Signaling for Growth Hormone Secretion: Lessons from the Dwarfs of Sindh*

Hiralal G. Maheshwari and Gerhard Baumann
Northwestern University Medical School, Chicago, Illinois

I. INTRODUCTION

Growth hormone [GH secretion is governed by GH-releasing hormone(GHRH)], somatostatin, and presumably, a third factor, as yet to be discovered, the putative natural ligand of the GH-releasing peptide (GHRP) receptor. Each of these hypophysiotropic factors interacts with one or more specific receptors of the seven-transmembrane domain, G–protein-coupled type. The relative contributions of the three factors to GH production and to the generation of a pulsatile secretion pattern remains somewhat unclear because of the difficulty in isolating their influences in the intact organism. Disturbances in any of the three can theoretically affect GH secretion. In this chapter we report a new type of genetic GH-deficient dwarfism that illustrates the critical necessity for GHRH signaling in the elaboration of GH.

II. AN EXPEDITION TO PAKISTAN

Following a newspaper article (1) reporting the discovery of several probably related dwarfs in two remote villages in Pakistan, we arranged a visit to study

---

*This work was presented in part at the 10th International Congress of Endocrinology, San Francisco, CA, 1996.

427

the phenomenon and its possible cause. Field studies quickly revealed the familial nature of this dwarfism and the absence of a general medical or nutritional reason for growth failure. Interestingly, dwarfism affected only the youngest generation, and parents were of apparently normal stature. We identified 18 dwarfs (15 male, 3 female), ranging in age from newborn to 28 years (Fig. 1). Except for the newborn, their height was 7–8 standard deviations below the normal mean (adult height averaged 130 cm for men and 114 cm for women). Dwarfism was proportionate, and patients had no deformities, no dysmorphic features, and no intellectual impairment. All were offspring of consanguineous marriages. Bone age was retarded by about 4–5 years. The physical phenotype was consistent with a defect in the GH–IGF axis. Insulin-like growth factor I (IGF-I) levels were extremely low (mean 5.2 ng/mL), IGF-binding protein 3 values were low (mean 0.42 μg/mL), and IGF-binding protein 2 values were elevated (mean 440 ng/mL). Serum GH did not respond to stimulation with GHRH, levo-dopa, or clonidine (peak levels < 0.3 ng/mL). The GH-binding

Figure 1   Six male dwarfs of Sindh (front row) with normal-stature persons (back row). The ages are—from left to right—20, 13, 15, 15, 21, and 25 years (one 15-year-old subject stands behind the other).

protein level was normal, and IGF-I levels rose normally after 5 days of exogenous GH treatment. Pituitary functions other than GH were intact, as were thyroid, adrenal, and gonadal hormones. General blood chemistry values were also normal.

We concluded that this syndrome was caused by a genetically isolated GH deficiency, inherited in an autosomal recessive manner. We obtained DNA and RNA from the patients and their relatives (approximately 50 subjects) to identify the faulty gene and the precise molecular defect. We considered several candidate genes that might be implicated in GH deficiency. The principal candidate gene, the GH-N gene, was amplified and fully sequenced, but no abnormality was found. We then used linkage analysis between the dwarf phenotype and chromosomal markers (microsatellites) near the loci of candidate genes to narrow the choice of candidates. Among several candidates, the only one showing linkage was the GHRH receptor gene on the short arm of chromosome 7 (maximum lod score 6.26). Stepwise amplification and sequencing of that gene revealed a G to T transversion in exon 3 of the GHRH receptor gene (2–4). This mutation converts the Glu^{50} codon to a stop codon. The result of this nonsense mutation is a GHRH receptor severely truncated in its extracellular domain. The mutant receptor can be predicted to be nonfunctional for a variety of reasons (disrupted binding site, lack of membrane anchoring, lack of G protein activation, or other). The mutation, in its homozygous form, was 100% concordant with GH deficiency and the dwarf phenotype in our kindred. Heterozygous family members showed only minimal or no height deficit, but had moderately decreased IGF-I and IGF-binding protein 3 levels. Of particular interest is the offspring of two homozygous, affected patients. The occurrence of a successful pregnancy demonstrates that both sexes are fertile in the absence of a functional GHRH receptor. This is an important finding in view of the expression of GHRH and its receptor in the gonads (5–7).

III. DISCUSSION

This new syndrome represents an unusual form of GH deficiency and illustrates the absolute requirement of GHRH signaling for GH secretion in humans. In its absence, it is clear that neither the somatostatin system nor the endogenous "GHRP" system can sustain GH production in physiologically meaningful quantities. (It is not yet known to what extent our patients would respond to exogenous GHRP, but we suspect that the GH response will be blunted or absent). The syndrome is the human equivalent of the little mouse (*lit/lit*) (8–9), which is also dwarfed and has a missense mutation (Asp^{38} → Gly) in the GHRH receptor gene. In the little mouse, the GHRH receptor is not truncated but, rather, is unable to bind ligand—with similar functional consequences. It is of interest that until now, only two other human cases of a defective GHRH receptor are known. They are also of consanguineous descent and have recently been reported in New York (10). Interestingly, their mutation is identical with the one we found in Pakistan, although there is no known blood relation between them and our kindred.

Familial isolated GH deficiency is usually caused by heterogeneous mutations in the GH-N gene (11). The mutated GHRH receptor gene we describe represents the second gene that is implicated in isolated GH deficiency. To date, in humans, no other genetic lesion causing isolated GH deficiency has been identified. The GHRH gene has been extensively investigated, but no mutation has yet been found in that gene (12). Mutations in the Pit-1 gene do cause GH-deficient dwarfism (13), but those patients also show thyrotropin (thyroid-stimulating hormone; TSH) and prolactin deficiency. Other potential genes include those involved in pituitary development (e.g., Prop-1 and others; 14), although they would likely affect multiple pituitary hormones. Our observation raises the possibility that less severely disabling mutations in the GHRH receptor gene may be responsible for milder cases of familial short stature.

ACKNOWLEDGMENTS

This work was supported by grants from the Genentech Foundation for Growth and Development, the Human Growth Foundation, and the Northwestern University Intramural Grant Program. We thank Dr. J. Dupuis for help with the linkage analysis, Drs. B. Silverman and L. Jameson for helpful discussions, and J. Hollands for expert technical assistance. We are grateful to Dr. D. Fisher, Corning-Nichols Laboratories, for performing IGF and IGFBP assays; Dr. J. Chipman, Eli Lilly Co. for providing human GH; and Prof. U. Heinrich, Heidelberg, for help in procuring GHRH.

REFERENCES

1. Anonymous. The Dwarfs of Sindh. DAWN (Karachi), February 21, 1994, p. 7.
2. Maheshwari HG, Baumann G. Genomic organization and structure of the human growth hormone releasing hormone receptor gene. Program 79th Meeting Endocrcine Society, Minneapolis, MN, 1997.
3. Mayo KE. Molecular cloning and expression of a pituitary-specific receptor for growth hormone-releasing hormone. Mol Endocrinol 1992; 6:1734–1744.
4. Gaylinn BD, Harrison JK, Zysk JR, Lyons CE, Lynch KR, Thorner MO. Molecular cloning and expression of a human anterior pituitary receptor for growth hormone-release hormone. Mol Endocrinol 1993; 7:77–84.

5. Berry SA, Srivastava CH, Rubin LR, Phipps WR, Pescovitz OH. Growth hormone releasing hormone-like messenger ribonucleic acid and immunoreactive peptide are present in human testis and placenta. J Clin Endocrinol Metab 1992; 75:281–284.

6. Bagnato A, Moretti C, Ohnishi J, Frajese G, Catt KJ. Expression of the growth hormone releasing hormone gene and its peptide product in the rat ovary. Endocrinology 1992; 130:1097–1102.

7. Matsubara S, Sato M, Mizobuchi M, Niimi M, Takahara J. Differential gene expression of growth hormone (GH)-releasing hormone (GRH) and GRH receptor in various rat tissues. Endocrinology 1995; 136:4147–4150.

8. Godfrey P, Rahal JO, Beamer WG, Copeland NG, Jenkins NA, Mayo KE. GHRH receptor of little mice contains a missense mutation in the extracellular domain that disrupts receptor function. Nature Genet 1993; 4:227–232.

9. Lin SC, Lin CR, Gukovsky I, Lusis AJ, Sawchenko PE, Rosenfeld MG. Molecular basis of the little mouse phenotype and implications for cell type-specific growth. Nature 1993; 364:208–213.

10. Wajnrajch MP, Gertner JM, Harbison MD, Chua SC Jr, Leibel RL. Nonsense mutation in the human growth hormone-releasing hormone receptor causes growth failure analogous to the little (lit) mouse. Nature Genet 1996; 12:88–90.

11. Phillips JA III, Cogan JD. Genetic basis of endocrine disease 6: molecular basis of familial growth hormone deficiency. J Clin Endocrinol Metab 1994; 78:11–16.

12. Perez Jurado LA, Phillips JA III, Francke U. Exclusion of growth hormone (GH)-releasing hormone gene mutations in familial isolated GH deficiency by linkage and single strand conformational analysis. J Clin Endocrinol Metab 1994; 78:622–628.

13. Parks JS, Kinoshita EI, Pfäffle RW. Pit-1 and hypopituitarism. Trends Endocrinol Metab 1993; 4:81–85.

14. Sornson MW, Wu W, Dasen JS, Flynn SE, Norman DJ, O'Connell SM, Gukovsky I, Carriére C, Ryan AK, Miler AP, Zuo L, Gleiberman AS, Andersen B, Beamer WG, Rosenfeld MG. Pituitary lineage determination by the Prophet of Pit -1 homeodomain factor defective in Ames dwarfism. Nature 1996; 384:327–333.

Index

Milton Keynes UK
Ingram Content Group UK Ltd.
UKHW020010071024
449327UK00031B/2724